Nondestructive Methods for Material Property Determination

Nondestructive Methods for Material Property Determination

Edited by

Clay Olaf Ruud

Materials Research Laboratory
Pennsylvania State University
University Park, Pennsylvania

and

Robert E. Green, Jr.

Materials Science and Engineering Department
The Johns Hopkins University
Baltimore, Maryland

PLENUM PRESS • NEW YORK AND LONDON

Library of Congress Cataloging in Publication Data

Symposium on Nondestructive Methods for Material Property Determination (1983:
Hershey, Pa.)
Nondestructive methods for material property determination.

"Proceedings of a Symposium on Nondestructive Methods for Material Property
Determination, held April 6–8, 1983, in Hershey, Pennsylvania"–T.p. verso.
Includes bibliographical references and index.
1. Non-destructive testing–Congresses. 2. Materials–Testing–Congresses. I.
Ruud, C. O. II. Green, Robert E. (Robert Edward), 1932–   . III. Title.
TA417.2.S938 1983   620.1'127   84-9941

Proceedings of a Symposium on
Nondestructive Methods for Material
Property Determination,
held April 6–8, 1983,
in Hershey, Pennsylvania

ISBN-13: 978-1-4684-4771-2   e-ISBN-13: 978-1-4684-4769-9
DOI: 10.1007/978-1-4684-4769-9

©1984 Plenum Press, New York
Softcover reprint of the hardcover 1st edition 1984
A Division of Plenum Publishing Corporation
233 Spring Street, New York, N.Y. 10013

PREFACE

The interest in materials property determination by nondestructive means is increasing especially for in-process and in-service inspection of structural and electronic materials and components. Such attention is due to several factors, including increased automation of manufacturing processes, the demand for greater reliability in consumer products and military hardware, and more severe demands on the performance of materials.

This book represents the proceedings for the Symposium on Nondestructive Methods for Material Property Determination held April 6 to 8, 1983, at the Hotel Hershey in Hershey, Pennsylvania. That symposium was one of the first meetings concerned specifically with nondestructive material property determination (characterization). Its purpose was to stimulate intercourse between researchers, engineers, and theoreticians so as to focus upon the multidiseiplinary problems of advancing the state of the art in this area.

The papers in the book are concerned mainly with acoustic (including ultrasonic), magnetic, electrical, and x-ray diffraction techniques and applications. Many of the papers describe well-developed technologies that are currently in practical application, while others discuss concepts which will never emerge from the laboratory but perhaps will provide the groundwork for more practical ideas.

The symposium was co-chaired by C.O. Ruud of The Pennsylvania State University and R.E. Green Jr. of The Johns Hopkins University. The Organizing Committee was as follows:

J. L. Parker, Lockheed Georgia Corporation
D. H. Pai, Foster Wheeler Corporation
J. R. Quinn, Electric Power Research Institute
H. Vanderveldt, Department of the Navy
R. N. Pangborn, The Pennsylvania State University
S. G. Wax, Defense Advance Research Projects Agency

The symposium was sponsored by the following:

The Pennsylvania State University
American Society of Nondestructive Testing (ASNT)
American Society of Mechanical Engineers (ASME)
Electric Power Research Institute (EPRI)
Department of the Navy
Defense Advance Research Projects Agency (DARPA)

Invited papers were presented by:

W. Lord, Colorado State University
J. R. Barton, Southwest Research Institute
V. Weiss, Syracuse University
E. P. Mueller, USFDA
K. Masubuchi, Massachusetts Institute of Technology
P. Höller, Fraunhofer-Institut für Zerstörungsfreie
  Prüfverfahren

Sesssion chairmen were as follows:

R. Hardy, Department of the Navy
R. N. Pangborn, The Pennsylvania State University
R. B. Thompson, Iowa State University
D. H. Pai, Foster Wheeler Corporation
R. E. Green Jr., The Johns Hopkins University

Administration of the Symposium was conducted by E. M. Hawk and R. L. Melchiorre of The Pennsylvania State University.

The editors are in debt to Lois Annechini-Moore of The Pennsylvania State University for her effort in organizing this book, preparing the index, table of contents, and list of contributors, and typing many of the manuscripts.

                                              C. O. Ruud

CONTENTS

ACOUSTO–ELASTIC EFFECT
OF STRESS IN ALUMINUM AND STEEL

X-RAY DIFFRACTION AND NEUTRON SCATTERING TECHNIQUES

X-RAYS FOR NON-DESTRUCTIVE CHARACTERIZATION OF MATERIALS PROPERTIES

Volker Weiss

Syracuse University
Syracuse, N.Y.

INTRODUCTION

X-rays found their first practical application, for radiography, almost immediately after their discovery by Roentgen in 1895. Crystal structure determination became possible in 1912 when Max von Laue derived the diffraction conditions for three-dimensional lattices. In 1914 C.G. Darwin published two papers, describing both the kinematical and the dynamical theory (1,2). The former is, of course, the basis for crystal structure determination. The need for the latter was rediscovered in the 1950's, when very perfect crystals became available.

Applications are based on the ABSORPTION, the EMISSION, and the DIFFRACTION characteristics. Exciting progress has recently been made in all three areas, sources have become more powerful (e.g. storage-ring synchrotrons), detectors more sensitive, and computer control and data analysis a common ingredient in x-ray instrumentation.

ABSORPTION TECHNIQUES

The absorption of x-rays, between absorption edges, is proportional to the third power of the atomic number of the absorber. The many medical and industrial uses of radiography are well known. Since the introduction of vidicon techniques, requiring smaller dosages and shorter exposure times, the development of "tomography" in the 1970's represents perhaps the most exciting innovation in radiography. The culmination of these methods of body-section radiography, computerized axial tomography (CAT), earned its developers McCormack and Hounsfield the 1979 Nobel Prize in medicine.

Linear tomography is schematically illustrated in Fig. 1. X-ray source and film move in opposite directions. Once the two velocities are chosen, a one-to-one correlation of representation on the film plane exists only for one plane (section) of the object. Density variations above and below the section of interest are smeared out. The same principle is employed for "transaxial tomography", where both object and film are rotated synchronously, Fig. 2.

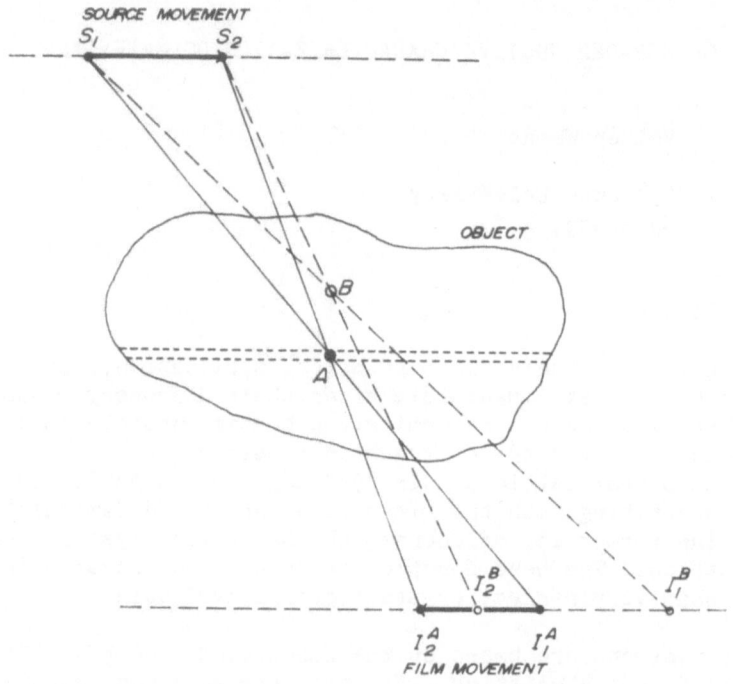

Fig. 1 Linear Tomography, schematic.

Computerized tomography, CT, and finally computerized axial tomography, CAT, were logical next steps (for more detailed discussions of these techniques see e.g. Ellingson and Berger (3) and Herman (4)). The principle of CAT is shown in Fig 3. Absorption information is obtained for many angular orientations (usually 180). The first scanners (in 1974) took about 15 minutes for one cross-section. Present state of the art equipment scanning times are in the order of seconds. In-plane resolution is about 2 mm. With further improvements in source and detection systems a resolution to 0.1 mm should become feasible in the near future. While most of the applications have been in medicine, Burstein, Mastronardi and Kirchner (5) have recently reported on the development of of an inspection system for Trident Rocket Motors and conclude that "CT for large rockets is not only feasible, but also yields inspection performance results superior to those offered by any competing technique or their projected future performance." The authors claim a resolution which allowed the detection of cracks as small as 0.08 mm (0.003 in).

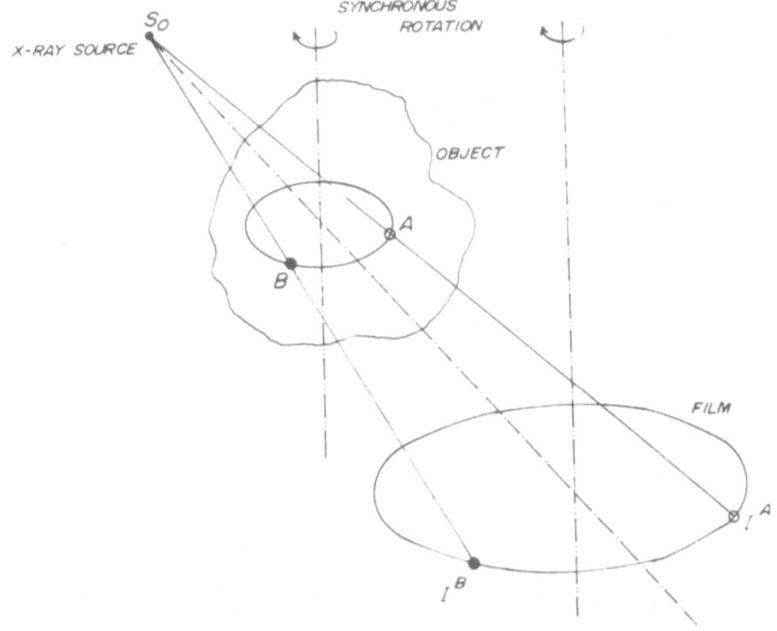

Fig. 2 Transaxial tomography, schematic.

Because the refractive index for x-rays is within a few parts per million of unity for all substances, it is not possible to construct lens systems to produce focussed images. Nevertheless, imaging systems have been developed using pinhole camera techniques. Strecker (6) described an x-ray scattering method for layer-by-layer imaging via a pinhole system, schematically illustrated in Fig. 4. For Al, a tube voltage of 200 kV, a .5 mm pinhole, and a 2 mm collimating slit required film exposure times of several hours. However with a commercial image intensifyer and a plumbicon TV-camera, real-time inspection of the image layer became possible.

An ingeneous use of the absorption properties of x-rays is made in the x-ray microscope (7). Depending on the film-to-sample distance the methods are referred to as contact or projection microradiography. The contact method is limited by the film grain size and diffraction effects to a resolution of 1 to 0.5 micrometers. For the projection method a microfocus x-ray tube of high stability is essential. In both instances proper choice of the radiation is important to develop the desired absorption contrast (7).

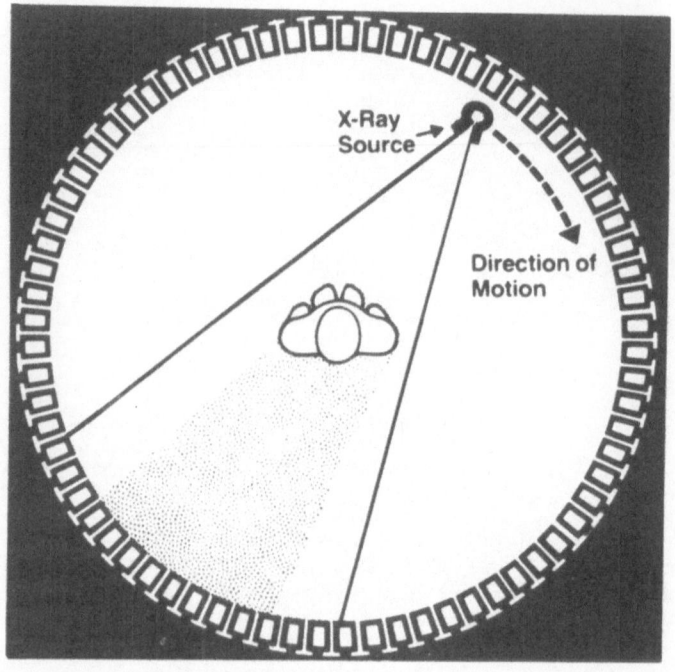

Fig. 3 Computerized axial tomography, source and detector assembly, schematic.

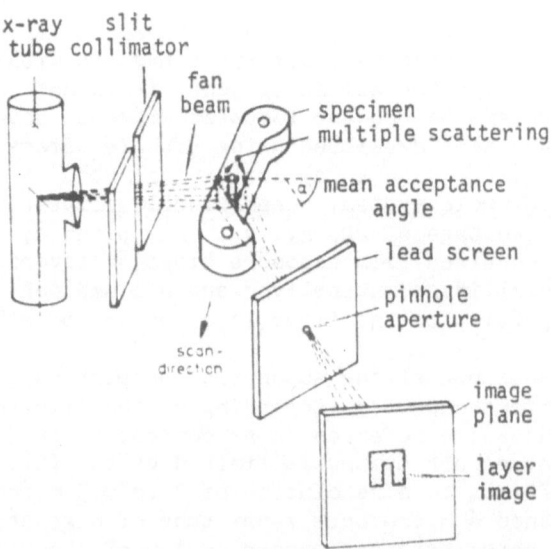

Fig. 4 Scatter imaging system, schematic. (From Strecker, ref. 6)

EMISSION METHODS

X-rays are generated through deceleration of electrons (Bremsstrahlung) and as high-energy level electron transitions within atoms (characteristic radiation). The latter provides us not only with a means of qualitative and quantitative analysis but also with information about the effects of electronic structure of atoms or compounds as gases, liquids or solids.

X-ray fluorescent analysis equipment has been available since the 1950's. Most chemical and materials laboratories have more or less advanced units for qualitative and quantitative analysis. In these units x-ray emission is stimulated by irradiation of the sample with high energy x-rays; the spectrum analysis is usually performed on a diffractometer using a Li crystal and a proportional counter.

The electron probe microanalyzer, now standard on most scanning electron microscopes (SEM), represents an exciting development for micro-chemical analysis. The sample in the SEM is bombarded by electrons. Back scattered and secondary electrons are used for image analysis. The x-ray emission, stimulated by the impinging electrons, is analyzed with a crystal detector system, or an x-ray energy analyzer (8, 9, 10). In this way "average" compositions of the surface area bombarded by the electrons can be obtained. It is also possible to determine the composition along a trace across the surface, as in elemental imaging of the field. The x-ray image quality, however, is inferior to that produced by secondary or back scattered electrons, because of the low number of photons generated per incident electron, (10^{-3} to 10^{-5}) and since x-rays come from much deeper regions of the sample.

ESCA, or electron spectroscopy for chemical analysis, represents one of the latest additions to the analytical tools for surface characterization. In one of the versions of ESCA photoelectron emission is excited by x-rays. The energies of the x-ray stimulated emissions, electrons and x-rays, are determined and information about the atom is obtained from

$$h\nu = E(binding) + E(kinetic)$$

Fig. 5 shows the schematic for such an instrument developed by DuPont, a photograph of the instrument and the representation of an experimental result.

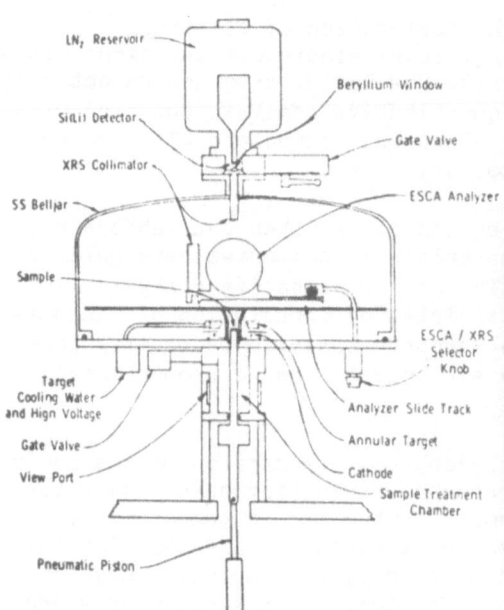

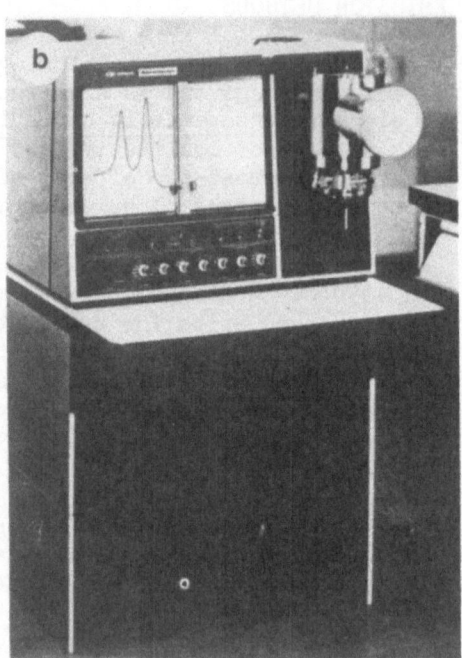

Fig. 5   a) Schematic, b) photograph and c) typical result of the DuPont ESCA - X-Ray analyzer.

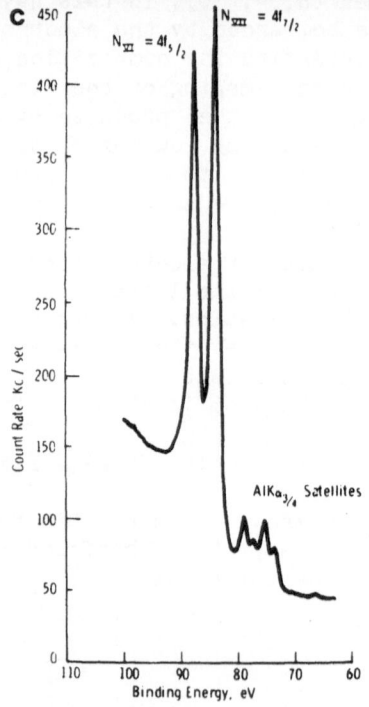

DIFFRACTION TECHNIQUES

It is hard to imagine today's state of our understanding of materials, of science, of nature, without the insights gained through x-ray diffraction. Confirmation of the structure of crystalline solids, structure determination , phase diagram studies, transformation studies, texture studies, stress measurement, fatigue an corrosion studies, particle size and shape studies in low angle scattering, electron density distributions, and structure determination for many non-crystalline substances (e.g. DNA) are among the many more where the use of x-ray diffraction has proven one, if not the essential ingredient. To do justice to only the recent developments in this field would go far beyond the scope of this brief overview. Moreover, excellent presentations on compound characterization, residual stress measurements, and topographic methods, all representing recent developments in non-destructive property characterization with x-rays, follow this overview. To set the stage brief reviews are presented here on a) the usefulness of phase composition characterization, b) Japanese developments for stress measurement and c) non-destructive characterization of fatigue damage.

A. PHASE COMPOSITION CHARACTERIZATION

Retained austenite measurements are a typical example of the determination of phase compositions by x-ray diffraction techniques. The composition fractions are calculated from the integrated intensities of the diffraction lines produced by the phases present. This is demonstrated below.

RETAINED AUSTENITE DETERMINATION

INTEGRATED INTENSITY =

$$I(hkl) = n_2^2 \; v \; m \; (LP) \; e^{-2m}(F.f)^2$$

```
n......number of cells in 1 cc.
v......volume exposed to x-ray beam
m......multiplicity of (hkl)
LP.....Lorentz polarization factor
e......Debye-Waller temperature factor
F......structure factor
f......atomic scattering factor
```

$$P = I(martensite)/I(austentite)$$
$$P = (Vol(m)/Vol(a))/G$$
$$Vol(austentite) = 1/(1 + G.P)$$
$$PCT(austentite) = (1 - carbide)/(1 + G.P)$$

$G = 2.50$ for Cr radiation and M(200), A(220) reflections.

Proper choice of radiation and diffraction lines is essential (11).
Difficulties can arise when the sample shows preferred orientation,
and when diffraction lines overlap. A lower detection limit of as
little as .2 percent retained austenite should be achievable with
most state-of-the art x-ray diffraction units. The equation given
above is, of course, applicable to any phase composition problem,
and certainly to the determination of martensite formed in stainless
steels, as a result of uniform or cyclic deformation. Fig 6. shows
the results of a study (12) in which the martensite content was measured during a fatigue test of AISI Type 304 stainless steel.
Martensite content appears to be a useful fatigue damage indicator.

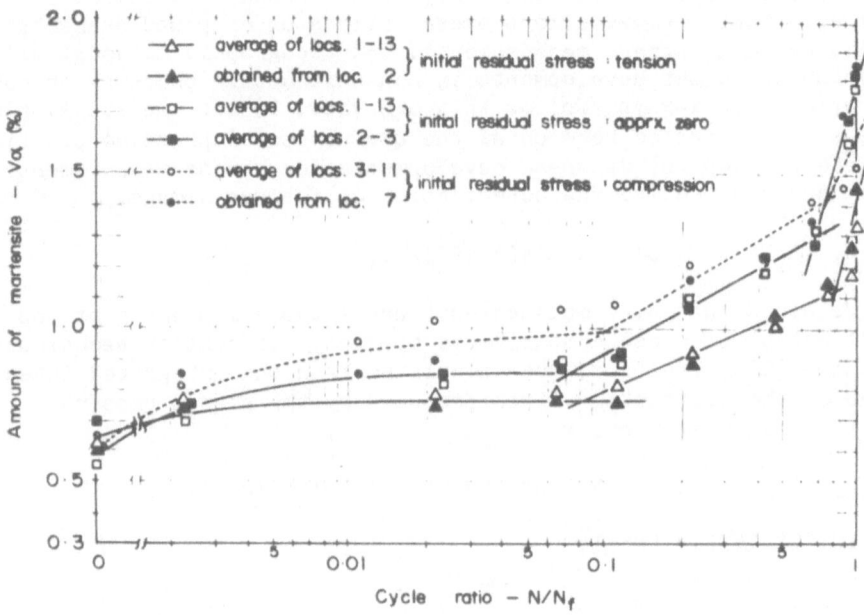

Fig. 6   Martensite content as a function of fatigue cycle ratio
(12).

B. STRESS MEASUREMENT

     The principle of stress measurement with x-rays is to measure
the lattice parameter in two or more directions and to calculate the
stress from these values in accordance to the theory of elasticity.
The well-known result is given below.

$$\sigma_x = \frac{E}{(1 + \nu)} \; \frac{1}{\sin^2 \psi} \cdot \frac{d_\psi - d_1}{d_1}$$

Potential problems involved in using these methods include elastic anisotropy, plastic anisotropy, diffraction line peak determination, texture, grain size, microstructure, surface condition and x-ray focusing geometry (e.g. 13). A "portable" Japanese unit, illustrated in Fig. 7, employs non-focusing parallel beam geometry which reduces the effect of positioning error considerably.

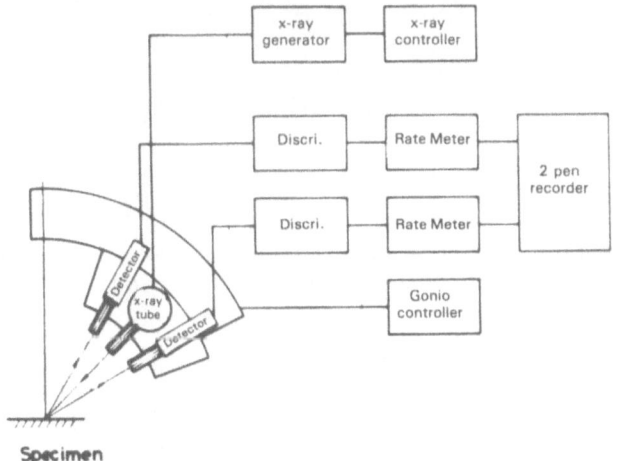

Fig. 7   Shimadzu x-ray stress analyzer.

Peak positions (references) are determined from the midpoint of the
half-intensity intercept (14). The introduction of position sensitive detectors coupled with computerized data reduction has led to
even faster and more versatile units (13).

C. FATIGUE DAMAGE CHARACTERIZATION

The desire to use x-ray diffraction techniques for fatigue damage characterization goes back to the 1930's when Regler studied
Austrian railroad bridges. Fig. 8 shows two of his diffraction patterns (15). Taira (16) devoted considerable efforts to this problem
during the past two decades. Most recently Kramer and Weissmann (17)
and our research group at Syracuse (12,18) have worked on this problem. While no unique, generally applicable method has been
developed, most researchers in the field agree that the potential
exists and that progress is being made. The most promising methods
are based on diffraction line shape analysis.

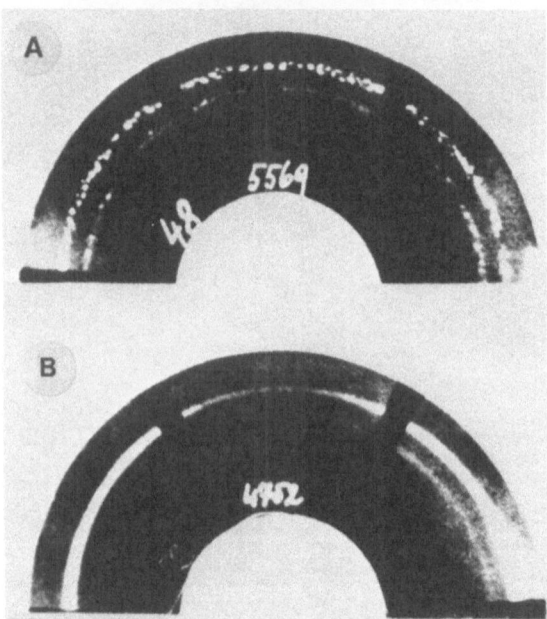

Fig. 8 Diffraction patterns obtained by Regler - A. undamaged sec-
        tion, B. fatigue damaged section.

Fig. 9 illustrates the typical line broadening observed with
increasing fatigue cycles. A plateau value is reached during the
early stages of cycling, with little change in half-width for most
of the life. Such a trend, observed on the surface, does not lend
itself as a predictor of failure. However, the level of the plateau
reached is related to the stress or strain amplitude and might well
serve as an estimator of the service stresses experienced.

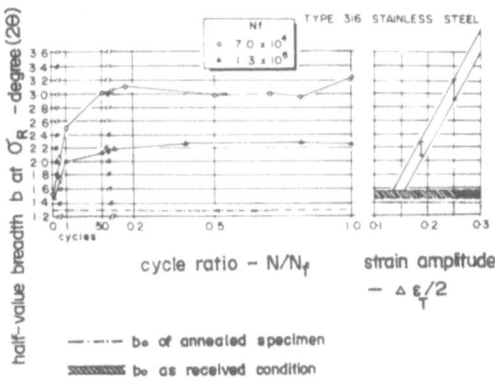

Fig. 9 Diffraction line width as a function of fatigue cycle ratio
and strain amplitude.

Fig. 10 shows this relationship between diffraction line width and
strain range for a group of samples subjected to different treatments. It appears that the effects of these treatments (prestrain-
ing and strain aging) disappear during the early fatigue cycles.

    Following these experiments, we are now conducting a detailed
diffraction line analysis study to measure fatigue damage in Al
7050. The experimental setup, utilizing a position sensitive detector and computerized Warren-Averbach data analysis, is illustrated
in Fig. 11. The results indicate that half-width is not as good a
damage indicator as the ratio of microstrain-to-particle size,
illustrated in Fig 12. This ratio, which can only be calculated
after the diffraction line broadening has been separated in its two
components, microstrain and particle size, can be shown to be proportional to the dislocation density (18). Moreover, the effect is
more pronounced for the (200) type reflections than for the (111)
type reflections. This appears to be another manifestation of the
selective nature of x-ray diffraction. For the focusing geometry

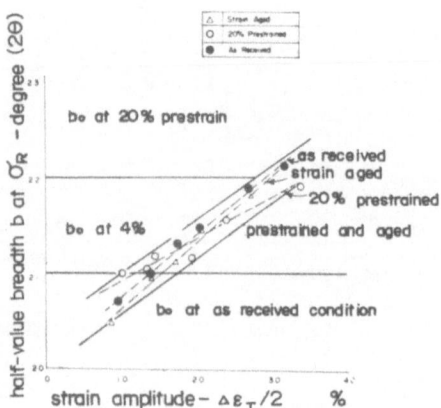

Fig. 10 Diffraction line width as a function of strain amplitude.

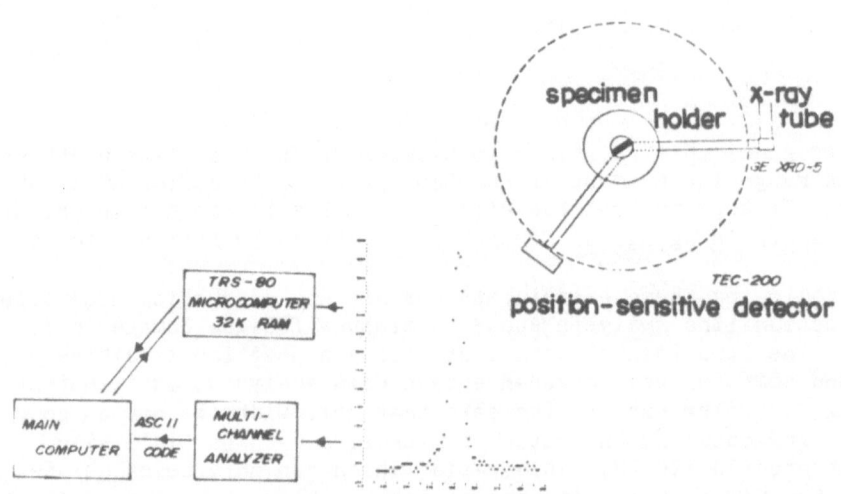

Fig. 11 Experimental setup for fatigue damage studies, schematic.

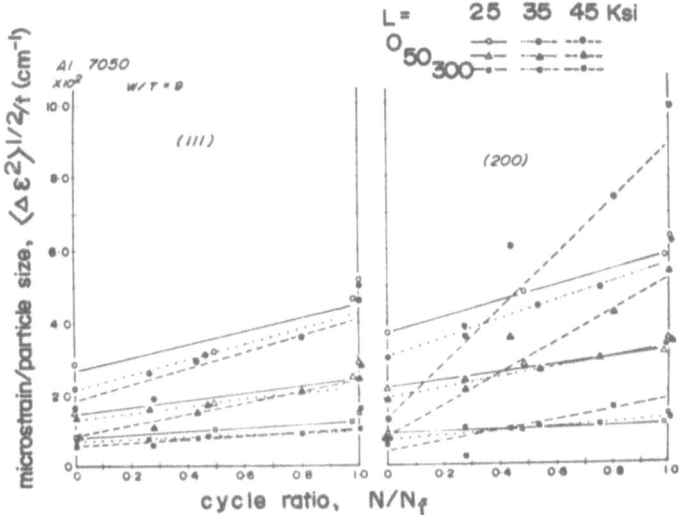

Fig. 12   Microstrain/particle-size as a function of cycle ratio
(surface).

used the (hkl) reflections are obtained from grains with (hkl)
planes parallel to the surface. It turns out that for the middle
region of wide bend specimens, plane strain conditions, the resolved
shear stress is approximately 50 percent higher in grains that have
their (200) planes parallel to the surface than for planes that have
their (111) planes parallel to the surface (18).

     The effect also varies with depth, as illustrated in Fig. 13.
The microstrain/particle size ratio decreases from the surface
inward, to a minimum followed by an increase to a plateau, the level
of which also depends on the amount of fatigue life exhausted, as
expressed by the cycle ratio N/N . Similar effects were observed by
Kramer and Weissmann on Al 2024 (17). These findings suggest that
the microstrain/particle size ratio obtained from some depth below
the surface should be a better damage indicator than its value
obtained from the surface. That is indeed the case, Fig 14. However, the results still seem to be somewhat dependent on the stress
amplitude. It appears that using the ratio of interior to surface
values of these microstrain/particle size ratios will reduce, if not
eliminate the stress amplitude dependency. Such a graph is presented
in Fig. 15. The initial value depends strongly on the surface prep-

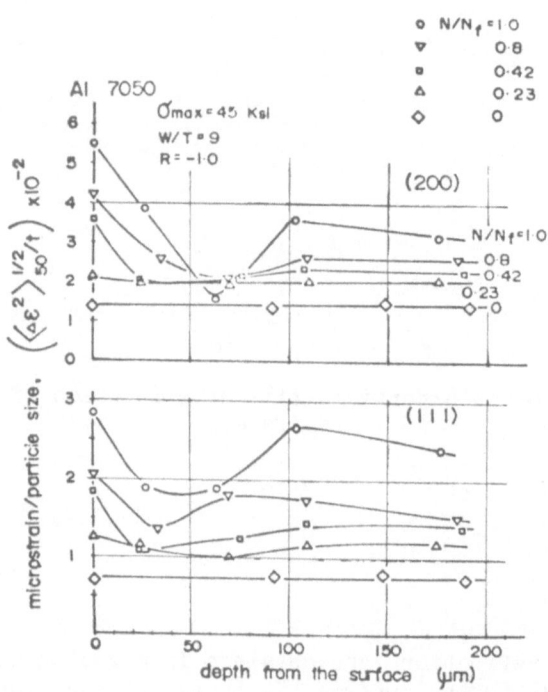

Fig. 13 Microstrain-to-particle-size ratio as a function of depth
below the surface, (200) and (111) reflections.

aration. It is expected to be near unity for an electropolished
surface, as is the case for our tests. The rising branch beyond a
cycle ratio of 0.5 appears to be general. The requirement to obtain
results from regions 0.1 to 0.2 mm below the surface can be met
non-destructively for Al and other low atomic number alloys by using
hard radiation (e.g. Mo). It is more difficult to achieve for ferrous materials. More work is required towards the development of a
practical system for fatigue damage characterization based on these
promising results.

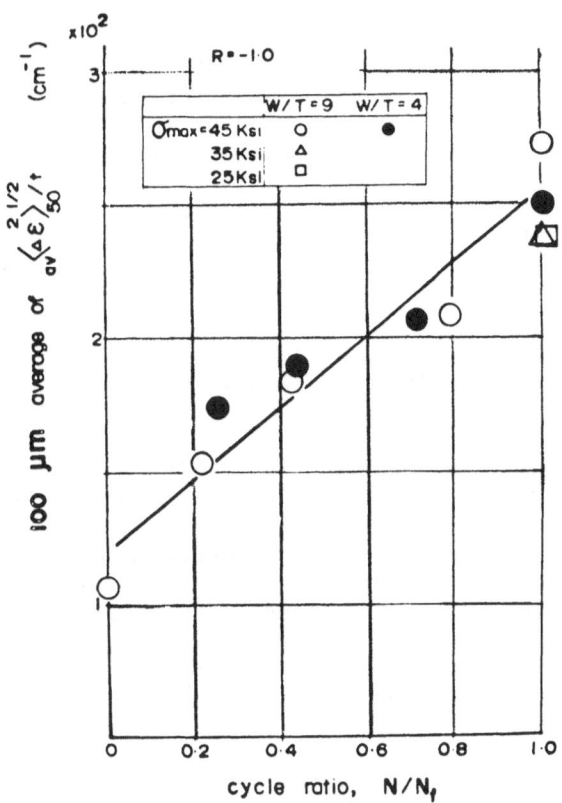

Fig. 14   Microstrain-to-particle-size ratio averaged over 100 μm as
a function of cycle ratio.

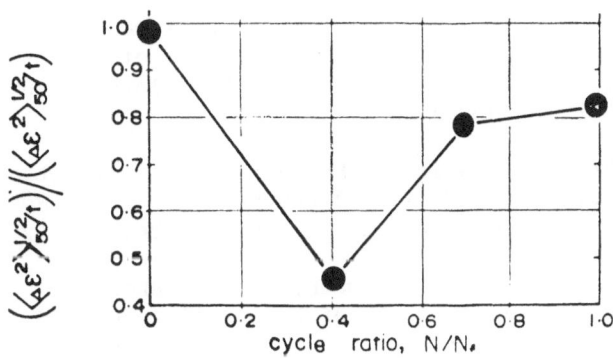

Fig. 15   Interior to surface ratio of microstrain-particle-size
ratio as a function of cycle ratio.

ACKNOWLEDGEMENTS

      The author wishes to thank Dr. Yoshiki Oshida for his assist-
ance with the manuscript preparation. Special thanks are due to Dr.
Irvin Kramer for his advice and guidance in connection with the
fatigue studies, the Naval Air Systems Command and the David Taylor
Naval Ship Research and Development Center for support of the
research program, and to Drs. Y. Oshida and A. Wu, who conducted the
fatigue damage studies.

REFERENCES

(1) C. G. Darwin: "The Theory of X-Ray Reflection". Phil. Mag., 27,
1914, pp.315-333.

(2) C. G. Darwin: "The Reflection of X-Rays from Imperfect
Crystals". Phil. Mag., 43, 1922, pp.800-829.

(3) W. A. Ellingson and H. Berger: "Three-dimensional Radiographic
Imaging". in "Research Techniques in Nondestructive Testing vol.IV"
ed. by R. S. Sharpe, Academic Press, 1980, pp.1-38.

(4) G. T. Herman: "Image Reconstructionfrom Projections - The Funda-
mentals of Computerized Tomography". Academic press, 1980, pp.26-39.

(5) P. Burstein, R. Mastronardi and T. Kirchner: "Computerized
Tomography Inspection of Trident Rocket Motors - A Capability Demon-
stration". Materials Evaluation, 40, 1982, pp.1280-1284.

(6) H. Strecker: "Scatter Imaging of Aluminum Castings Using an
X-Ray Fan Beam and a Pinhole Camera". Materials Evaluation, 40,
1982, pp.1050-1056.

(7) P. S. Ong: "X-Ray Microscopy". in "Techniques of Metals Research
vol.II, Techniques for the Direct Observation of Structure and
Imperfections, Part 1" ed. by R. F. Bunshah, Interscience
Publishers, 1968, pp.159-196.

(8) J. J. Burke and V. Weiss: "Characterization of Materials in
Research - Ceramics and Polymers" proceedings of the 20th Sagamore
Army Materials Research Conference, Syracuse University Press, 1975.

(9) K. F. J. Heinrich, D. E. Newbury and H. Yakowitz: "New Techniques for the Surface Analysis of Nonmetallic Solids". Ref.(8),
1975, pp.73-102.

(10) H. K. Herglotz: "Characterization of Polymers by Unconventional
X-Ray Techniques". Ref.(8), 1975, pp.103-136.

(11) J. Durnin and K. A. Ridal: "Determination of Retained Austenite Steel by X-Ray Diffraction". J. of Iron and Steel Inst., 1968, pp.60-67.

(12) V. Weiss, Y. Oshida and A. Wu: "A Note on Fatigue Damage Assessment by X-Ray Diffraction Techniques for a 304L Stainless Steel Specimen". J. of Nondestructive Evaluation, 1 (3), 1980, pp.207-213.

(13) E. Kula and V. Weiss: "Residual Stress and Stress Relaxation". Proceedings of the 28th Sagamore Army Materials Research Conference, Plenum Press, 1982.

(14) J. J. Burke and V. Weiss: "Nondestructive Evaluation of Materials" Proceedings of the 23rd Sagamore Army Materials Research Conference, Plenum Press, 1979.

(15) H. K. Herglotz: "A Historical Example of Fatigue Damage". Ref.(14), 1979, pp.55-67.

(16) S. Taira and K. Kamachi: "Detection of Fatigue Damage by X-Rays". Ref.(14), 1979, pp.21-54.

(17) R. N. Pangborn, S. Weissman and I. R. Kramer: "Determination of Prefacture Fatigue Damage". David W. Taylor Naval Ship Research and Development Center, DTNSRDC - 80/006, Jan. 1980.

(18) A. Wu: "Accumulated Fatigue Damage Characterization by Nondestructive Methods". Ph.D. Dissertation, Syracuse University, Jan. 1983.

APPLICATION OF A POSITION SENSITIVE SCINTILLATION DETECTOR TO NONDESTRUCTIVE X-RAY DIFFRACTION CHARACTERIZATION OF METALLIC COMPONENTS

Clay Olaf Ruud

Materials Research Laboratory
The Pennsylvania State University
University Park, Pennsylvania

ABSTRACT

A unique instrument for the x-ray diffraction (XRD) characterization of crystalline materials has been developed and is being applied to a number of studies of metallic components. The instrument is based upon a unique position sensitive scintillation detector (PSSD) for x-rays. The methodology for the application of this instrument to the single-exposure technique (SET) of x-ray stress measurement is described. A number of applications of the device to residual stress measurement and qualitative cold work damage assessment in metallic components are discussed and data from these applications presented. Also, the application of the PSSD instrument to texture measurement is described.

INTRODUCTION

X-Ray Diffraction (XRD) characterization of polycrystalline materials relies upon the ability of this method to accurately measure the interatomic planar spacing of the material under study. In the most elementary form the Bragg equation

$$n\lambda = 2d \sin\theta$$

describes the relationship between the various parameters, where n is normally unity, λ is the wavelength of the interrogating radiation, d is the interatomic spacing, and θ is the Bragg angle. Characterization in XRD is usually accomplished by selecting a monochromatic x-radiation and measuring θ and thus obtaining the interplanar spacing, d. The mean value of d may be used for the

21

measurement of elastic strain, and therefore applied and residual
stress [1,2]; preferred orientation, or texture [2,3]; phase
identification [4]; and quantitative phase analysis, including
retained austenite determination [3,4]. The distribution of the d
values about the mean is indicative of cold work damage [2];
crystallographic lattice parameter variation, including elemental
variation [4]; and particle size [4].

In most cases each of the aforementioned characteristics may
be measured independently, although in certain instances the
isolation of a single property may require rather sophisticated
analysis [5].

BACKGROUND

Extensive industrial application of XRD as a characterization
tool has been realized with respect to phase identification and
phase analysis, as is manifest in the thousands of computer-
automated x-ray diffractometers which are being used in routine
analysis. However, the application of XRD to residual stress,
texture, cold-work damage, and in some cases phase analysis has
been greatly hampered by restrictions on specimen size and shape
dictated by the instrumentation available.

There are several basic types of instrumentation used for
material analyses and characterization by XRD, but the most popular
by far is the Bragg-Brentano para-focussing goniometer. Several
manufacturers in the free world are marketing diffractometers based
upon this design which has remained fundamentally the same for
several decades. Essentially this XRPD diffractometer consists of
a mechanical turntable capable of positioning a powder specimen at
various known θ angles with respect to an incident x-ray beam, and
a detector scanning at twice the angle that the specimen is
rotating. The scanning is accomplished in such a way that the
detector is in a position to intercept the diffracted x-ray beam
whenever the Bragg condition is met. There are several excellent
descriptions of this device, including Barrett [3], Cullity [2],
and Klug and Alexander [4]. This scanning detector device has been
the workhorse of XRD analysis since the late 1940's and presently
the most modern versions are computer automated.

In the early 1970's however the technology for x-ray detectors
which did not have to be moved, i.e., mechanically scanned, to
detect angle of Bragg x-ray diffraction evolved to produce reliable
devices with sufficient spatial resolution to be suitable for
application to XRPD. These so-called position sensitive detectors
(PSD) found many applications where they were substituted for the
x-ray detector normally used on the scanning diffractometers [6].
By and large these detectors have been based upon the same principle

as proportional x-ray detectors and thus were named position
sensitive proportional detectors (PSPD). Attempts were also made
to use these detectors in XRPD geometries other than the Bragg-
Brentano diffractometer [7,8]. However, although the attempted
designs were innovative and scientifically sound, many practical
shortcomings prevented their general acceptance and the PSPD has
essentially remained as an accessory to the Bragg-Brentano scanning
goniometer.

Recognizing the potential of the position sensitive detector
in XRD and the shortcomings of the PSPD to industrial applications
of materials characterization, this author and Professor C.S.
Barrett of the University of Denver set out to design a more
versatile position sensitive detector for certain XRD applications.
The goal was to develop a device that could be made spatially
efficient in the vicinity of the x-ray beam and specimen and that
possessed more than a single x-ray sensitive surface. Secondary
goals were to provide better x-ray peak position resolution, a
broader x-ray flux detection range, and better reliability in
industrial environments. The result was the Ruud-Barrett position
sensitive scintillation detector (PSSD). Although the fundamental
design was conceived in the early 1970's, it was not until 1980
that an adequate device was built and proven reliable. The
following describes the principles and advantages of this x-ray
detector and cites examples of its application to XRD materials
characterization.

DESCRIPTION OF INSTRUMENTATION

Principle

The principle of the operation of the PSSD is based upon the
coherent conversion of the diffracted x-ray pattern into an
optical signal, i.e., light; the conduction of this light signal,
over several linear centimeters of two coherent, flexible fiber
optic bundles; the amplification of this signal by electro-optical
image intensification; the electronic conversion of the signal; and
the transfer of the electronic signal to a computer for refinement
and interpretation, see Fig. 1. In the present PSSD detector
design that is being marketed,* a film of scintillation material
has been adhered to one end of each of two flexible fiber optic
bundles, 3 by 13 mm in cross-section. This scintillation coated
end of the fiber bundles is the x-ray sensitive area which is
located in the vicinity of the x-ray source and specimen. The
length of the bundles vary from tens of centimeters to two meters,

---

*The PSSD is being manufactured and marketed by Denver X-Ray
 Instruments Inc.

depending upon the x-ray diffraction application. The other,
non-x-ray sensitive cross-sectional end of the flexible fiber
optic bundles are optically coupled to a proximity focussed image
intensifier (II) which amplifies the visible light version of the
x-ray pattern. The light signal then having been amplified several
thousand times is transferred, via a rigid fiber optic couple, to
two parallel self-scanning, linear silicon diode arrays where the
light pattern is converted to an analogue electronic signal. This
signal is then sent to an analogue-to-digital converter, then on to
a PDP 11/03 computer where it is stored, refined, and interpreted.
The image intensifier and the diode arrays are enclosed in a cube
which is approximately 50 x 50 x 50 mm. This enclosure is hermetically sealed and the temperature sensitive diode arrays are cooled
by a solid state cooler also contained in the electronics cube. A
recent paper describes the PSSD in more detail [9].

Advantages

     The present version of the Ruud-Barrett PSSD shows a linear
(x-ray peak position) resolution of approximately fifty micrometers
over its entire length along with a photon conversion efficiency of
about fifteen percent. This compares well with very best
resolution of about sixty micrometers at the center of the detector
for a few PSPD's and one to two hundred for most of those used in
x-ray diffraction. Furthermore, the PSPD's usually show poorer
resolution away from the center of their x-ray sensitive area.
What is meant by linear resolution is the detector's ability to
accurately identify the point along its x-ray detection length at
which an x-ray photon is incident. The resolution value given is
essentially the degree of uncertainty associated with the reported
position of an x-ray peak. The fact that the x-ray sensitive
surface is a scintillation coated glass bundle provides for the
option that the surface shape is not restricted to a long, thin
plane. It can be ground and polished to other configurations,
e.g., a circle to conform to the radius of the focussing circle in
a special x-ray diffraction application. Also, since the x-ray
sensitive surface is a simple fiber end, means to conform the PSSD
to any number of x-ray focussing geometries are easily fabricated.

     Proportional detectors, whether they are position sensitive or
not, inherently are capable of processing only one x-ray photon at
a time and there is a time delay, or dead time, during which no
other photon can be processed. The dead time is usually on the
order of a few microseconds and the maximum x-ray flux in which a
detector is functional is a few ten-thousand counts per second over
its entire x-ray sensitive surface. This means that under conditions of application where rapid data accumulation requires very
high x-ray fluxes, a proportional detector would be ineffective.
The scintillation detector described herein has no dead time period
and therefore is effective over a very broad range of x-ray flux

intensities. For example, it has been shown to be applicable to
very faint diffracted x-ray peaks as might be provided with a one
to two hundred watt x-ray tube source; on the other hand, it has
been shown to be operable in the direct x-ray beam of an x-ray
source operating at over a kilowatt.

     The Ruud-Barrett PSSD is inherently rugged. Since the
electronics cube is hermetically sealed and the flexible fiber
optics enclosed in a heavy plastic envelope, the PSSD is operable
in dusty, humid industrial environments. Furthermore, temperature
ranges between 4 and 43°C (40 and 110°F) have little effect on the
detector's operation, and it shows no loss of linear resolution or
stability under moderate mechanical vibration. However, the
computer to which the detector is interfaced is subject to the
normal environmental limits of such instrumentation, but the
computer can be operated remotely from the detector cube. It
should also be noted that the cube can be placed remotely from the
x-ray source since the flexible fiber optics can be two meters
long; this provides for application of the PSSD in high electrical
and/or magnetic fields.

     The detector previously described has been interfaced with at
least three different configurations of x-ray sources, all filament
gun tubes, in its application in the author's laboratory. These
have included copper and chromium target Dunlee, Amprex, and Kevex
x-ray tubes. The latter provided an extremely compact measurement
head several times smaller than that provided with the Amprex tube
with the same PSSD, and at least an order of magnitude smaller than
any other type of portable XRD device presently marked, see Fig. 2.
This miniature XRD instrument is described in a recent publication
[10]. This small device promises unprecedented portability and
versatility, especially in industrial application of XRD materials
characterization.

APPLICATIONS

Residual Stress Techniques

     The position sensitive scintillation detector (PSSD) described
in this paper has been applied mainly to x-ray residual stress
measurement thus far. This is not because its application to other
x-ray diffraction uses is more difficult, but simply because that
is the first application which was attempted and there has been a
great demand for stress data generation. There are three basic
techniques for obtaining stress (strain) readings from a polycrystalline solid by the XRD method. The three are called the sin-
square-psi technique ($\sin^2\psi$), the double-exposure technique (DET)
and the single-exposure technique (SET). It must be noted that in
order for the x-ray method to be used to measure absolute residual

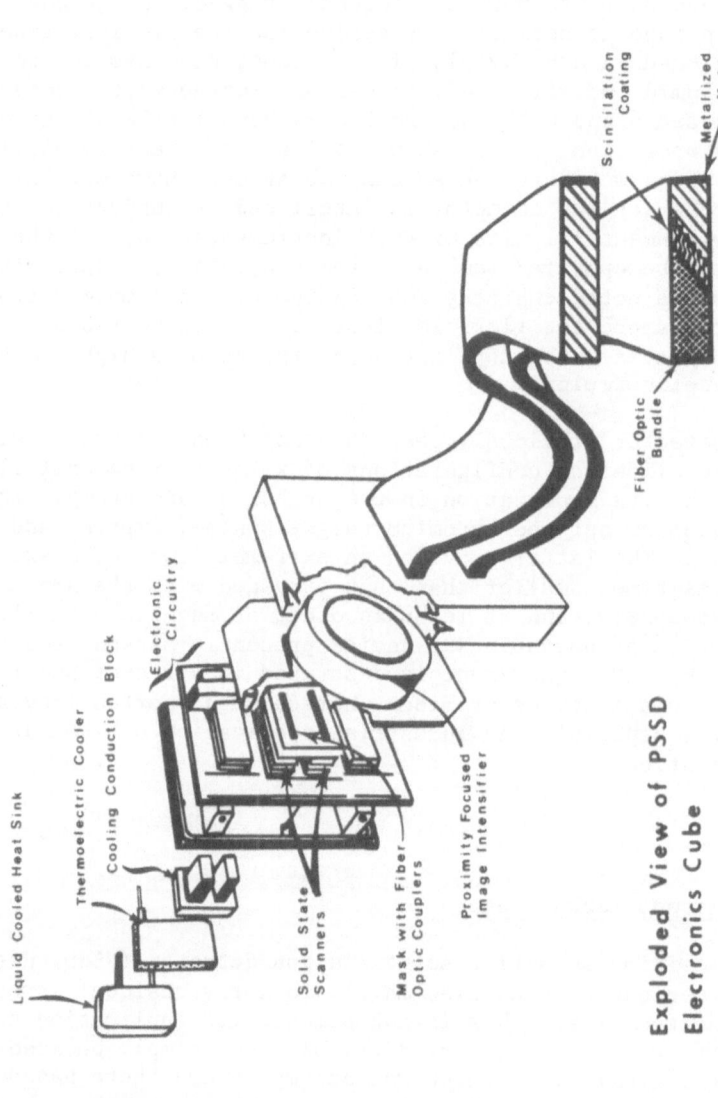

Liquid Cooled Heat Sink

Thermoelectric Cooler

Cooling Conduction Block

Electronic Circuitry

Solid State Scanners

Mask with Fiber Optic Couplers

Proximity Focused Image Intensifier

Scintillation Coating

Metallized Mylar

Fiber Optic Bundle

Exploded View of PSSD Electronics Cube

Fig. 1 Exploded view of Ruud–Barrett position-sensitive scintillation detector.

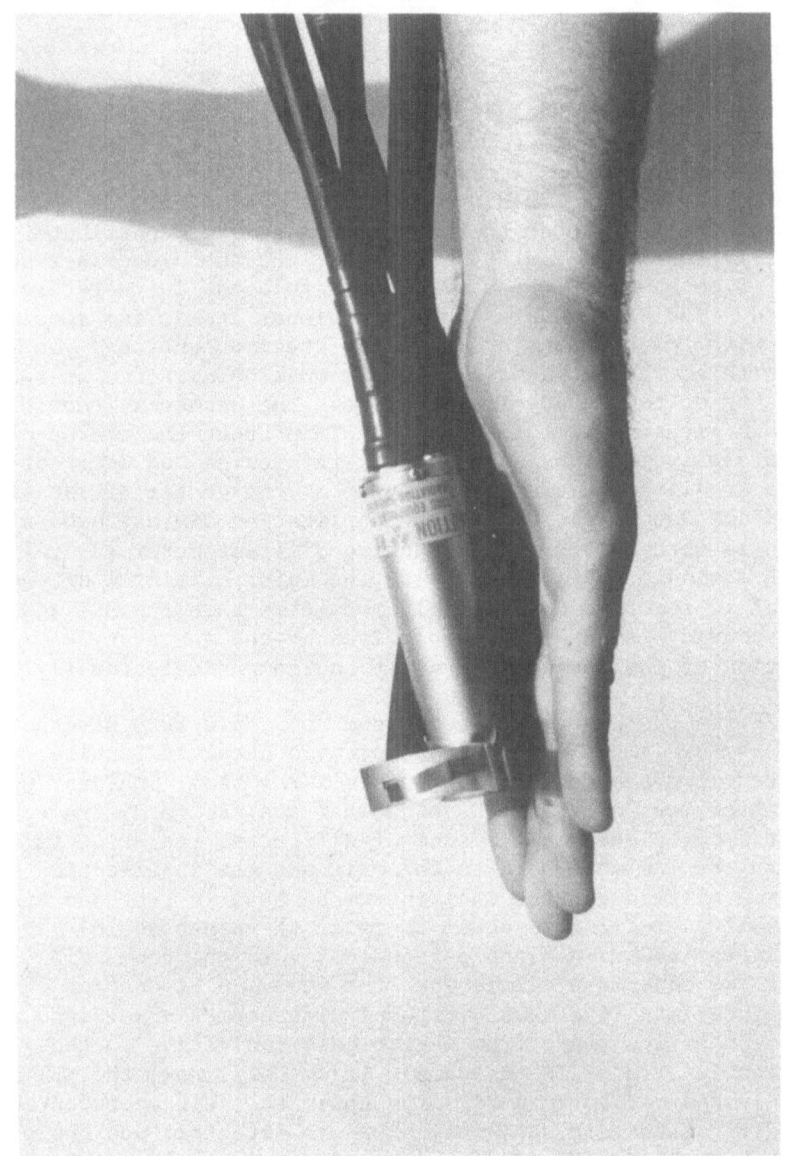

Fig. 2 Photograph of miniature x-ray stress measurement head, including the PSSD detection surfaces mounted in the single-exposure technique configuration.

or applied stress, the d spacing of planes of the same crystallographic character (i.e., Miller indicies) must be measured at two
different orientations with the polycrystalline solid's surface.
What is meant by absolute measurement is that no previous or
subsequent measurement of that metal piece need have been obtained
in a zero-stress, or some other known stress, condition for
calibration.

The sin-square-psi technique is usually recommended where
certain characteristics (e.g., texture, severe stress gradients)
are likely to degrade the accuracy of the x-ray stress measurement.
Furthermore, the several readings at a series of different
attitudes of the x-ray beam to the specimen surface contribute to
statistically superior accuracy. However, the technique increases
the time for each stress measurement many fold and is therefore
seldom used. The DET measures the interplanar atomic (d) spacing
of planes with two orientations with the specimen surface. To
accomplish this, the incident x-ray beam must be directed at two
different angles to the specimen surface. The necessary reorientation is done either by rotating the specimen about the center of
the area of analysis or by moving the x-ray source and detector
through an arc with the area of analysis at the center of rotation.
This technique was developed for the application of instruments
with a single detector, e.g., a scanning diffractometer [2,3,4], or
those with a single PSD surface [8]. The major criticism of the
DET is that it requires considerably more time than the SET and
that its accuracy is severely compromised by the necessary
reorientation of the specimen or x-ray source and detector [1,11].

The SET is the preferable technique for rapid XRPD stress
measurements that are convenient and provide accurate results. The
SET is based upon the fact that a single x-ray beam, incident upon
at least a few tens of thousands of powder grains, is diffracted at
a constant θ angle such that a cone of diffracted radiation is
formed with the incident beam as the axis and the intersection of
the specimen surface and the incident x-ray beam is the cone apex.
The cone half angle, 2η, is equal to twice the Bragg angle
subtracted from 180°. A plane perpendicular to the cone axis
intercepts the cone in a circle when the specimen is unstressed;
but if the specimen is stressed, the circle becomes an ellipse and
the deviation of its shape from a circle is indicative of the
stress (elastic strain) in the sample [2]. Until now, the SET has
been largely restricted to x-ray film apparatus, since conventional
scanning diffractometers have only a single detector; and the size
of the position sensitive proportional detectors (PSPD) circumvents
more than one of them being conveniently located near the x-ray
source [8]. However, the position sensitive scintillation detector
described herein is well suited to SET application, the major
reasons being that it has two detection surfaces that are spatially
efficient since the active detection surfaces cover over 95% of its

fiber optics cross-section, see Fig. 3. The ability of the
Ruud-Barrett PSSD to apply the SET then provides a great advantage
for its use in x-ray stress measurement over other detectors and
diffractometers. However, this PSSD can also be applied to the
sin-square-psi and double-exposure techniques, usually with more
proficiency and rapidity of data collection and equal accuracy as
other XRD instrumentation.

Stress Measurement Application

     Two x-ray stress measurement instruments have been built
utilizing the PSSD described herein. One has been mounted on a
horizontal rack with a carriage capable of translating the stress
measuring head in two orthogonal directions. This instrument is
used for mapping surface stresses in large specimens. It has been
applied to investigating stress corrosion cracking problems in
bronze naval ship propellors and Inconel tubing, mapping residual
stress fields in heavy steel plate weldments, and evaluating
surface treatments for fatigue life improvement of martensitic
stainless steel. Calibration results indicate that it is accurate
to less than 7 MPa (1 KSI) with a standard deviation of 7 MPa
(1 KSI) for ferritic steels. Data collection times for ferritic
steels range from 1 to 10 seconds and the time required for one
complete reading, i.e., from data collection through stress
calculation and hard copy printout, is about 90 seconds. This
indicates that upwards of 30 readings can be made in one hour. Its
application to heavy plate weldments (450 to 500 kg, 1000 to 2000
lbs) has demonstrated its use on large specimens, and its use on
nickel-base tubing has demonstrated its unique ability to resolve
stress gradients over very small areas with unprecedented data
collection speed and accuracy.

     A second PSSD was incorporated in a residual stress analyzer
designed to be inserted into pipe with inside diameters as small as
230 mm (9 inches) [12]. This instrument has been applied to
nondestructive residual stress measurement on the inside of girth
welded austenitic stainless steel piping. No other XRD stress
measurement instrumentation approach could provide an instrument
capable of being inserted into the piping; and the accuracy and
precision which has been demonstrated are unprecedented for XRD
stress measurement on stainless steel. For example, calibration
curves of applied stress versus XRD measured stress show standard
deviations of less than 20 MPa (3 KSI).

Cold Work Assessment Application

     A number of other applications of the two PSSD based
instruments described above have been investigated. One is to
measure the amount of cold work induced in metallic components by
shot peening. Cold work in metals is indicated by x-ray peak

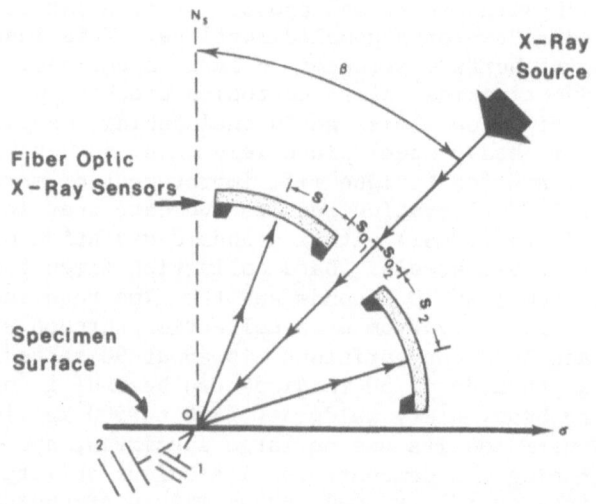

Fig. 3 Schematic drawing of single-exposure technique (SET)
showing location of the PSSD x-ray sensing surfaces. N_S is
the specimen surface normal; β is the angle between the
incident x-ray beam and N_S; S_1, S_{O1}, S_{O2} and S_2 are
related to the angle of diffraction from planes 1 and 2,
respectively; O is the intersection of the incident x-ray
beam and the specimen surface.

broadening, as well as residual stress inducement, and these two quantities can be measured simultaneously with the PSSD instrument. However, the change in residual stress in the surface of the material due to shot peening does not provide an adequate measure of shot peen effectiveness. This is because the stresses increase to near the maximum stress induced by an infinite amount of shot peening almost immediately, then remains essentially constant through continued shot peen treatment. On the other hand, the breadth of the x-ray peak is indicative of microstresses, dislocation density, etc. and increases gradually with the degree of shot peening, see Fig. 4. This peak breadth parameter could well provide an accurate tool for the measurement of the degree of shot peening, and data accumulation with the PSSD is on the order of a few seconds.

Peak breadth also has been used in conjunction with residual stress measurement to ascertain the degree of annealing afforded a cold worked surface during welding. Figure 5 shows two pairs of peaks representing single-exposure technique (SET) diffraction of Chromium K-beta radiation from the (311) planes of austenitic stainless steel [13]. The left pair is broad due to cold work induced by machining of the stainless steel surface before welding. The right pair shows narrower peaks resulting from the annealing afforded by the heat of welding. Peak breadth provided an indication for the thermal effect of welding on this stainless steel material.

Texture Evaluation Application

In most practical applications of XRD measurement of residual stresses, it can be demonstrated that a plot of d or θ versus the square of the sin of an angle related to the orientation of the measured crystallographic planes to the specimen surface, usually designated psi, will provide a straight line [1]. However, it has been noted that often when preferred orientation (texture) is evident in the specimen under study, this sin-square-psi plot does not provide a straight line [14]. Thus, such a plot is often used to detect the presence or degree of texture in a material, especially in conjunction with XRD residual stress measurement.

Figures 6 and 7 show sin-square-psi plots from two Inconel 600 tubes, reportedly manufactured by the same process. Little deviation from linearity is evident in the data from the slightly textured tube shown in Fig. 6, while the Fig. 7 data shows an extreme degree of texture. The data shown in each of these figures was acquired in less than an hour, with little or no special fixturing using the PSSD described. This same task, using a conventional diffractometer, would have required at least a day.

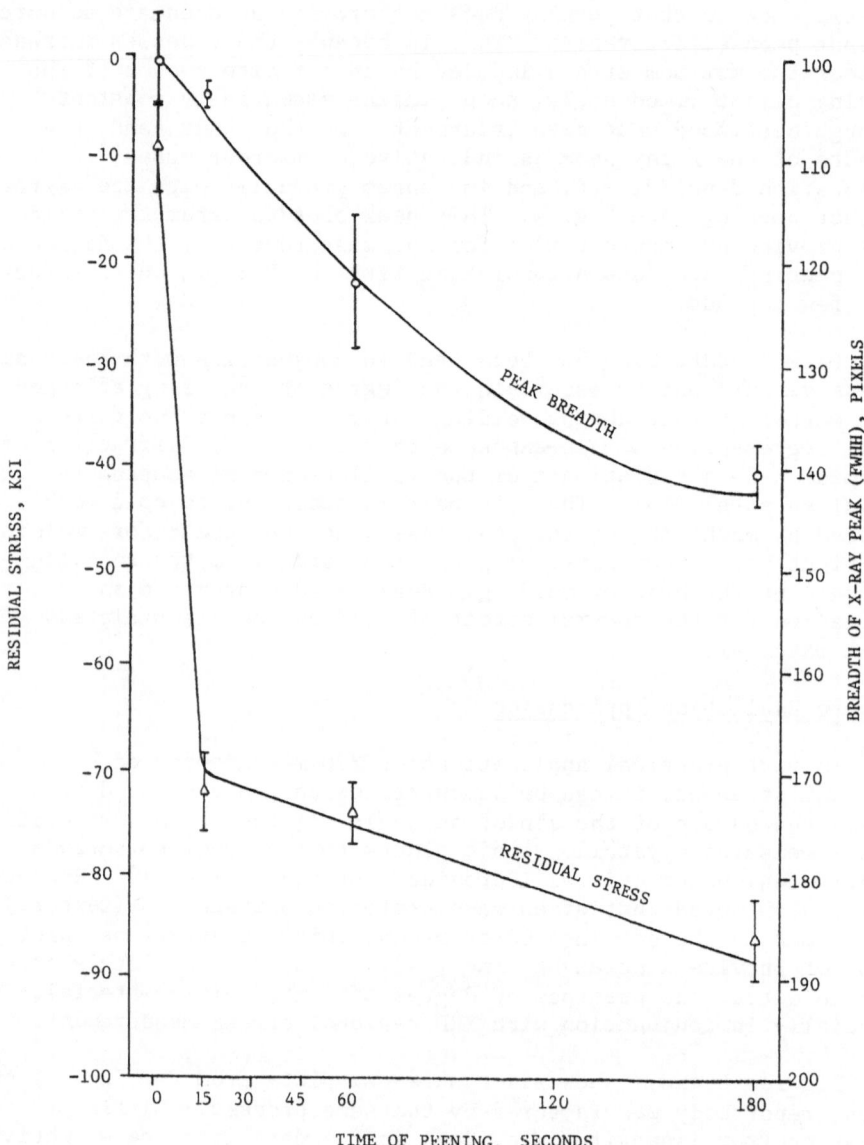

Fig. 4 Plot of residual stress, or breadth of the x-ray peak,
versus the time (or degree) of shot peening for high
strength steel gears.

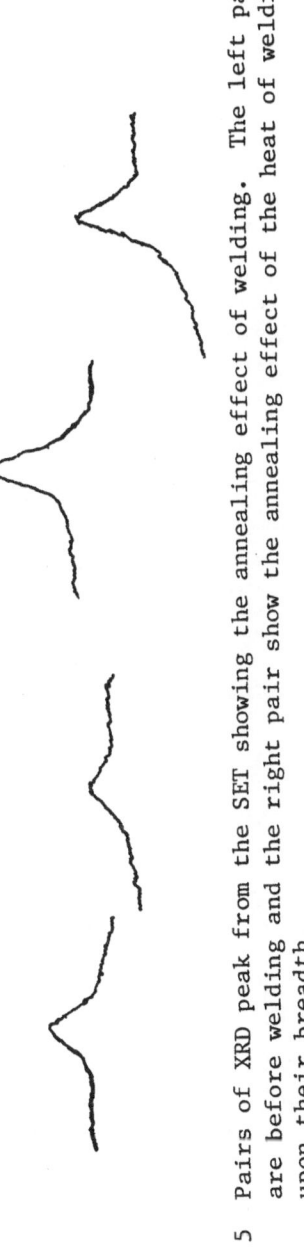

Fig. 5   Pairs of XRD peak from the SET showing the annealing effect of welding. The left pair are before welding and the right pair show the annealing effect of the heat of welding upon their breadth.

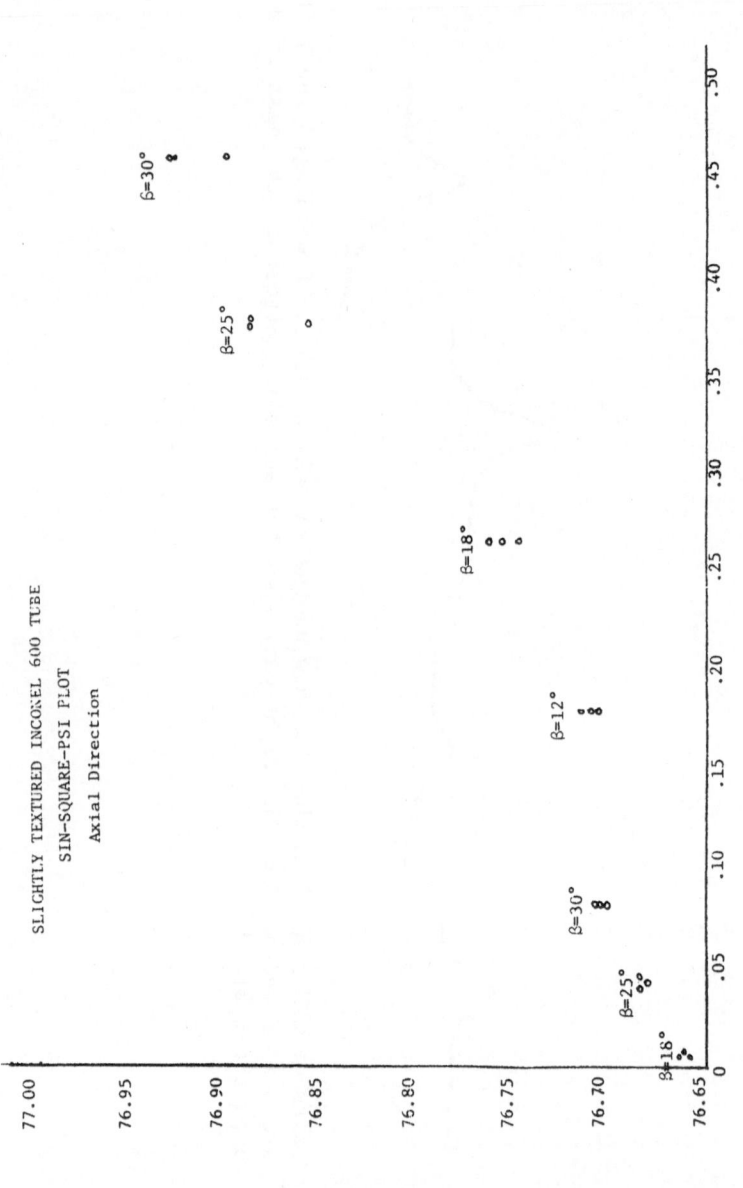

Fig. 6 Sin-square-psi versus Bragg angle plots for a slightly textured Inconel 600 tube.

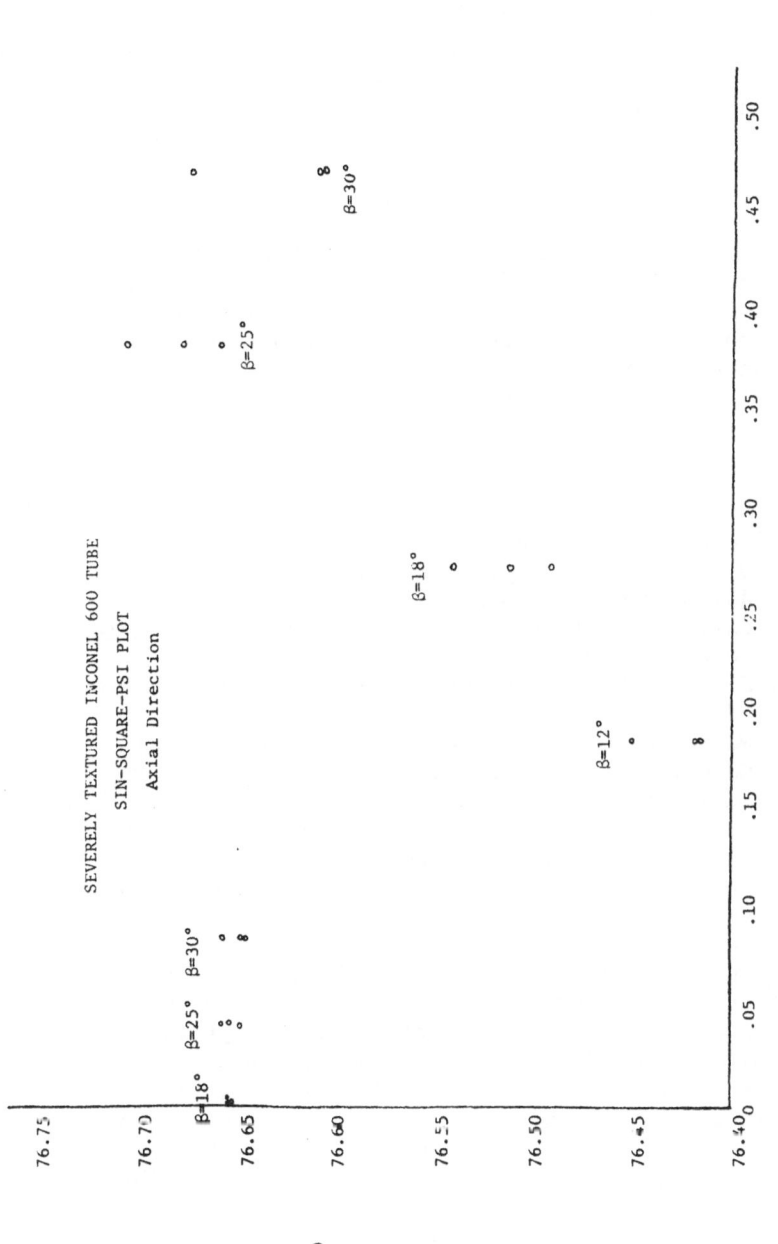

Fig. 7 Sin-square-psi versus Bragg angle plots for a severely textured Inconel 600 tube.

SUMMARY

Most important in the consideration of the applications described above is that the measurements can be done rapidly enough to be used to characterize material on line in manufacturing processes. Further, the instrument described is compact and rugged enough to be deployed in a manufacturing environment or be used for inservice inspection.

CONCLUSION

The advanced instrument described has been shown to be capable of providing several types of materials characterization measurements rapidly, accurately, and with sufficient ease that these measurements could be performed in industrial manufacturing and/or inspection processes.

REFERENCES

1. SAE, "Residual Stress Measurement by X-Ray Diffraction: SAE J784a," Soc. of Auto. Eng. Inc., Warrendale, PA.
2. B. D. Cullity, Elements of X-Ray Diffraction, 2nd Ed., Addison-Wesley, Reading, MA 1978.
3. C. S. Barrett and T. B. Masalski, Structure of Metals, 3rd Ed., J. Wiley and Sons, NY, 1966.
4. H. P. Klug and L. E. Alexander, X-Ray Diffraction Procedures, 2nd Ed., J. Wiley and Sons, NY, 1974.
5. B. E. Warren and B. L. Averbach, J. Appl. Phys. 21, 595 (1950).
6. H. E. Göbel, "A New Method For Fast XRPD Using a Position Sensitive Detector," Adv. in X-Ray Anal., Vol. 22, pp. 255-265, Plenum Press, NY, 1979.
7. S. K. Byram, B. Han, G. H. Rothbort, R. N. Samdahl, and R. A. Sparks, "A Novel X-Ray Powder Diffractometer Detector System," Adv. in X-Ray Anal., Vol. 20, pp. 529-545, Plenum Press, NY, 1977.
8. M. James and J. B. Cohen, "PARS: A Portable X-Ray Analyzer for Residual Stresses," JTEVA, Vol. 6, pp. 91-97, March 1978.
9. C. O. Ruud, "Position-Sensitive Detector Improves X-Ray Powder Diffraction," Ind. Res. and Dev., January 1983.
10. C. O. Ruud, P. S. DiMascio, and D. J. Snoha, "A Miniature Instrument for Residual Stress Measurement," Adv. in X-Ray Anal., Vol. 27, 1984 (to be published).
11. J. T. Norton, "X-Ray Stress Measurement by the Single-Exposure Technique," Adv. in X-Ray Anal., Vol. 11, pp. 401-410, Plenum Press, NY, 1968.
12. C. O. Ruud, P. S. DiMascio, and D. M. Melcher, "Application of a Position Sensitive Scintillation Detector for Nondestructive Residual Stress Measurements Inside

Stainless Steel Pipe," Adv. in X-Ray Anal., Vol. 26, pp. 233-243, 1983.

13. C. O. Ruud and C. S. Barrett, "Use of Cr K-Beta X-Rays and Position Sensitive Detector for Residual Stress Measurement in Stainless Steel Pipe," Adv. in X-Ray Anal., Vol. 22, pp. 247-249, 1979.

14. E. Marcherauch, "Lattice Strain Measurement on Deformed FCC Metals," Adv. in X-Ray Anal., Vol. 9, pp. 103-114, 1966.

RESIDUAL STRESS MEASUREMENTS IN Ti-6Al-4V and α-BRASS WITH SUB-

SURFACE FLAWS USING X-RAY DIFFRACTION TECHNIQUES

Vijay Alreja and John M. Liu

Department of Materials Science & Engineering
State University of New York at Stony Brook
Stony Brook, N.Y. 11794

ABSTRACT

This paper describes a method of characterizing sub-surface
flaws by the measurement of residual stresses associated with
these flaws. The residual stresses were estimated using the two
angle x-ray diffraction method. The flaws were artificially formed
within the material using mechanical and metallurgical (diffusion
bonding) methods. Specimens with sub-surface flaws were loaded in
tension and subsequently unloaded. A clear indication of spatial
dependence of residual stress in the vicinity of the flaw was observed. Ti-6Al-4V and α-Brass are the two alloys used in this
investigation. The origin of the residual stress is macroscopic
in case of α-Brass, which is a single phase alloy as compared to
the microscopic origin in case of Ti-6Al-4V, which is a two-phase
alloy. Ti-6Al-4V specimens have residual stresses opposite in
nature to the applied stress. Residual stress pattern around the
drilled hole specimens is opposite in sign to other specimens of
the same material and the residual stress distribution is insensitive to the flaw in case of the diffusion bonded specimens.

INTRODUCTION

This paper characterizes sub-surface flaws using residual
stress analysis. Ti-6Al-4V and α-Brass were chosen for this investigation as they have differing mechanical properties with respect to yield strength, ductility and hardness. The fact that
the two alloys have differing lattice structures, and are extensively applied, makes them good candidates for comparison. It is
important to remember that Ti-6Al-4V is a two-phase alloy and α-
Brass is a single-phase alloy.

39

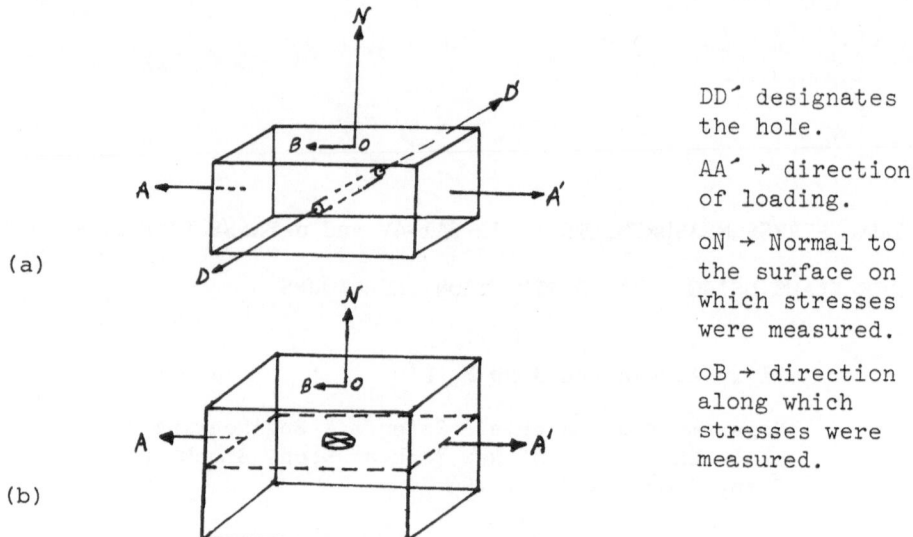

DD´ designates the hole.

AA´ → direction of loading.

oN → Normal to the surface on which stresses were measured.

oB → direction along which stresses were measured.

Fig. 1. Sub-surface flaws. (a) drilled hole specimen;
(b) diffusion bonded specimen depicting internal flaw.

        Artificial flaws were generated using a combination of hot
press bonding (1) and mechanical methods. This simulates actual
flaws generated during the various processes of fabrication. The
two kinds of flaws used are depicted in Fig. 1. These flaws are
characterized by spatial dependence of residual stresses in the
vicinity of the deformed region. The residual stresses are measured using 'the two-angle' x-ray diffraction method (2,3).

EXPERIMENTAL

        A Siemens diffractometer was used to measure the peak posi-
tions in the fixed time mode. The diffracted beam intensity was
recorded as counts, and the peak positions estimated using the
three point parabola method (4). LP, LPA and background correction
factors were employed in calculating the peak positions (2,5).
The residual stresses were estimated using these 2θ values and

$$\sigma\phi = K(2\theta_p - 2\theta_\psi) \qquad\qquad (1)$$

wherin $\sigma\phi$ = stress value in the ϕ direction,
    $2\theta_p$ = Peak value for ψ = 0 degree position,
    $2\theta_\psi$ = Peak value for ψ = 45 degree position, and

$$K = \frac{E \cot \theta}{(1+\nu) \ 2 \ \sin^2\psi} \cdot \frac{\pi}{180} \tag{2}$$

The values of 'K' used in this work were obtained from (6) for the
(213) peak of Ti-6A1-4V and (7) for the (420) peak of α-Brass.
The x-ray wavelength used for this investigation is 1.542Å and
the nominal values of the 2θ angles are 141.7 degrees for Ti-6A1-
4V and 138.9 degrees for α-Brass. The α phase of Ti-6A1-4V was
used for residual stress analysis. The various specimens used for
investigating the spatial dependence of residual stresses were
masked according to the procedure depicted in Fig. 2.

RESULTS

Residual Stresses in Ti-6A1-4V

    Three types of Ti-6A1-4V specimens were investigated for
residual stresses; as-received, diffusion bonded and drilled hole
specimens. In case of the as-received specimens, as seen in Fig.
3, the residual stress is compressive initially, and becomes more
compressive as the specimen is pulled in tension and released.
The diffusion bonded specimens with internal flaws, have residual
tensile stress initially and this becomes compressive as the specimen is pulled in tension and released. This is depicted in Fig.
4. In case of the drilled hole specimens of Ti-6A1-4V, the specimens have uniform residual compressive stress across the length
of the specimen initially (see Fig. 5). As the hole is drilled,
and the specimen pulled in tension and released, the residual
stress near the center becomes less compressive and towards the
ends the residual stress becomes more compressive, for higher
levels of applied stress.

Residual Stresses in α-Brass

    As-received and drilled hole specimens of α-Brass were ex-
amined for residual stresses. As-received specimens have residual
compressive stress initially (see Fig. 6). This residual stress
becomes tensile as the specimens are pulled in tension and released.
The drilled hole specimens of α-Brass exhibit uniform residual compressive stress across the length of the specimen initially. As
these specimens are pulled in tension and released, the residual
stress pattern is as seen in Fig. 7, i.e. the residual stresses
near the hole become more compressive and towards the side the
residual stress is tensile.

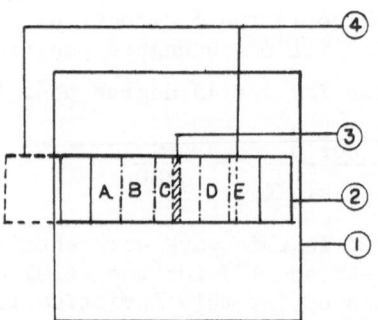

Fig. 2. Schematic drawing depicting the masking procedures.
1) Specimen holder; 2) Specimen; 3) Flaw. ABCDE identify
the five areas in which stress measurements were made.
To facilitate the measurement of stress in the areas on
either side, i.e. A,B,D,E, the specimen was shifted as
indicated by the dotted lines (4).

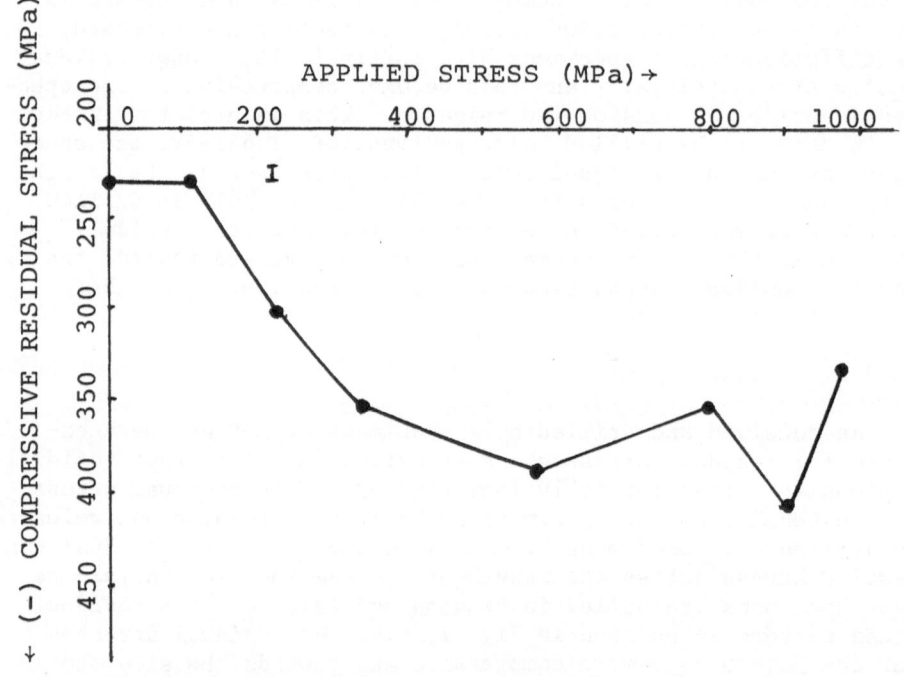

Fig. 3. Residual stress as a function of applied stress for
Ti-6Al-4V As-Received Specimen.

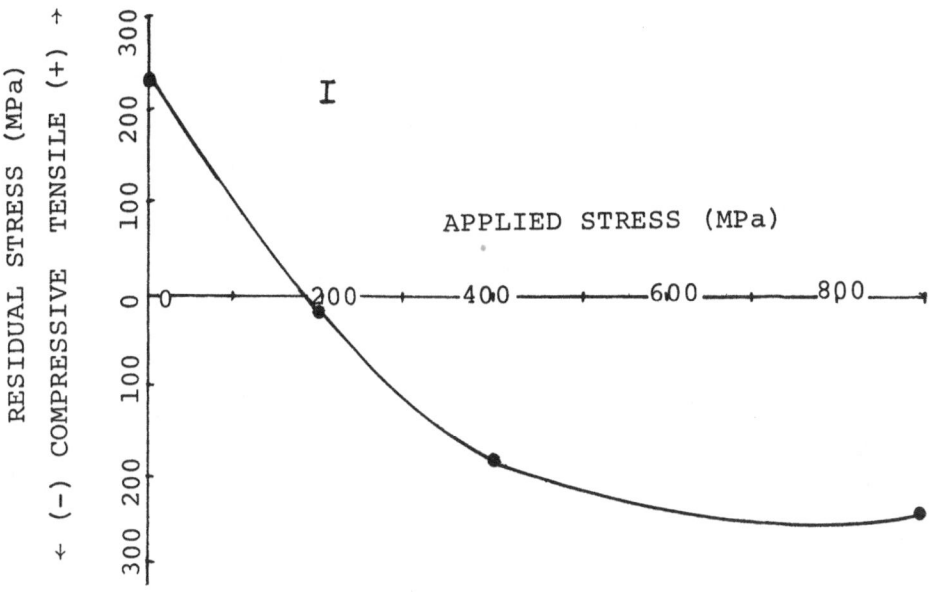

Fig. 4. Residual stress as a function of Applied Stress for
         Ti-6Al-4V diffusion bonded specimen

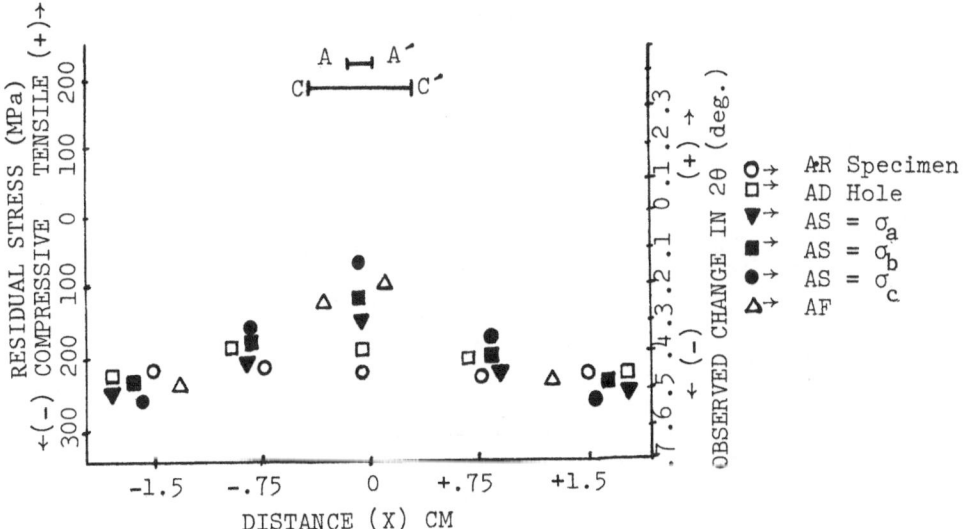

Fig. 5. Spatial dependence of residual stress for Ti-6Al-4V.
         AA´ → size of hole; CC´ → size of unmasked area exposed
         to the x-ray bean, AR-As-Received; AD-After Drilling;
         AS-Applied Stress; AF-After Fracture. $(\sigma_c > \sigma_b > \sigma_a)$.

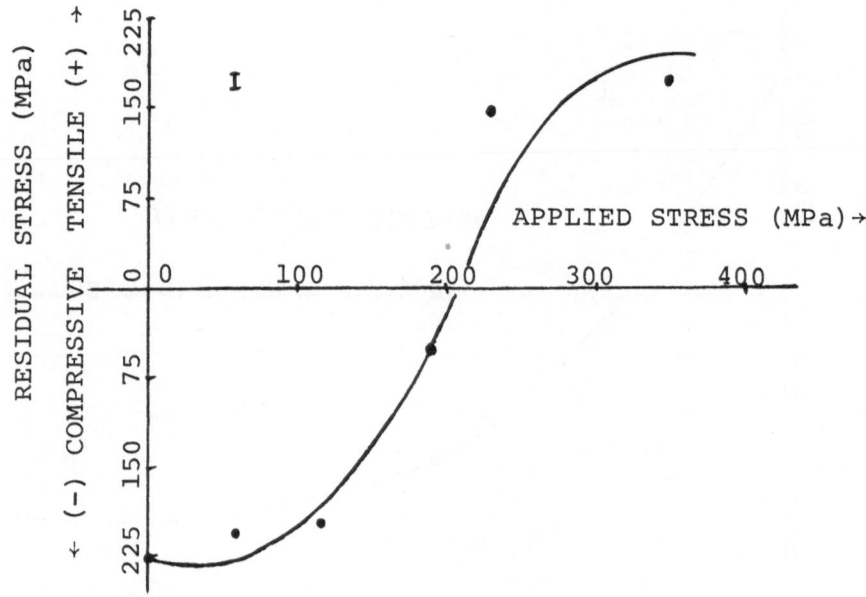

Fig. 6. Residual stress as a function of applied stress for
         α-Brass as-received specimen.

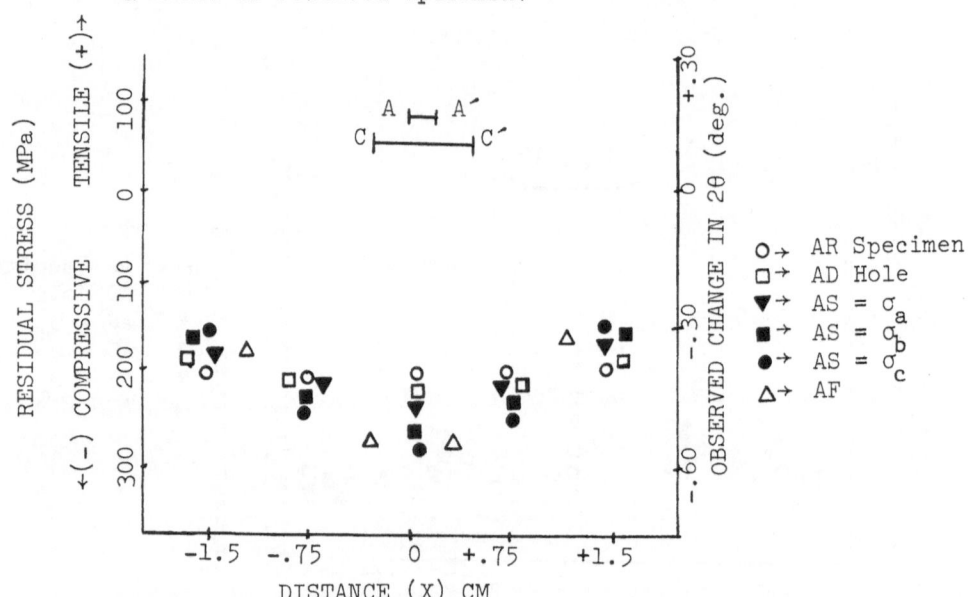

Fig. 7. Spatial dependence of residual stress for α-Brass.
         AA′ → size of hole; CC′ → size of unmasked area exposed
         to the x-ray beam. AR-As-Received; AD-After Drilling;
         AS-Applied Stress; AF-After Fracture. ($\sigma_c > \sigma_b > \sigma_a$).

DISCUSSION

    As-received specimens of both Ti-6Al-4V and α-Brass have com-
pressive residual stress initially. This initial residual stress
is a function of the prior history of the specimens. It could be
a result of operations like heat treatment and/or mechanical
working. As the specimens are pulled in tension and released, the
residual stress becomes tensile in case of α-Brass and more compressive for Ti-6Al-4V specimens. As tensile stress is applied
to α-Brass specimens, they retain a part of this stress due to
plastic deformation as residual tensile stress. This retained
residual stress in Ti-6Al-4V and α-Brass is a fraction (approximately 1/3 - 2/3) of the yield stress of the material.

    The anomalous residual stress behavior observed in case of
Ti-6Al-4V specimens is termed pseudo-macro stress (PMS) (8). It
is important to note that Ti-6Al-4V is a two-phase alloy, and
studies made on two-phase Ti alloys by Saleh and Margolin (9) for
Bauschinger effect did reveal that plastic deformation in β occurs
at much higher values as compared to α.

    The differences in behavior between the two-phase Ti alloy
and single-phase α-Brass, as well as that between the specimens
with drilled holes and those without such holes, can be explained
by considering the effects of inhomogeneous plastic deformation in
a material in which constraints due to geometry and microstructure
exist. In single-phase α-Brass, the deformed structure consisting
of dislocation arrays apparently distributes quite uniformly over
large regions of the specimen. As a result, most of the applied
tensile strain remained after unloading due to the locking of
strain in a uniform dislocation distribution. On the other hand,
in the α-β Titanium alloy, the β-phase has a yield stress higher
as compared to the α-phase. During tensile loading, the α-phase
goes through elastic and plastic deformation, while the β-phase
remains in the elastic range. Upon unloading, the dislocations
stored in the α-phase do not allow the β-phase to completely return
to its initial unstrained state. This results in residual tensile
stress in the β-phase and compressive residual stress in the α-
phase. Since the volume fraction of α-phase is substantially
larger than β-phase in our Ti-6Al-4V specimens, the X-ray measurements showed a change in lattice parameter consistent with the
existence of compressive stress when the specimens were unloaded.
(The β-phase, which is low in volume, only manages to scatter X-
rays into the tails of the diffraction line or into the general
background without substantially altering the line position) (8).
Thus, as shown in Fig. 5, the residual compressive stress of Ti-
6Al-4V, in close proximity of the flaw, becomes less compressive
as the specimens were loaded in tension and released, and finally
it becomes slightly more compressive after fracture.

The nature of the observed behavior for the specimens with
drilled holes can also be explained following a similar line of
reasoning. In case of α-Brass, since the residual stress was tensile in the vicinity of the plastically deformed region near the
hole, the residual stress far removed from the hole, i.e. near the
outside surface of the specimen, must be compressive in order to
maintain the macroscopic conditions of force equilibrium. The x-
ray beam (which measures the lattice spacing 'd' near the surface)
sensed this material in compression even though the material
inside is in tension. The same explanation holds for the specimen
of Ti-6Al-4V with drilled holes. Thus, the apparent opposite
response in specimens of both types of material with drilled holes,
while comparing them to unflawed specimens of the same material,
is another reflection of the effects of geometrical constraint.
Here, the effect is a macroscopic one, in contrast to the microscopic origin of the effect in the two-phase α-β Ti alloy.

In case of the diffusion bonded specimens as seen in Fig. 4,
the residual stress is tensile to start with. This is again a
phenomenon of PMS, because in bonding these specimens, they were
subjected to compressive loading and ended up having residual
tensile stress. No spatial dependence of residual stresses for
the flaws introduced inside these specimens was observed. This is
so, possibly because the flaws introduced are relatively small and
deep inside the material. The X-ray beam has very limited penetration and is not sensitive to small artifacts seeded deep within
the material.

REFERENCES

1. J. H. Ashton, Z. R. Wolanski, J. S. Kunselman, C. L. Veech,
    B. O. McCauley, J. A. Regalbuto, "Solid State Diffusion
    Bonding III. Vacuum-Hot Press Parameter Study", General
    Dynamics, FZM-12-13108 (1970).
2. B. D. Cullity, "Elements of X-ray Diffraction", Addison Wesley
    Pub. Co., Reading, Mass., Ch. 16, (1970).
3. P. H. Klug and E. L. Alexander, "X-ray Diffraction Procedures",
    John Wiley and Sons, New York, Chapter 10, (1962).
4. D. P. Koistinen and R. E. Margurger, ASM Transactions, Vol. 51,
    pp. 537, (1959).
5. C. Kim, Advances in X-ray Analysis, Vol. 25, pp. 343, (1981).
6. P. S. Prevey, Advances in X-ray Analysis, Vol. 20, pp. 345,
    (1977).
7. W. Wallance and T. Terada, Advances in X-ray Analysis, Vol. 14,
    pp. 389, (1970).
8. B. D. Cullity, Advances in X-ray Analysis, Vol. 20, pp. 259,
    (1977).
9. Y. Saleh and H. Margolin, Acta Met., Vol. 27, pp. 535, (1964).

DETERMINATION OF RESIDUAL STRESS FROM

TWO-DIMENSIONAL DIFFRACTION PATTERNS

Giancarlo M. Borgonovi

Science Applications, Inc
1200 Prospect Street
La Jolla, California 92075

INTRODUCTION

The need for a reliable nondestructive technique for measuring residual stress, especially for field applications, is emphasized by the number of physical properties and effects which have been investigated as possible indications of residual stress. Most of these properties and effects correlate well with stress, but depend strongly on the microstructure of the material and require, to be useful, a good characterization of the material. Of the methods used for determining residual stress, x-ray diffraction has some distinct advantages. It is reliable and it has been long established. One limitation of the technique is that it provides surface stress measurements only, however this limitation is largely offset by the present capabilities of correlating, through calculations, the surface stresses to the bulk stresses.

Considerable progress has been made in residual stress measurements by x-ray diffraction with the application of position sensitive detectors, either proportional[1] or solid state types.[2] All existing techniques are one-dimensional and measure the shift of one or more diffraction peaks in a scattering plane. It appears that additional progress can be made by using two-dimensional detectors, which allow one to determine the coordinates of a large fraction of a diffraction ring. In stress measurements by x-ray diffraction, the results are very sensitive to small changes in the sample-detector distance.* This distance needs to be controlled very accurately, a

---

* Personal communication with C.O. Ruud at The Pennsylvania State University (PSU) indicated that in recent work, to be published, distance variations of greater than ±1 mm are tolerable with an x-ray stress analyzer developed at PSU.

task not easily accomplished, especially under field conditions. The distance sample detector can, however, be inferred by analysis of the shape of the diffraction rings, if two-dimensional detectors are used.[3]

If the distance sample-detector changes, the size of the diffraction ring changes, in a known way. If the components of the stress change, the shape and curvature of the ring changes, in a different way. A least squares analysis, in two dimensions, of the shape of the diffraction ring, allows one to discriminate between the effect of stress and the effect of distance variation.

The approach described has three potential advantages over one-dimensional techniques:

(a) It allows one to remove errors due to very small changes in sample-detector distance.
(b) It permits measurements on coarser grain materials, which produce spotty diffraction rings, because it uses a larger part of the diffraction cone.
(c) It permits the simultaneous determination of both stress components parallel to the sample surface.

THEORETICAL APPROACH

Figure 1 shows the geometry for conventional measurements of residual stress by x-ray diffraction. Figure 2 shows the geometry for the approach described here. The approach is based on the analysis of the shape of the diffraction ring which results from the intersection of the diffraction cone with a flat detector. With no stress applied, the locus of reciprocal lattice vectors in a polycrystalline sample is a sphere and the diffraction cone is a circular right cone, illustrated in Figure 3. With stress present, the locus of reciprocal lattice vectors becomes an ellipsoid, and the diffraction cone is a distorted one, resulting from the inter-section of the ellipsoid with the sphere of reflection as shown in Figure 4.

The theoretical approach consists first of determining the equation of the distorted diffraction cone. This requires knowledge of the orientation of the stress ellipsoid and stress components. The equation is a relationship, in implicit form, between two angles defining the direction of scattering. The relationship is then put in explicit form by use of a perturbation solution. Once the equation of the diffraction cone is available, one can then calculate the coordinates of the intersection of the diffraction cone with a flat detector. These coordinates are the variables that will be measured from the detector output. Finally, a least squares method that fits the function derived above and provides estimates of the distance sample-detector and of the stress components is developed.

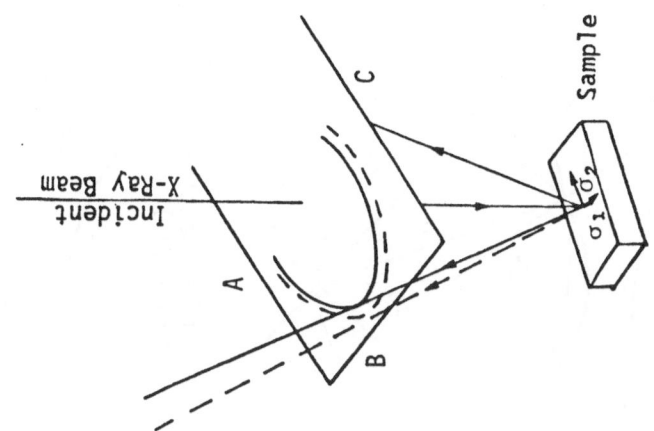

Figure 2. Two-dimensional geometry for stress measurements by x-ray diffraction.

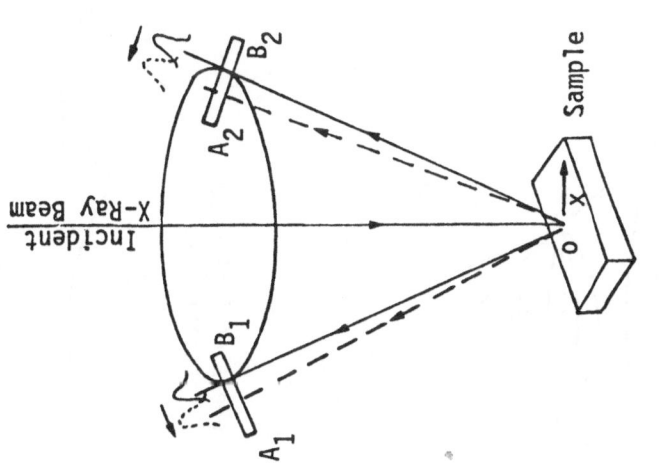

Figure 1. Geometry for conventional stress measurements by x-ray diffraction.

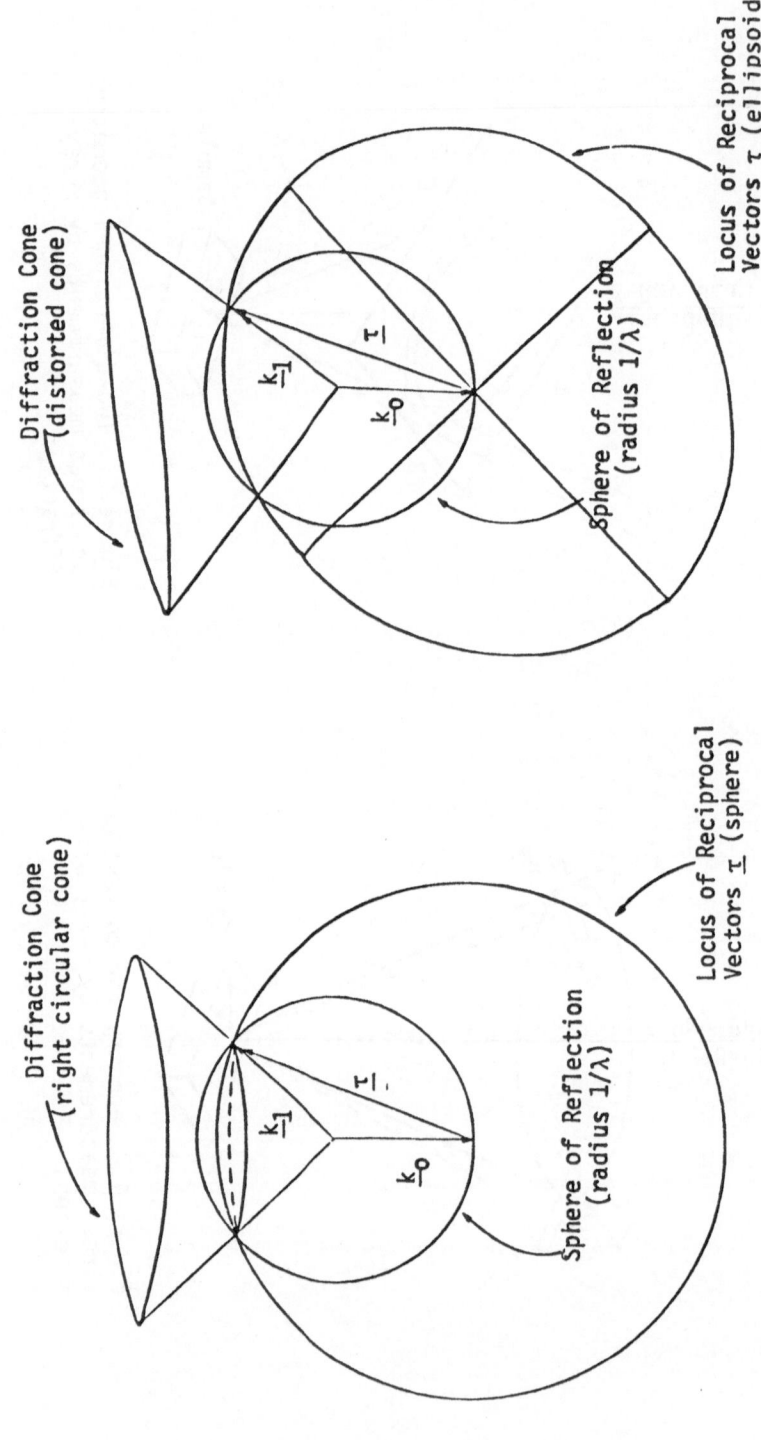

Figure 4. Diffraction Geometry With Stress.

Figure 3. Diffraction Geometry With No Stress.

Equation of Diffraction Cone

The first step is to define the reference systems used. Figure 5 shows the laboratory system OXYZ, with the direction of the incident x-ray beam coinciding with the negative direction of the Z axis. Figure 5 also shows the sample system Oxyz. The axes x and y are taken to be coincident with the direction of the principal stresses σ_1, and σ_2, and z is normal to the surface of the sample. The orientation of the sample system with respect to the laboratory system is defined by the three Eulerian angles ε_1, ψ_1, and ϕ_1.

The components τ_i of a vector in the sample system are related to its components τ_i^* in the laboratory system as follows:

$$\tau_i^* = \sum_{i=1}^{3} M_{ik}\, \tau_k \tag{1}$$

where

$$
\begin{aligned}
M_{11} &= \cos\psi_1 \cos\phi_1 + \sin\psi_1 \sin\phi_1 \cos\varepsilon_1 \\
M_{12} &= -\cos\psi_1 \sin\phi_1 + \sin\psi_1 \cos\phi_1 \cos\varepsilon_1 \\
M_{13} &= \sin\psi_1 \sin\varepsilon_1 \\
M_{21} &= -\sin\psi_1 \cos\phi_1 + \cos\psi_1 \sin\phi_1 \cos\varepsilon_1 \\
M_{22} &= \sin\psi_1 \sin\phi_1 + \cos\psi_1 \cos\phi_1 \cos\varepsilon_1 \\
M_{23} &= \cos\psi_1 \sin\varepsilon_1 \\
M_{31} &= -\sin\phi_1 \sin\varepsilon_1 \\
M_{32} &= -\cos\phi_1 \sin\varepsilon_1 \\
M_{33} &= \cos\varepsilon_1
\end{aligned}
\tag{2}
$$

A reciprocal lattice vector $\underline{\tau}$ which diffracts the incident beam is identified by the angles ϕ^* and ψ^* in the laboratory system, and by the angles ϕ and ψ in the sample system. For zero stress the angle ψ^* is constant for different ϕ^*. This constant value is given by ψ_o^*, such that

$$\cos\psi_o^* = \lambda/2d_o \tag{3}$$

where λ is the wavelength and d_o the unstressed spacing of the diffracting plane.

The diffraction condition for the vector $\underline{\tau}$ can be expressed as:

$$\tau = \frac{2}{\lambda} \cos\psi^* \tag{4}$$

In the presence of a stress field characterized by the two principal stresses σ_1 and σ_2 (σ_3 is assumed to be zero) the length of the

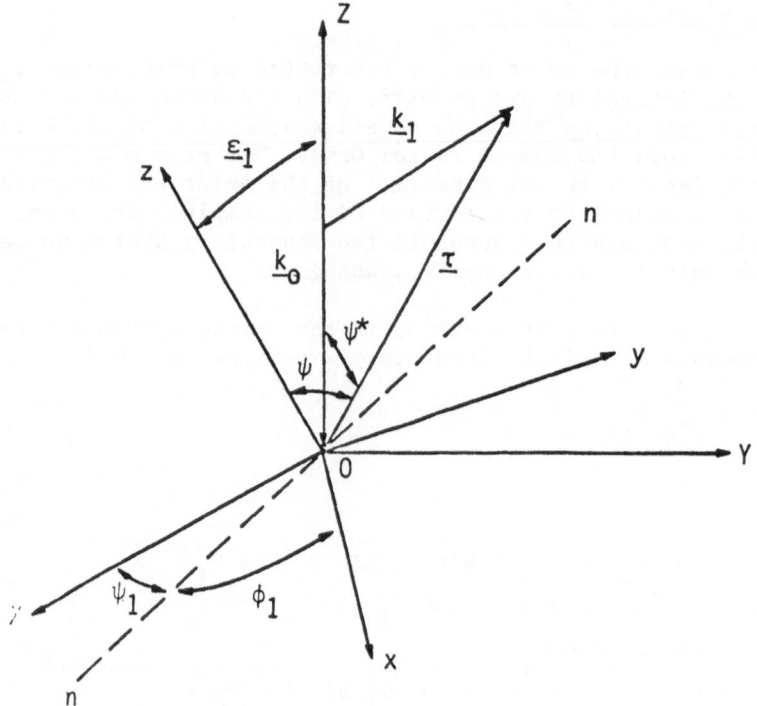

Figure 5.   Reference Systems.

vector $\underline{\tau}$ is

$$\tau = \cfrac{1}{d_o \left[1 + \cfrac{1+\nu}{E} \sigma_\phi \sin^2 \psi - \cfrac{\nu}{E} (\sigma_1 + \sigma_2) \right]} \tag{5}$$

where $\sigma_\phi = \sigma_1 \cos^2 \phi + \sigma_2 \sin^2 \phi$, E is the Young's modulus and ν the Poisson's ratio.

Combining equations (4) and (5) one obtains a relationship among ψ^*, ϕ, and ψ. However, ϕ and ψ can be expressed in terms of ϕ^* and ψ^*, so that one obtains a relationship between ψ^* and ϕ^*. Its derivation requires use of equations (1) and (2) and only the final result is reported here. The implicit form of the relationship between ψ^* and ϕ^* is:

$$\cos \psi^* \left| H + Z_1 \sin^2 \psi^* + Z_2 \cos^2 \psi^* + Z_3 \sin \psi^* \cos \psi^* \right| = \frac{\lambda}{2d_o} \tag{6}$$

where

$$H = 1 - \frac{\nu}{E} (\sigma_1 + \sigma_2)$$

$$Z_1 = \frac{1+\nu}{E} (\sigma_1 R_1{}^2 + \sigma_2 R_2{}^2) \qquad\qquad S_1 = M_{31}$$

$$Z_2 = \frac{1+\nu}{E} (\sigma_1 S_1{}^2 + \sigma_2 S_2{}^2) \qquad\qquad S_2 = M_{32}$$

$$Z_3 = \frac{2(1+\nu)}{E} (\sigma_1 R_1 S_1 + \sigma_2 R_2 S_2) \qquad\qquad S_3 = M_{33} \qquad\qquad (7)$$

$$R_1 = M_{11} \cos \phi* + M_{21} \sin \phi*$$

$$R_2 = M_{12} \cos \phi* + M_{22} \sin \phi*$$

$$R_3 = M_{13} \cos \phi* + M_{23} \sin \phi*$$

Perturbation Solution

For a given value of $\phi*$, equation (6) is a transcendental equation in $\psi*$. If $\cos \psi*$ and $\sin \psi*$ are expressed in terms of $m = tg\psi*/2$, one obtains a sixth degree equation in m, which would be cumbersome to solve. Since the stress effect is a small one, it is more convenient to look for a perturbation solution of the form:

$$\psi* = \psi_o* + \Delta\psi* \qquad\qquad (8)$$

When (8) is inserted into (6) and second and higher degree terms in $\Delta\psi*$ are neglected one obtains the solution:

$$\Delta\psi* = \frac{\cos \psi_o* (1-Q_1)}{Q_2 \cos \psi_o* - Q_1 \sin \psi_o*} \qquad\qquad (9)$$

where

$$Q_1 = H + Z_2 + (Z_1-Z_2) \sin^2 \psi_o* + Z_3 \sin \psi_o* \cos \psi_o*$$

$$Q_2 = 2(Z_1-Z_2) \sin \psi_o* \cos \psi_o* + Z_3 (\cos^2\psi_o* - \sin^2\psi_o*) \qquad (10)$$

Calculation of Coordinates

The detector is identified by a vector $\underline{D}$ normal to the detector plane, from the origin of the laboratory system to the detector plane, and by two unit vectors in the plane of the detector. The vector $\underline{D}$ can be identified by its modulus d and by two angles ψ_D and ϕ_D, as shown in Figure 6.

A diffracted beam due to the reciprocal vector $\underline{\tau}$ identified by the angles $\phi*$, $\psi*$ intersects the detector plane at a point P having coordinates ξ and η. The calculation of ξ and η from $\phi*$, $\psi*$ and the geometry parameters is straightforward. The relationships derived allow one to calculate the shape of the distorted diffraction ring for any orientation of the sample, and for any orientation of the detector.

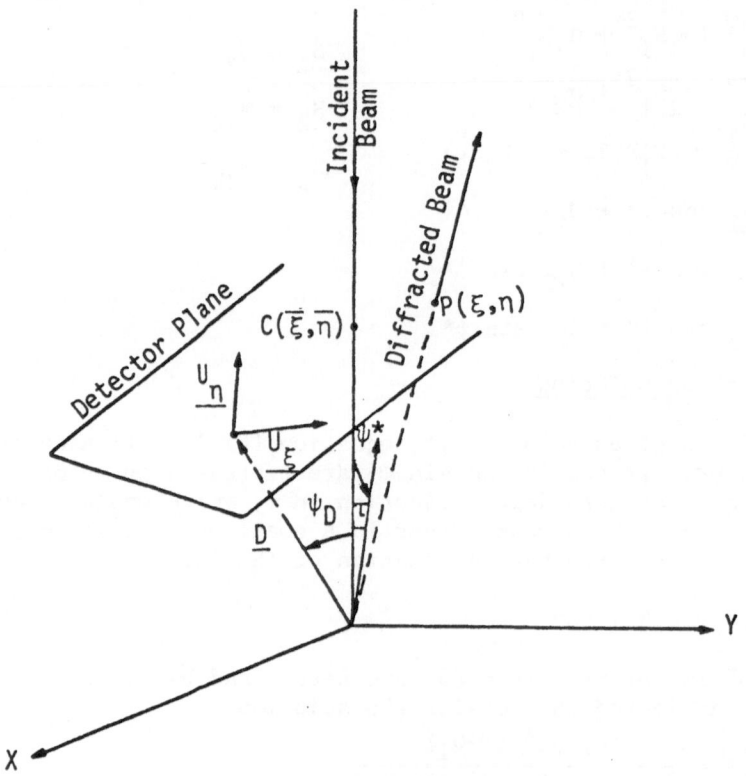

Figure 6.   Detector Geometry.

Least Squares Analysis

The previous considerations have indicated that a functional
relationship exists between the coordinates ξ and η of a point on the
diffraction ring on the detector plane. This relationship depends on
the parameters d, σ_1, σ_2, and ϕ_1. Thus, if one measures these coordinates, one can use a nonlinear least squares method to estimate the
sample-detector distance and the stress components and orientation.
The relationship between the coordinates and the parameters is a relatively complex one, so it is simpler to do the least squares fit on
the angles ψ^*, ϕ^*, derived from the experimentally measured coordinates ξ and η for a mechanically or otherwise determined value of the
distance d. One can then iterate over the detector distance and determine the stress values and orientation by nonlinear least squares.

For an assumed value of d we can then calculate, from pairs of coordinates ξ_i, η_i the pairs of angles ϕ_i*, ψ_i* that will be input to the least squares, using equations (8) and (9) as the analytic form for the fit. The derivatives with respect to σ_1, σ_2, and ϕ_1, which are required for the nonlinear least squares method, can be calculated from equations (9) and (10). The estimates obtained can be used to recalculate the values of ψ* and of the derivatives. One keeps iterating until the residual sum remains constant. This provides an estimate of σ_1 and σ_2 for the assumed distance d. One can repeat the least squares calculation for several values of d and select the one which gives the minimum residual value. With this search-iteration procedure one can converge on the best value of the distance and stress components.

RESULTS OF CALCULATIONS

A small number of simulations has been carried out to test the above ideas. The simulations have assumed a wave length of 2.085 Å and a spacing of 1.085 Å, very close to the values for the Chromium K_β radiation and the 311 plane of austenitic stainless steel. The distance of the sample from the detector was assumed to be 60 mm, the orientation of the sample with respect to the laboratory (instrument) system was assumed to be defined by the three Euler angles $\epsilon_1 = 30°$, $\psi_1 = 70°$, $\phi_1 = 25°$, the stress components at the surface of the sample were assumed to be $\sigma_1 = 30$ ksi and $\sigma_2 = 60$ ksi respectively, the Young's modulus and Poisson's ratio of the sample were assumed to be 3.10^4 ksi and 0.2806 respectively. With these assumptions the coordinates of a number of points on the distorted ring on the detector planes were calculated. A random variable uniformly distributed between $\pm$ an input value was then added to the coordinates, to simulate the experimental error in the determination of the coordinates. The data so generated were then analyzed with the iterative least squares technique, providing estimates of the stress components and orientation of the main stress axes. The results of the analysis depend on the fraction of the diffraction ring analyzed and on the size of the random error attached to the data. Table 1 summarizes the results obtained. When no random error is added to the data, the least squares values converge on the values of the parameters used to generate the data. When random errors with standard deviation equal to 0.01 and 0.02 mm are added, the analysis using one fourth (90°) of the diffraction rings produced considerably inaccurate results (different from the input values used to generate the data). However, when the fraction of the ring observed and analyzed is increased to 50 percent and more, the accuracy of the method increases considerably. The sensitivity could most likely be improved by proper choice of the orientation parameters. The size of the random error attached to the calculated coordinates was comparable to the accuracy of some present day position sensitive detectors. The results presented and other simulations

Table 1. Example of Analysis of Simulated Data.

INPUT VALUES

σ_1 = 30 ksi σ_2 = 60 ksi ϕ_1 = 25°

FRACTION OF DIFFRACTION RING					
	90°	180°	300°	360°	STANDARD DEVIATION ON x,y COORDINATES (mm)
σ_1	30.	30.	30.	30.	0
σ_2	60.	60.	60.	60.	
ϕ_1	25.1	25.2	25.3	25.3	
σ_1	61.2	27.9	26.9	29.7	.01
σ_2	10.5	59.1	57.9	57.2	
ϕ_1	119	25.0	26.	24.2	
σ_1	113	25.8	23.8	29.3	.02
σ_2	-59.5	58.1	55.9	54.5	
ϕ_1	21.1	24.8	26.6	22.9	

carried out indicate that, with proper choice of the detector and
sample inclination and reasonable experimental errors, enough sensitivity should be attained to compensate for distance variations
and determine the stress components with sufficient accuracy.

CONCEPTUAL DESIGN OF APPARATUS

Figure 7 shows a conceptual design for implementing the two-
dimensional stress analyzer. The design is based on a small x-ray
tube (available from Kevex Corporation) and a position sensitive
detector (available from Surface Science Laboratories). The x-ray
tube is small enough that, in the configuration shown, a large
fraction of the diffracted radiation cone will reach the plane sensitive surface. The sensitive surface is a thin scintillator, covered by a thin layer opaque to light, in contact with a specially
tapered fiber optic bundle. The small side of the fiber optic
bundle is in optical contact with the faceplate of the detector,
which contains a photocathode, a microchannel plate and a
resistive anode plate. X-ray photons impinging on the scintillator produce pulses of light which are converted to electrons.
The electrons are multiplied in the microchannel plate and

piped down to the resistive anode. The pulses collected at each corner of the resistive anode allow one to determine the position coordinates of the initial event.

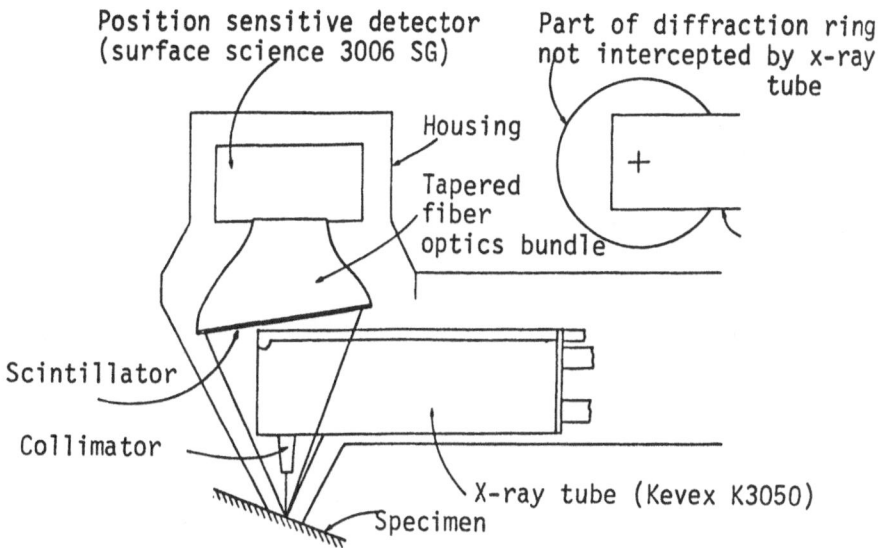

Figure 7. Conceptual design for implementation of two-dimensional stress analyzer.

The pulses, which carry the position information, will be processed by the electronic unit (commercially available in conjunction with the detector) and will increment locations in a digital memory. It may be desirable to reduce the size of this memory by preprocessing the pulses and storing only those that fall on locations near the Debye-Scherrer ring. A minicomputer will be necessary to perform the least squares analysis and provide the estimates of the stress components. The detector has low noise, generates individual pulses for each x-ray photon, and does not require cooling.

REFERENCES

1. M. James and J. B. Cohen, "PARS - A Portable X-Ray Analyzer for Residual Stresses", Journal of Testing and Evaluation, 6, 91, 1978.
2. C. O. Ruud and C. S. Barrett, "Use of Cr K-Beta X-Rays and a Position Sensitive Detector for Residual Stress Measurement in Stainless Steel Pipe", Advances in X-Ray Analysis, Vol. 22, 247, 1979.
3. G. Borgonovi, D. Epperson, G. Houghton and J. Orphan, "Technical Feasibility of a Borehole Probe for In-Situ X-Ray Diffraction Analysis", Advances in X-Ray Analysis, Vol. 24, 197, 1981.

SMALL ANGLE NEUTRON SCATTERING

AN NDE TOOL

C. T. Oien

Acceptance Technology Division
Sandia National Laboratories
Livermore, CA 94550

ABSTRACT

Small angle neutron scattering (SANS) is a method where the
very small angle neutron scattering ($2\theta \approx 2°$ to $5°$) profile from
test specimens is recorded and analyzed. The analysis of the
scattering profile is very similar to that already developed
extensively for x-ray scattering. The method provides microstructure information on test samples which correlates very well
with traditional microstucture analysis methods like TEM and
SEM. The main difference and advantage of SANS is that it is a
bulk measurement as opposed to thin film or surface. Recent
developments in the application of this powerful analytic tool
have provided the potential to use this technique as an NDE
tool. This paper presents a brief overview of the SANS
technique, the developments which may allow the application of
this technique to NDE problems, and a few examples of how this
technique is being utilized at Sandia National Laboratories
(SNLL) and elsewhere for microstructure investigation.

POTENTIAL MATERIALS EVALUATION

Introduction

There are several materials being used and proposed for use
at SNLL and in the industry which are potential candidates for
evaluation using neutrons of selected wavelength as the materials
probe. Information is available from both the transmitted neutron
signal as well as the scattered neutron profile. Some advantages
of using neutrons as a material probe are; 1) neutrons can
differentiate between isotopes of the same species, 2) neutrons

59

can differentiate between materials which are very similar chemically (similar electronic structure) and 3) neutrons can easily
differentiate between gases and structural materials. This is due,
primarily, to the fact that neutrons interact with the atomic
nucleus itself and not the electron cloud as do the TEM, SEM and
x-ray probes. Finally, particularly for nondestructive evaluations, neutrons can penetrate several centimeters of structural
material and still yield excellent counting statistics. The
disadvantages associated with neutron techniques for materials
evaluation are: 1) the relatively low fluxes of neutrons available
for measurements at selected neutron wavelengths and 2) the interpretation of, particularly, the scattered neutron profile. The
following are a few potential materials evaluation problems that
lend themselves to a study using neutrons of selected wavelength.

Gases in Metals

The gases, hydrogen, deuterium, tritium, oxygen, and
helium-3, in metals can be readily detectable with neutrons of
thermal energy (or colder). This is due to the relatively large
difference in neutron cross sections for metals and these gases.
This difference is in general enhanced as the neutron energy is
lowered. Total neutron attenuation measurements yield the
concentration of a gas in the metal. If the gas is considered a
defect, the small angle neutron scattering profile can determine
the defect size distribution, shape, volume and texture. Neutron
diffraction studies could yield strain (stress) as a function of
defect type and/or concentration. There has been a little work in
this area although most has been done on pure metals.

Metals Evaluation

Many structural materials (metals) in use could be evaluated
as well with small angle neutron scattering techniques. SANS
evaluations can determine, for example, precipitate size
distribution, dislocation concentrations and the amount of a
particular phase of material present. Much work is being done
currently on structural materials to evaluate the neutron induced
damage to the metal lattice by high fluxes of neutrons. Other
applications include the evaluation of parent materials versus
welded material for determination of the extent of the heat
affected zone or the difference in the amount of a particular phase
present in the welded materials versus the parent material. There
has been a good deal of work done in this area already.[2,4,5]

Powders Evaluation

Much of the analysis of powders is done with TEM and SEM
methods. Neutrons can be used to make very similar measurements
but with a major difference--that is, that the measurements can

be done on bulk samples and/or encapsulated bulk samples. Neutron
attenuation mesurements can determine, for example, the extent of
hydriding taking place and the concentration of gases or other
species in the powders. The SANS evaluations can yield information on specifics associated with various hydrides (e.g., void
volumes) and what configuration (atomic) they assume. Also
information on the differences between the various isotopes of
hydrogen and how they form hydrides can be evaluated since the
scattering lengths for these isotopes are different. Limited
SANS work with powders has been done to date.

Solid State Bonds

One area of metals joining has historically been very
difficult to evaluate nondestructively. This is the area of
solid state bonding. Evaluating nondestructively the very small
bond region associated with resistant forge welds, plug welds and
pinch welds is very difficult. SANS, which is very sensitive to
conditions which change the neutron scattering length, could
possibly be used to evaluate this type of metal bonding. There
has been some work on evaluating the heat affected zone in
weldments.[2]

SANS TECHNIQUE

Theory

Small angle neutron scattering refers to the coherent
scattering of neutrons at small angles ($2\theta \approx 5°$). The coherent
scattering of neutrons is a consequence of the interaction of
neutrons with the nuclei throughout the material. The secondary
rays scatter in all directions but because phase relationships
are preserved on coherent scattering, interference effects with
rays from adjacent centers will occur so that the particular
scattering profile observed is a function of the spatial
arrangement of the nuclei in the test specimen. Generally, there
are three types of scattering phenomena which can be observed at
small angles. First, discontinuous, where scattering is favored
at specific angles (small angle diffraction) and is usually a
consequence of definite, large lattice spacings in the sample.
Second, beam broadening arising from neutron refraction when the
direct beam passes through regions with different refractive
indices. This effect is small but can be observed at small
scattering angles. The third type of small angle neutron
scattering is continuous, where the scattering intensity varies
monotomically with angle. Continuous scattering arises from
particles or regions more or less randomly distributed throughout
the material. The term "particle" or "region" is used to denote
any area where the neutron scattering length is different than

that of the surrounding region. Neutrons with wavelengths
approximately twice the lattice spacing are used to avoid double
Bragg type reflections which make the profile analysis difficult.

The primary purpose of a SANS experiment is to determine the
angular distribution of the scattered intensity. Once this is
available there are a variety of analytical tools available,
primarily based on the comprehensive analytical work done for
x-ray scattering, which can be used to evaluate parameters such as
size, shape and distribution of inhomogeneities.[1,2]

The main elements of a SANS system include:

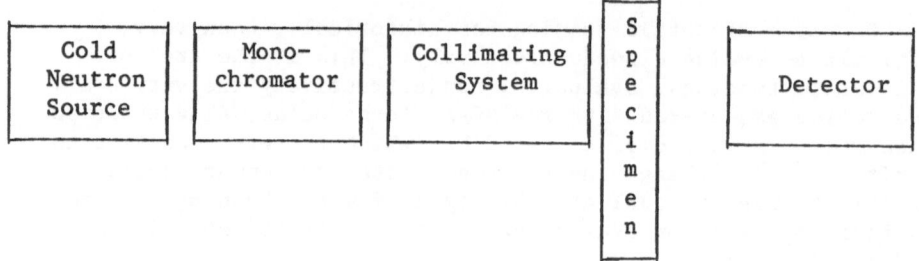

The major source of neutrons is a steady state reactor. Cold
neutrons are obtained by moderation in a refrigerator or by heavy
filtering with beryllium and/or bismuth. Collimation is accomplished with critical angle deflections and/or absorbing slits.
Monochromatization is accomplished with critical angle deflection,
crystal deflection, and/or mechanical selection. The scattered
beam is detected by a modified film method, position counters or
area detectors. The scattered intensity profiles are then
produced by appropriate integration of computer hardware to the
detector system. The introduction of area detectors for SANS work
has greatly increased the speed of the data acquisition.

Facilities

There are several facilities in the USA where one can now
obtain SANS profiles. One is at Brookhaven National Laboratory,
another is the National Center for Small Angle Scattering Research
(NCSASR) located at the Oak Ridge National Laboratory, and finally
the SANS Instrument at the National Bureau of Standards (NBS). The
BNL instrument is essentially a laboratory tool and is not suited
specifically to outside users. The NCSASR and NBS facilities are
more flexible, from an outside user suitability standpoint.   The
The European community has been extensively involved in SANS work
for about the last fifteen years and has many excellent SANS
facilities.

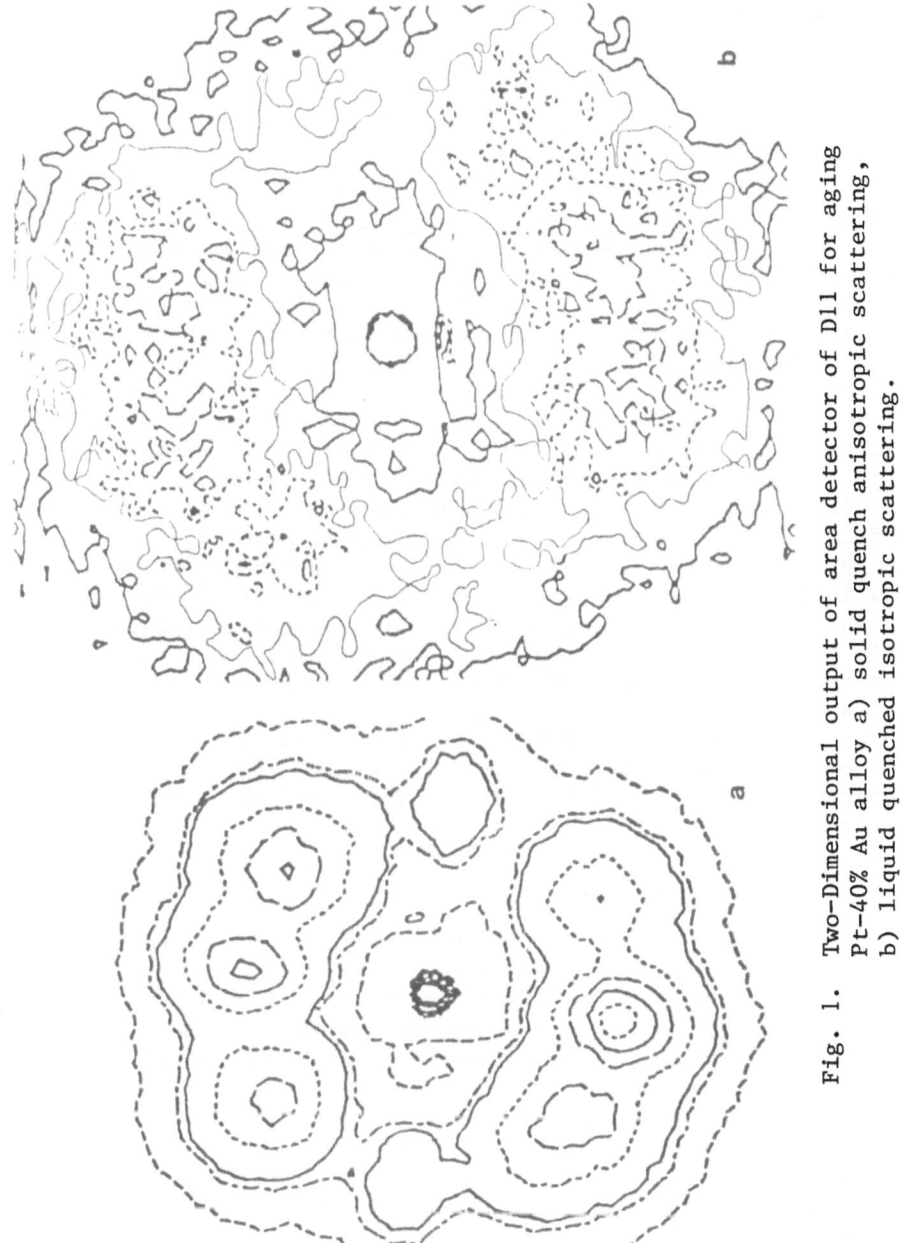

Fig. 1.   Two-Dimensional output of area detector of D11 for aging
          Pt-40% Au alloy a) solid quench anisotropic scattering,
          b) liquid quenched isotropic scattering.

SANS APPLICATIONS

The following presents two of the ways in which SANS has been employed to study technological materials perform NDE on alloys-in-use.

Thermal Treatments

Thermomechanical and precipitation strengthening heat treatment of a wide range of ferrous and non-ferrous materials makes these materials excellent candidates for SANS studies. The shortcomings of using x-ray probes to study solid state precipitation are 1) the limited techniques of analysis that yield meaningful structural information, 2) the absorption of x-rays is very high requiring samples to be very thin (e.g. .003 inch for aluminum alloys), 3) in order to detect clusters using x-rays requires that the respective atomic numbers differ considerably. This is due to the fact that small angle scattering of x-rays is caused by the extra-nuclear electrons. Unless the number of extra-nuclear electrons is considerably different for the respective clusters, there will be little difference in scattering intensity. Neutrons on the other hand scatter with strengths unrelated to atomic number. Neutrons can traverse considerably thicker sections of materials, and of particular importance to the data analysis, neutrons can be slowed sufficiently to yield a reasonable number of neutrons with long wavelengths beyond the Bragg cutoff which eliminates the confusing double Bragg relations in the data. An example, shown in Figure 1, shows the different scattering profiles produced by different heat treatments.[6] These profiles are typical outputs of an area detector for a SANS instrument. The lines shown are iso-intensity lines of neutron intensity versus solid angle. Figure 1-A shows an anisotropic scattering profile while Figure 1-B shows an isotropic scattering profile.

Gases in Metals

Hydrogen is known to be highly soluble in technological materials. SNLL has had some particular experiences with the effects of hydrogen in metals and now there is growing concern for other gases in metals. Particulary, the presence of He-3 has been shown to have a significant degrading effect on metals of interest to SNLL. The detailed role of gases in metals and how they contribute to failure is anything but certain. SANS can contribute in this area in several ways: 1) it can determine the location and concentration of the gas in a metal section, 2) it can determine some information on how the gas is trapped in the lattice—for example, if the hydrogen has formed voids or if the He-3 is clustering and what the void or cluster size(s) are, and 3) it can be used as an NDE tool to help predict failure even in the

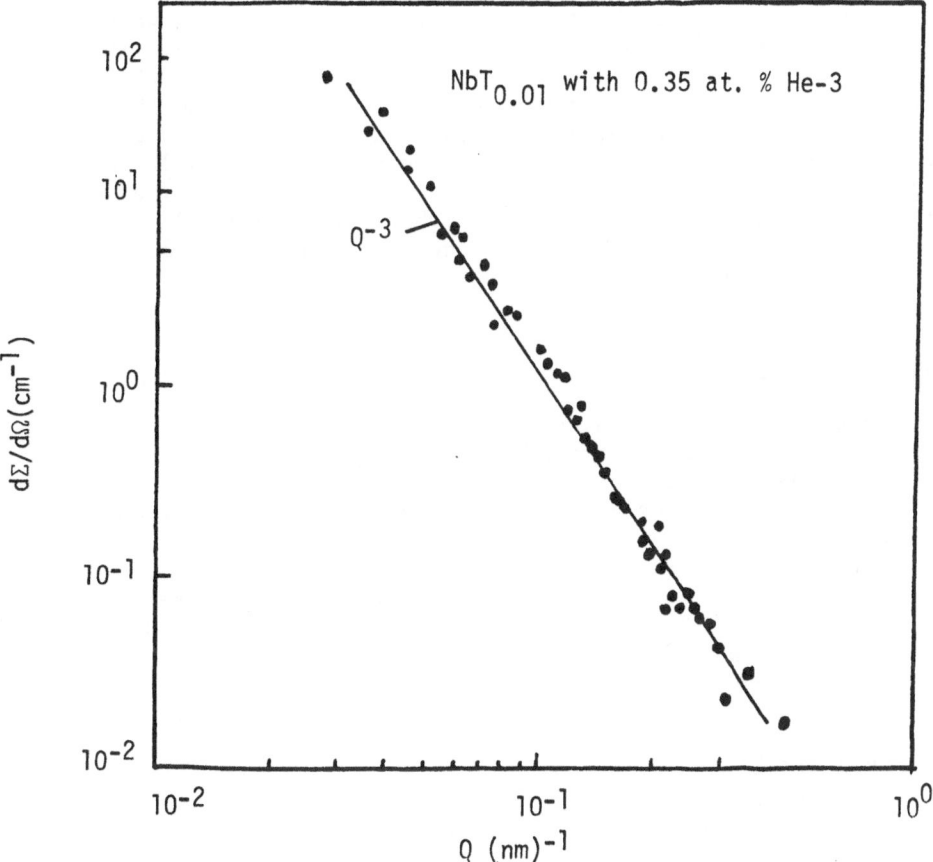

Fig. 2. SANS profile from He-3 charged niobium.

absence of detailed understanding. As an example, Figure 2 shows
the scattering profile obtained from a He-3 loaded niobium
sample. This particular sample is unique in that the He-3 present
is due to the decay of tritium introduced into the sample several
years ago, resulting in very high concentrations of He-3 without
any implantation damage. The scattering profile exhibits a very
strong Q^{-3} dependence on differential scattering cross-section
$(d\Sigma/d\Omega)$ versus scattering angle (Q). The differential Σ scattering
cross-section is the scattered neutron intensity per solid angle,
I(Q), where Q is given by $Q = 4\pi Sino/\lambda$. Here λ is the wavelength
of the incident neutrons and 2o is the scattering angle. This is
indicative of dislocation scattering. Commonly used scattering
cross-sections and their asymptote form can be found in the
literature.[2,5] Preliminary evaluations indicate enhanced
scattering from dislocations due to the presence of He-3.[7]
Below is a list which outlines other potential applications of SANS.
The list is not intended to be comprehensive but to indicate the
amount and variety of work being done with SANS to stress this
potentially significant NDE and materials research tool.

Applications of SANS (2)

        Precipitation Hardening Systems
                Optimizing Thermal Treatments
                Quality Control
                Property Degradation (e.g., through coarsening and phase
                            change)
                (Examples: Steels, Superalloys, Al-Alloys
                            Al-Mg, Al-Si, Al-Cu, Al-Zn, Al-Mg-Zn)

        Gases in Metals
                Hydrogen
                Radiation Damage

        Radiation Effects
                Voids and Swelling
                Radiation-Induced Transformations

        Microvoids in Ceramics
                Sintering
                Porosity
                Microcracks (Si_3N_4....)
                Cermets
                Oxide Coatings
                Cement

Welding and Fabrication
    HAZ
    Hydrogen

Preferred Orientation
    Two-Phase System with Area Detector

Surface Effects
    Corrosion
    Stress Corrosion
    Coatings
    Surface Roughness

Degradation due to:
    Fatigue (High and Low Temperature)
    Creep (and in combination with fatigue)
    Plastic Deformation

Composite Materials

Fine Particles (e.g. Catalysts, Lubricant Analysis)

CONCLUSION

SANS has been shown to be a viable NDE technique with application to a wide range of materials characterization problems. Areas requiring further effort are scattering profile interpretation and work improving the neutron beam flux, quality and detection.

REFERENCES

1.   A. Guinier and G. Fournet, "Small Angle Scattering of X-Rays," Wiley, New York (1955).
2.   H. Herman, "Non-Destructive Evaluation of Materials with Cold Neutron Beams," Report to Naval Air System Command, Washington, D.C., Contract No. N00119-77-M-0418, AD A053073 (1977).
3.   Handbook of Materials Testing Reactors and Associate Hot Laboratories in the European Community," D. Reidel Publishing Co., Boston (1981).
4.   C. Jantzen, D. Schwahn, J. Schelten, H. Herman, "X-Ray and Neutron Small-Angle Scattering Study of Phase Decomposition of Amorphous SiO_2-Al_2O_3," 4th International Small Angle Scattering Conference, Gatlinberg, Tenn. (1977).

5.   P. Pizzi, "Application of Small-Angle Neutron Scattering
         to NDI of Materials and Manufactured Components,"
         AGARD Meeting, Structures and Materials Panel, "NDI
         Relationships to Aircraft Design and Materials"
         (1977).
6.   S. P. Singhal, Thesis, Personal Communication, National
         Bureau of Standards.
7.   G. J. Thomas, Sandia National Laboratories, Livermore,
         Personal Communication.

EDDY CURRENT AND ELECTRICAL CONDUCTIVITY APPLICATIONS

PREDICTION OF LASER IRRADIATED SURFACE PROPERTIES

BY EDDY CURRENT METHOD

R. Palanisamy
Ames Laboratory, USDOE

P. A. Molian
Department of Mechanical Engineering
Iowa State University
Ames, IA   50011

ABSTRACT

The laser surface modification poses two significant problems, namely, a) accurate determination of melt depth as the metallurgical and mechanical properties of the diffusion layer change with this depth and b) detection and characterization of defects within the modified surface. The objective of this work is to determine theoretically if eddy current method of nondestructive testing (NDT) is sensitive enough to determine changes in material properties, melt depth and defect parameters in surfaces modified by high energy laser beams. Using a finite element numerical analysis technique eddy current signals were predicted (absolute and differential probes) for different fusion zone geometries produced by a single laser pass. The results show that eddy current NDT techniques can be employed to determine the changes in properties and cracks in laser modified surfaces.

INTRODUCTION .

The recent applications of directed energy sources, such as lasers, to materials processing include surface melting, alloying, cladding and heat treating of crankshafts, valve seats, piston ring grooves, cutting tools, etc. The beneficial effects of laser surface treatment are minimum distortion, development of compressive residual stresses, formation of nonequilibrium crystalline and amorphous phases[1], refinement of grains, homogenization of microstructures, increased solid solubilities of alloying elements, and

modification of segregation patterns. Consequently the mechanical
and physical properties such as the resistance to wear, corrosion
and fatigue are improved which enhance the life, reliability and
performance of engineering components.

The key process variables that control the superior microstructures and mechanical properties of laser irradiated surfaces
are the fusion and heat-affected zone depths, and the composition
of fusion zone. In spite of the significant advantages, laser
processing, like conventional welding processes, introduces
porosity, cracks and irregular boundaries in both fusion and heat
affected zones which impair the surface properties of the irradiated parts considerably. At present for want of nondestructive
examination techniques, information concerning melt depth and
subsurface flaws are obtained by destructive means. The data
obtained by destructive methods become inaccurate as the true
surface characteristics and part or whole of melt depth are lost
in sample preparation process. The lesser the melt depth, the
more unreliable the inspection results are.

EXPERIMENTAL PROCEDURE

Cold rolled iron and carbon steels were electrodeposited with
chromium and irradiated with a 1500 watts continuous wave CO_2 gas
transport laser to generate Fe-Cr-C alloys to melt depths 100 to
1500 μm. The fusion zone shape, composition, flaws, microstructures and mechanical properties were determined by destructive
methods[2].

EXPERIMENTAL RESULTS

Fig. 1 depicts the definition of laser surface alloying
process. The three typical profiles of fusion zone, observed in
laser surface alloying process, are presented in Fig. 2. The
"hour-glass" fusion zone, shown in Fig. 2a, is characterized by a
deep penetration (melt depth > 700 μm) but usually accompanied by
loss of material through vaporization, low cooling rates, shrinkage
cracks, 'key-hole' porosity and highly non-uniform composition.
Fig. 2c shows a semi-circular fusion zone and this has the advantages of very high cooling rates (an important parameter to produce
nonequilibrium and amorphous phases), uniform composition and
minimum defects. However it is limited by a shallow melt depth
(10-300 μm). The triangular fusion zone (Fig. 2b) has the intermediate characteristics of the above two.

Figures 3 to 6 show the presence of cracks and pores in fusion
zones. The centerline cracks are identified as solidification
shrinkage cracks and are due to inadequate feeding of molten metal.

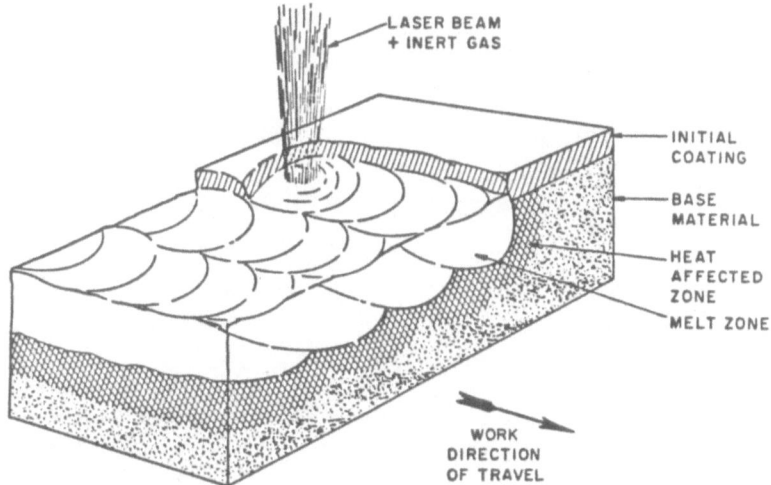

Fig. 1. A schematic of laser surface alloying process.

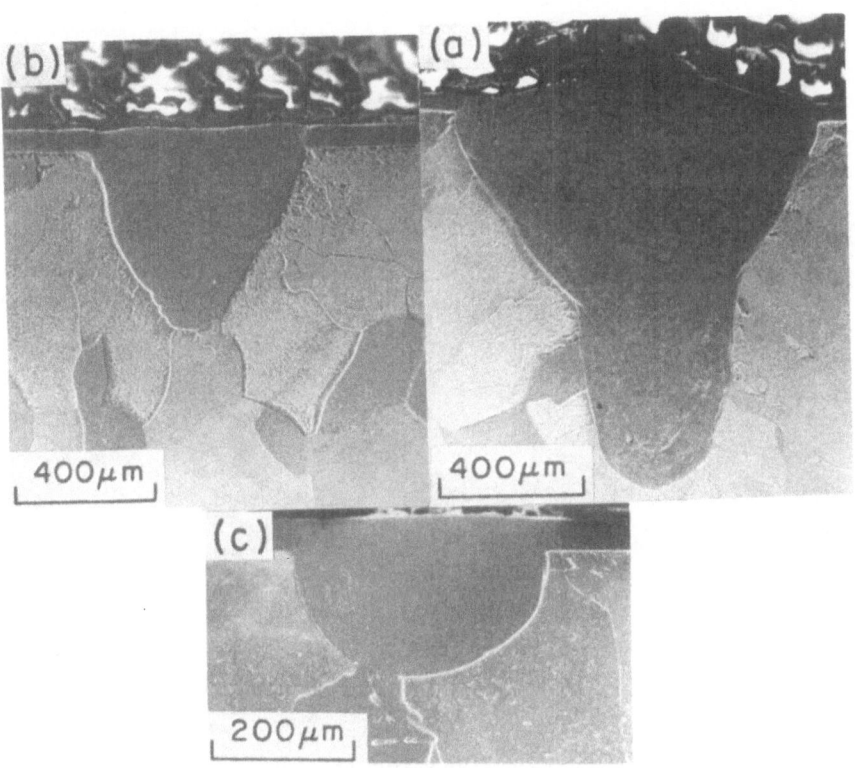

Fig. 2. Transverse sections of laser alloyed fusion zones.

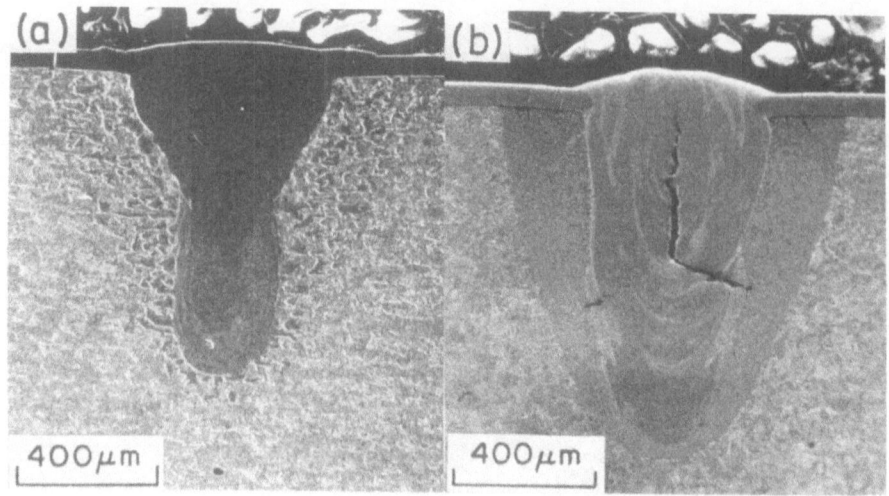

Fig. 3.   Transverse sections of laser alloyed fusion zones showing
          the effect of carbon on cracking.   (a) 0.2% C steel;
          (b) 1.0% C steel.

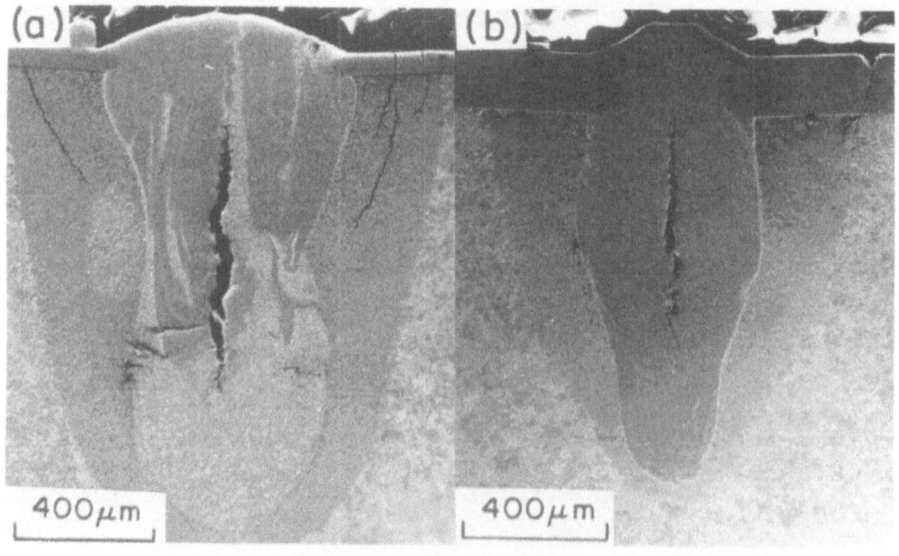

Fig. 4.   Transverse sections showing the centerline cracks.

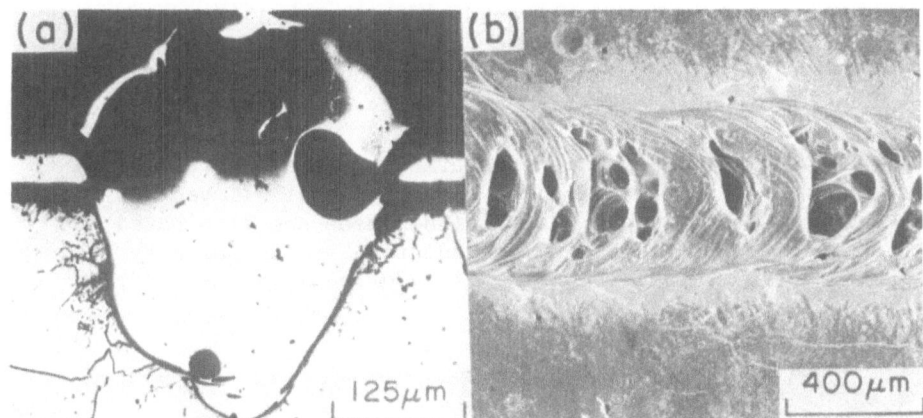

Fig. 5. Top views of laser alloyed surfaces. (a) with trapped
gases; (b) without trapped gases.

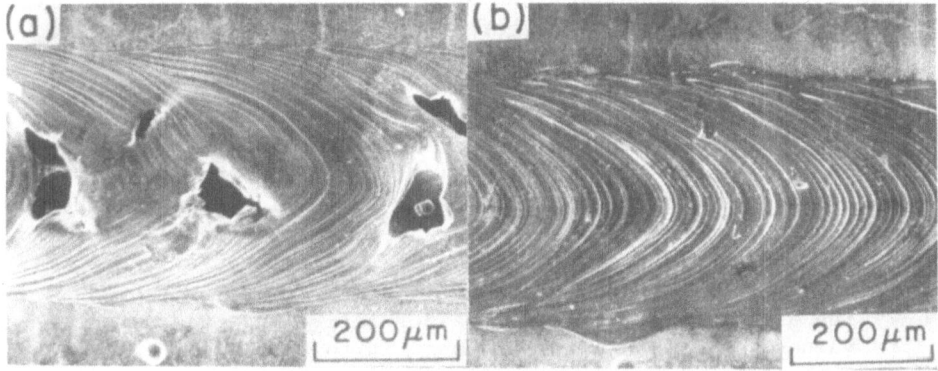

Fig. 6. Transverse and top views of laser alloyed surfaces
showing the pores.

The susceptibility to this type of cracking is a function of both the fusion zone geometry and composition. The fusion zones containing 0.2% C and 1.0% C are shown in Fig. 3 and are produced under identical laser alloying conditions. This illustrates the effect of carbon in promoting the crack formation. A comparison between Figs. 4a and 4b indicate that the chromium content of the fusion zone does not reduce the propensity of cracking.

The gases entrapped in the materials generate pores during laser melting. Figures 5a and 5b show the laser alloyed surfaces with and without entrapped gases. (The gases are eliminated by vacuum degassing prior to laser irradiation.) The porosity is also aggravated by rapid scanning of laser beam on the material surface (Figs. 6a and 6b).

EDDY CURRENT CHARACTERIZATION

In the regular production line, one should be able to characterize the laser irradiated surface properties and geometry by non-destructive means. The possibility of applying eddy current NDT method to this purpose is dealt in the following sections.

Method of solving eddy current problems employing finite element numerical technique, and modeling of eddy current NDT phenomena using this technique are well described in references 3 and 4, respectively. Knowing the test geometry and material properties (electrical conductivity and magnetic permeability), the impedance of an eddy current coil can be calculated numerically. Change in material properties adjacent to the coil changes the coil impedance. The sensitiveness of an eddy current coil to the changes in surface properties can be determined by calculating the difference in impedance values before and after laser irradiation.

An absolute coil encircling a carbon steel rod (1" O.D.) is shown in Fig. 7a. By passing a laser beam around the rod once, a triangular fusion zone is created similar to the one shown in Fig. 2b. The complex impedance (Z) is calculated using the finite element model for different positions as the coil moves past the fusion zone. The probe sensitivity is defined as $|(Z-Z_0)/Z_0|$, where Z_0 is the impedance obtained before laser irradiation. Commercial eddy current instruments have sensitivity value around 10^{-3}. The electrical conductivity and relative magnetic permeability values for carbon steel are assumed to be 5×10^6 (ohm-m)$^{-1}$ and 100, respectively. Corresponding values for fusion zone are 1×10^6 and 1. The predicted results of a differential eddy current probe for the fusion zone shapes shown in Fig. 2 are presented in Figs. 8, 9 and 10. The change in Z introduced by a shrinkage crack (Fig. 4b) can be noticed in Fig. 10.

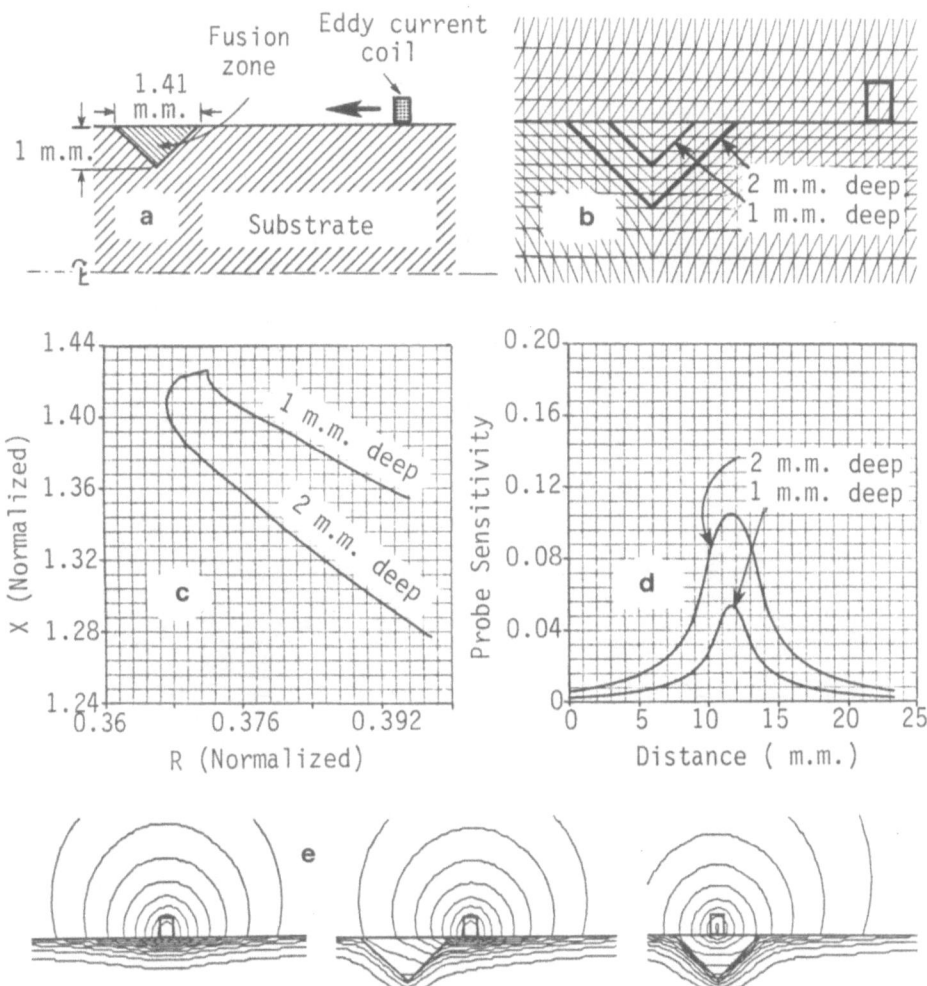

Fig. 7. (a) Absolute eddy current coil encircling a circular rod
         having a triangular fusion zone; (b) part of finite
         element discretization; (c) complex impedance plane
         trajectories (normalized imaginary part of coil impedance
         X vs. normalized real part R); (d) predicted sensitivity
         as a function of probe position; and (e) field plots for
         different probe positions. Frequency: 3 kHz.

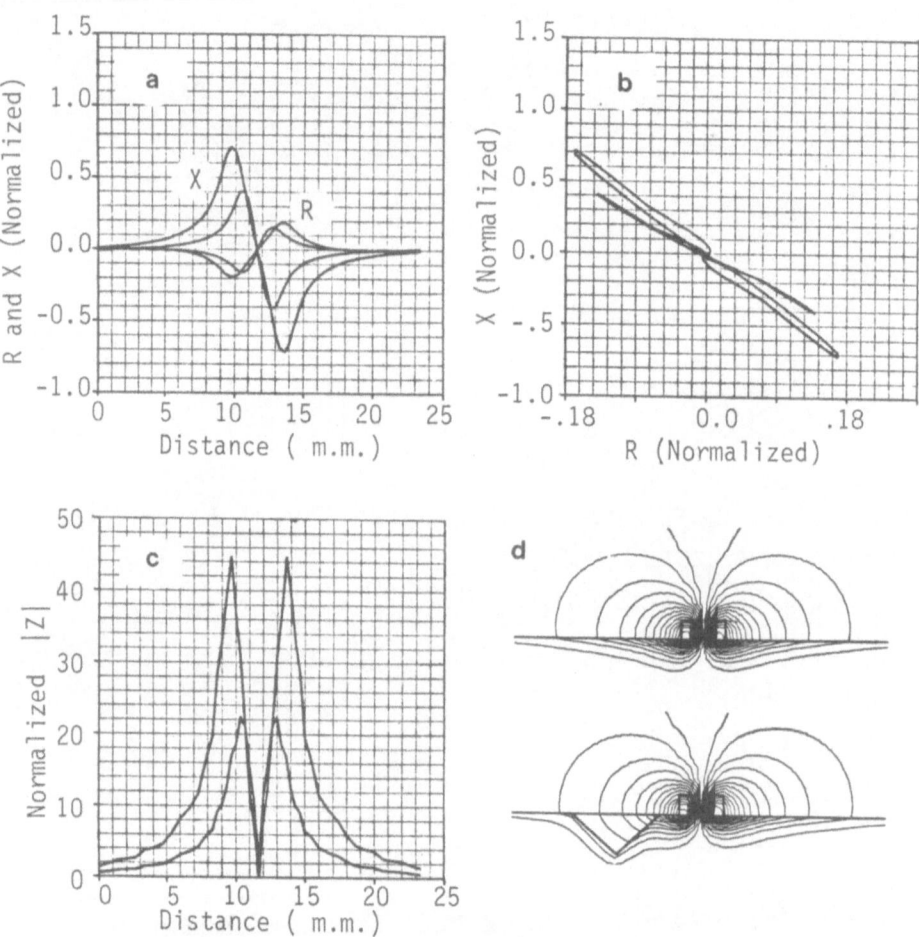

Fig. 8. (a) Normalized real (R) and imaginary (X) parts of differential eddy current probe impedance as a function of
probe location corresponding to Fig. 7a; (b) complex
impedance plane trajectories; (c) normalized impedance |Z|
vs. probe position; and (d) field plots. Larger and
smaller signals correspond to 2 and 1 mm deep triangular
fusion zones, respectively. Frequency: 3 kHz.

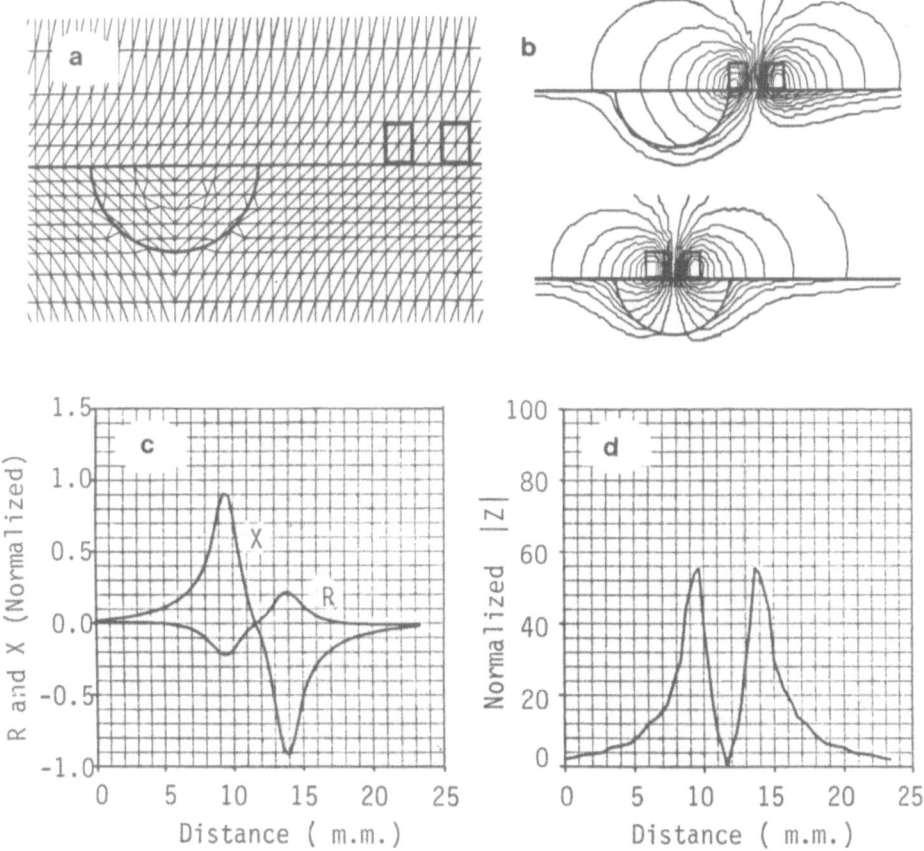

Fig. 9.   (a) Part of finite element mesh for analyzing 2 mm deep
          semi-circular fusion zone; (b) field plots; (c) normalized
          differential impedance components as a function of probe
          position; and (d) normalized impedance |Z| vs. probe
          position. Frequency: 3 kHz.

CONCLUSION

     The theoretical results presented in this paper are from an
axisymmetric finite element code. In the absence of a perfect
circular symmetry, a three-dimensional code should be employed.
Application of this model requires a knowledge of electrical and
magnetic properties of all material media involved in the analysis.
Though the predicted eddy current results are promising, they are

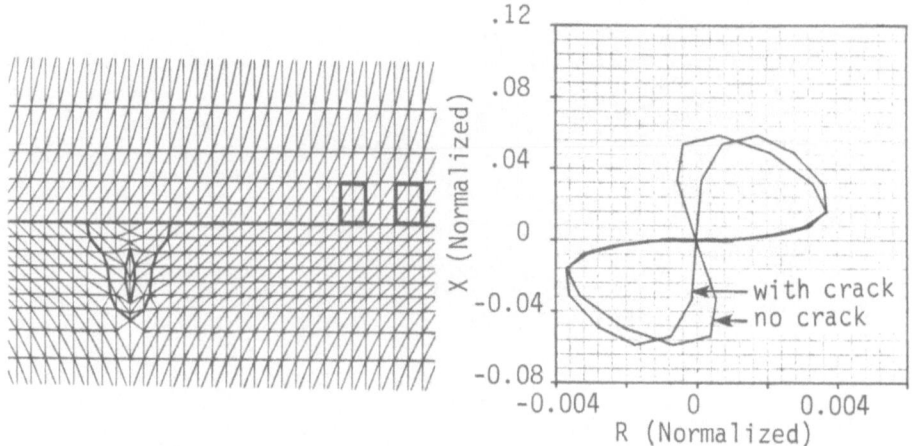

Fig. 10. Finite element predicted differential eddy current
          signals for a hourglass shaped fusion zone with and
          without crack. Frequency: 3 kHz.

yet to be verified experimentally. Since the electrical and
magnetic properties of an amorphous layer differ drastically from
those of crystalline substrate, eddy current inspection technique
may serve as a viable tool for characterizing such surface layers.

REFERENCES

1. H. W. Bergmann and B. Mordike, "Laser and electron beam melted
       amorphous layers," Journal of Materials Science, V. 16,
       1981, pp. 863-869.
2. P. A. Molian, "Laser Surface Alloying of Chromium and Nickel
       on Iron and Carbon Steels," Ph.D. Thesis, Oregon Graduate
       Center, Beaverton, OR 97006, 1982.
3. M. V. K. Chari, "Finite element solution of the eddy current
       problem in magnetic structures," IEEE Transactions on Power
       Apparatus and Systems, Vol. PAS-93, No. 1, January-February
       1974, pp. 62-72.
4. W. Lord and R. Palanisamy, "Development of theoretical models
       for NDT eddy current phenomena," Ed. G. Birnbaum and
       G. Free, ASTM Pub. Code No. (PCN) 04-722000-22, 1981,
       pp. 5-21.

SOLVING FOR THERMAL HISTORY AND METALLURGICAL
CONDITION OF A QUENCHED AND AGED ALUMINUM ALLOY
THROUGH CONDUCTIVITY AND HARDNESS PROPERTIES

Russell A. Chihoski

Martin Marietta-Denver Aerospace
Denver, Colorado

SUMMARY

For any heat-treatable aluminum alloy reporting strength or
hardness (H) with electrical conductivity (C), the accumulations of
data fall into regressing scatterbands which trend down in H and up
in C. Conductivity is useful for sensing anomalies in products
made with these alloys, but the relationship is too dispersed to
find usefully accurate hardness or strength values. The vagueness
is commonly attributed to conductivity measurement inaccuracies.

For a program studying C vs H relationships in the 2219 alloy,
specimens were produced with a variety of thermal histories: that
is, combinations of quenching time and aging time. Each specimen
was found to have a unique C vs H coordinate position. Plotted on
a C vs H format, these points fill a fan-like envelope. The
envelope can be divided into progressive stages of quenching time
(Qt) and into progressive stages of aging time (At). In principle,
a point defined by values of H and C will reveal the values for Qt
and At. Either pair of these dimensions will identify the metallurgical status.

With slower quench rates H decreases and C increases to mark
an HC locus which does lie parallel to the familiar axis of
regression. Moving down this locus marks degradation by the
increase of the coarse θ precipitate (which forms at higher
temperatures). In contrast, aging a quenched specimen at lower
temperatures causes an HC locus which traces a path across this
regression axis to give breadth to the envelope. This records the
precipitation of the fine θ' and θ" strengthing precipitates.

The envelope of practical possibilities makes it clear that any collection of pieces with the same nominal strengths actually can have a wide range of conductivities. With the results of an organized thermal history variety, it is possible to demonstrate the metallurgical differences that account for this.

This work creates a model that coherently and quantitatively ties two nondestructively acquired properties to thermal history and metallurgical condition. This model demonstrates that strength cannot be specified from conductivity alone. However, it shows that the combination of conductivity and hardness can be used to solve for the particular metallurgical condition of an alloy piece. From this known condition we should be able to know all the properties associated with it.

DISCUSSION

When 2219 panels are cooled from their solution heat treatment temperature at different rates, different metallurgical states result. Seven panels whose quenching time (Qt) ranged from very fast to very slow (see Fig. 1) were made. Their condition was noted by their hardness (H) and electrical conductivity (C) after quenching. On an HC format the seven HC coordinate points delineate a locus which parallels the well-known sense that

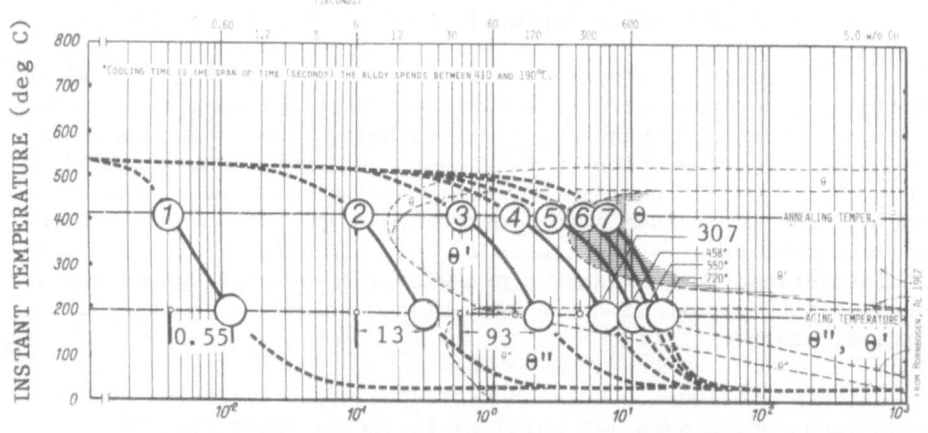

ELAPSED TIME (minutes)

Fig. 1. Graph indicating the severity of quench of several 2219 aluminum alloy specimens plotted as instant temperature vs elapsed time from the initiation of quench. The seven delayed paths cross deeper into nucleation fields leaving, after stabilization, seven graduated metallurgical states.

increasing C accompanies decreasing strength. This locus is the A
line in Fig. 2. The best quenched piece is panel #1, found to have
the HC coordinate value marked by 1A. At 1A Cu, the major alloying
constituent is essentially in solid solution in the alloy matrix
(α). The material is 2219 with 5.8% Cu.

When this 1A material is raised to 190°C and held, its
hardness increases rapidly (1B, 1C, 1D) while its conductivity
remains practically unchanged. After 135 minutes (1F) the aging
course of 1A suddenly reverses to a slow decrease in H with a rapid
increase in C (1F, 1G, 1H). Movement of the HC point slows until
after 2000 minutes it practically stops (1I, 1J). Panels like #3
start at A from a lower-H and higher-C position (3A). Each of the
courses after #1 are characterized by a progressively shorter
H-rise during the aging phase and a shorter C-rise during
overaging phase.

The HC locus of all quenched panels after extended aging is
marked by the J line (1J, 2J, 3J, etc.). What is apparent now is
an envelope (from A to J) that shows all the HC combinations
possible as a result of the widest range of heat treatments. In
this it can be seen how a room full of pieces with a narrow range
of strengths can have a disturbing range of conductivities. It is
clearly possible for metallurgical, instead of measurement, reasons.

Transmission Electric Microscopy (TEM) examination of
specimens taken along the #1 aging course shows that this H-rise
corresponds to the precipitation of very fine θ'' from the matrix
(α). The C-rise corresponds to the nucleation and growth of θ'
when the θ'' disappears. The later panels (#3, etc.) spend enough
time at high temperature during quenching to allow the formation of
a large widely spaced blocky θ ($CuAl_2$) which takes Cu from the
matrix without itself contributing to strength. The reduction of
Cu decreases α's ability to precipitate θ'' and θ'. By #7 so much
θ is formed during quenching, locking up so much Cu that θ'' and θ'
precipitation is actually too small or infrequent to be seen by
electron microscopy. Still their minute formation is seen by the
H-rise and the subsequent C-rise during the aging experience.

Quenching time turns out to be the most important parameter
to know about a heat-treatable alloy produced for aerospace or
vehicular use. After 1 second the increases in Qt decrease
mechanical properties first gradually, then swiftly. Lack of
attention given to exact execution of quenching (plates or
forgings) results in variable quenching times. Those differences
have to be factored against the alloy's quench sensitivity to
deduce consequences. For 2219, Fig. 2 shows the effect of 0.5, 13,
93 seconds, etc. on strength. The change in morphology and
properties here is continuous.

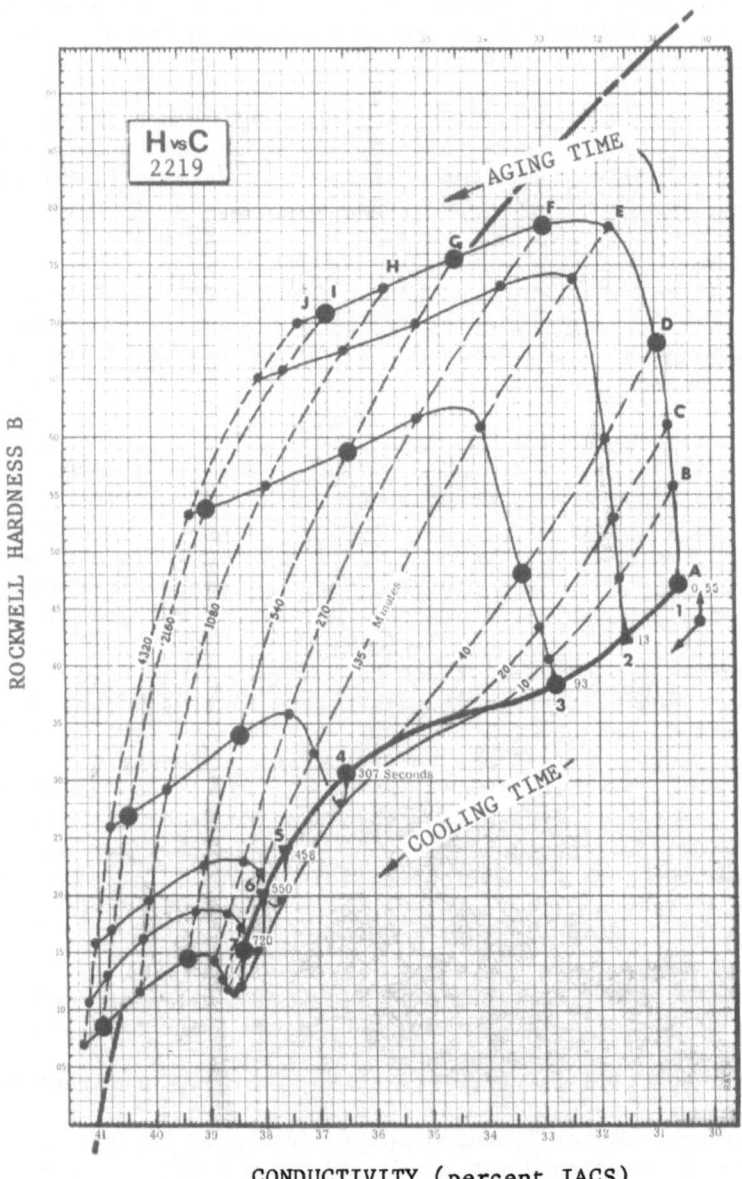

CONDUCTIVITY (percent IACS)

Fig. 2. Plot of Conductivity vs Hardness for an array of
         specimens made by a range of graduated quenching and
         aging times. A wide envelope of possible H vs C
         properties is revealed.

There are two quite different branches of this subject which are interesting and important in several respects. The first extends nucleation and precipitation matters; the second pictures the statistical consequences of Qt and At variations.

Nucleation diagrams illustrate the time of appearance of different precipitates at chosen temperatures. The start boundaries are normally found by taking specimens put into solid solution at a sub-solidus temperature (such as 540°C) and quenching them. They are then raised to an assigned temperature and held. Specimens are removed in time and examined under TEM. This is how Hornbogen's Diagram[1] shown in Fig. 3 was found. His boundaries, shown as dashed lines outlining shaded areas, mark nucleation starts for θ'', θ', and θ. The black dots mark specimens removed for a TEM determination of the precipitate existing at those moments in time.

On this format from reference 1 the aging course of the #1 panel has been superimposed and identified as A through H. From the reference fields θ'' nucleation should start after A around 0.45 minutes. This should start to turn to θ' between 2 minutes and 40 minutes. The end of θ'' is not distinguishable.

The HC format of Fig. 2 revealed these precipitate appearances for 2219 with superior clarity. TEM examinations confirmed what the HC locus of Fig. 2 predicted: θ'' precipitation is in progress

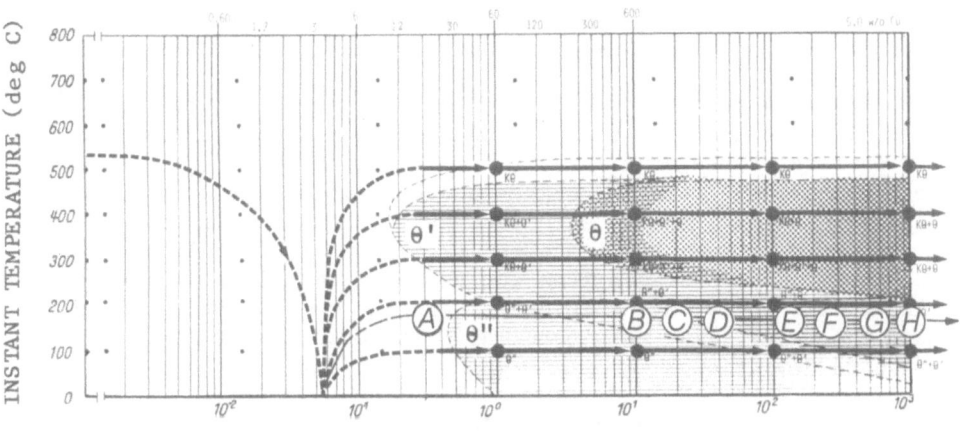

ELAPSED TIME (minutes)

Fig. 3.   Graph indicating how nucleation diagrams are made here by removing 5% Cu specimens at the specified times from the different isothermal holds for a TEM search for the onset of certain precipitates. A through H have been added here to represent the aging of pieces described in this work.

at A and before artificial aging although after 96 hours at room
temperature. It proceeds for 135 minutes (E) whereupon it quickly
disappears in favor of θ' and is gone after 270 minutes (F). This
can be translated to the nucleation diagram, Fig. 3. The first
nose at 190°C (light parallel lines) is earlier than Ref. 1 for
this condition of 2219. The θ' start nose is just before 135
minutes (E) and the last significant θ" will be seen at 270
minutes (F). Hardness or conductivity properties have been used
to show the degree of an undetailed transformation. It appears
possible here to make distinctions between the formations of θ",
θ' and θ, and actually draw nucleation boundaries from the
relative changes of H and C.

Future work will test this prediction. It is interesting to
recognize how these simple techniques can provide access to
sophisticated knowledge of the microscopic interior of a
material. This can put such insight onto a factory floor.

Manufactured products experience cooling from solution
treatment temperatures more like one of the paths shown in Fig. 1.
When practices vary so as to use different cooling paths (cooling
times), the quenched alloy finds itself on different positions on
the A line, Fig. 2. These positions, of course, represent
differences in mechanical, as well as physical, properties. These
variations, either the output of a particular mill or of the
entire industry, show up expressed merely in the form of a
statistical distribution. It is not sufficient to merely
characterize the probable spread of the product. One should want
to know more about the nuances of production practices that lead
to scatter from which design strength standards are calculated.

Figure 4 represents an 11 by 14 foot panel that was produced
several years ago by a common production and inspection system.
The pattern of the constant strength areas, mapped from converted,
portable Brinell reading, suggests quench-water variations.
Consider that if tensile specimens were taken from one corner, the
plate would be accepted and used; from another it would have been
rejected. Coupons taken from any corner from a large number of
like plates would have shaped the Gaussian curve that would be
said to characterize this material. It would have a large
standard deviation.

The nondestructive evaluation discussed here truly is one of
those methods that determines the nature and properties of sound
material without consuming any part of it.

The work led to an assessment of the sensitivity of strength
to manufacturing variations and an appreciation of cooling time as
highest in a rank of influential variables. This comes from a
sense of relative inconsistency of this manufacturing parameter

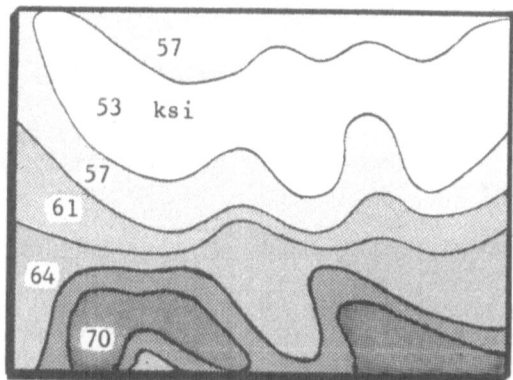

Fig. 4.   Constant strength areas mapped from hardness measurements
         taken on a production plate of 2219.

and the gain that the time has on the consequences. It leads to
questions about the effectiveness of sampling systems on a product
so easily altered.

CONCLUSION

      Hardness and conductivity coordinates can be used to deduce
the quenching and aging times, and the precipitate status of
alloys like 2219. Here is the basis of a system that can
certainly prevent the shipment and use of anomalous products.
Through specified NDE devices a means exists for defining and
culling outstandingly improper pieces. But more important to
consistency and productivity, we have an instrument for preventing
their manufacture.

      The result of implementation will be the shrinking of
standard deviations and the raising of design allowables; at the
same time guaranteeing that no failure-prone sound substance ever
is included in a DOD, aerospace, or civilian vehicle or structure.
The extraordinary significance of this should not be overlooked.

      Of course all studies of all NDE phenomena hope to eventually
accomplish this much. Whether the mere provision of metallurgical,
process, and property answers from combinations of physical
phenomena, as has been done here with hardness and conductivity,
will actually evolve to influence processes and products in a
question whose answer is yet to be seen.

      More information on this subject may be found in recent
publications: References 2 and 3.

REFERENCES

1.   E. Hornbogen, Aluminum, Volume 43, 1967, Parts I-IV.

2.   R.A. Chihoski, Metal Progress, Conductivity-Hardness
     Reveal Heat Treat History of Aluminum Alloys, Part I,
     May 1983, pp. 27-32.

3.   R.A. Chihoski, Metal Progress, Part II, July 1983, pp.
     27-34.

EDDY CURRENT MEASUREMENT OF RESIDUAL STRESSES

IN A HEAVY SECTION STEEL WELDMENT

W. G. Clark, Jr. and W. R. Junker

Westinghouse R&D Center
Materials Engineering Department
Pittsburgh, Pennsylvania 15235

ABSTRACT

A preliminary investigation was conducted to explore the applicability of an eddy current nondestructive evaluation technique to the characterization of residual stresses in a low alloy, heavy section steel weldment. Simulated weld repair samples were examined with a hand held eddy current probe and the results correlated with strain gage measurements of residual surface stresses. Eddy current measurement profiles were used to estimate residual stress contours. The advantages and potential problems associated with an eddy current approach to the measurement of residual stresses are discussed.

1. INTRODUCTION

The accurate characterization of residual stresses in structural components is critical to the development of stress analyses required for design assurance and hardware qualification. An area of particular concern is heavy section steel weldments where thermal stress relief treatments are not practical or of questionable value (for example, in-field weld repairs on large structures). Presently, two techniques are widely used to characterize residual stresses: x-ray diffraction techniques and hole drilling (metal removal) methods combined with strain gage measurements. The x-ray technique is essentially nondestructive in nature and problems with portability of equipment have recently been resolved with some unique developments.[1] However, this technique is limited to the measurement of very shallow surface-stresses on the order of a few microns deep. The hole drilling-strain gage method on the other hand can measure subsurface stresses but is a destructive test which leaves small diameter holes in the hardware being evaluated. The significant nature of these limitations is reflected by the fact that it is common design practice to forego measurements and to assume residual stresses equivalent to the yield point of the material involved. Obviously, this assumption can yield an overly conservative analysis. An improved method of residual stress measurement could resolve this situation and lead to improved structural integrity analyses.

Recent experimental work has demonstrated that conventional eddy current (EC) nondestructive evaluation techniques can be used to characterize applied and residual surface stresses in low-alloy steels.[2] The technique involves the use of a simple hand-held probe (coil) which measures changes in electromagnetic properties caused by the application of stress (strain). A calibration curve of eddy current response versus applied stress for the material of concern is required to correlate test results with stress conditions. The technique is particularly useful for mapping surface stress contours and changes in test frequency can be used to control the depth of surface penetration yielding depth profiles of the stress field.

In view of the success encountered to date, work is underway to further explore the applicability of the eddy current characterization of residual stresses and to develop optimum techniques. This paper presents the results of preliminary work associated with the characterization of residual stresses in a heavy section ASTM A533 steel weldment. Specifically, eddy current measurements made on simulated weld repair specimens are compared with residual stress data determined by the more conventional metal removal-strain gage technique. A brief review of the principles of eddy current testing is included along with a summary of results developed to establish calibration data. Recommendations for further work and potential problem areas are discussed.

2. PRINCIPLES OF EC TESTING

The basic concept underlying the use of eddy current techniques to characterize the properties of materials is the relationship between material structure (microstructure, texture, etc.) and electromagnetic behavior.

More specifically, eddy current testing relies on the electro-magnetic interaction between a coil driven by an alternating electric current and the material under test. This interaction is governed by Faraday's law which says that an electrical conducting loop (coil) placed in a changing magnetic field has a voltage generated across the ends of the loop which is proportional to the time rate of change of the field enclosed by the loop. If the loop is closed, a current flows in the loop in the direction opposite to the change in the magnetic field. If the loop is replaced with a conducting plate, the changing magnetic field produces a current in the plate which flows in closed loops and is referred to as an eddy current field. Eddy current testing uses as the source of the changing magnetic field an inspection coil which is driven by an alternating electric current. The voltage across the coil is proportional to the time rate of change of the magnetic flux generated by the coil. However, as the coil is brought near a metallic surface the total magnetic flux seen by the coil is changed by the currents generated in the surface. In turn, the coil voltage is altered.

With the coil driven by a sinusoidal varying current, the relationship between the voltage and current is given by the steady state equivalent of Ohms law, $V = IZ$ where V is the voltage across the coil, I is the current in the coil and Z is the impedance of the coil. In general, the impedance is composed of two orthogonal components. One is associated with the losses in the coil and the other associated with the inductance of the coil. The effect of bringing the coil near a metallic surface is to alter the coil impedance by introducing changes in both the loss and inductive components. For a particular coil-material interaction, the exact value of the coil impedance will depend on: (1) the coil geometry, (2) the spacing between the coil and the material, (3) the electrical conductivity of the material, (4) the magnetic permeability of the material and (5) the frequency at which the coil is excited. In general, the characterization of materials with a conventional eddy current NDE system involves the measurement of impedance changes rather than impedance alone. Specifically, the presence of a surface flaw can produce a significant change in impedance as compared to the unflawed material yielding an effective flaw detector. All eddy current data reported in this investigation involve the measurement of impedance changes.

In order to measure applied or residual stresses with a conventional eddy current approach, the application of stress must change either the conductivity or permeability of the material such

that a detectable change in test coil impedance occurs. For
ferromagnetic materials it is well known that the application of
stress (strain) will alter both the electrical and magnetic
properties. This phenomenon provides the basic concept underlying
the possible use of eddy current techniques to measure applied and
residual stresses in structural steels.

3. EC MEASUREMENT OF STRESSES

The preliminary evaluation of the eddy current approach to the
measurement of stresses was conducted with simple tension and
bending specimens under known load conditions.[2] Specifically, an
eddy current probe was held against the specimen (in the gage
length) under test and the eddy current response (change in coil
impedance) recorded during loading and unloading. Figure 1 shows
the eddy current response versus applied stress data for an ASTM
type A533 steel (65 ksi, 449 MPa yield strength) specimen subjected
to tensile loading and unloading. Similar results have also been
developed for an ASTM A471 steel (111 ksi, 766 MPa yield strength)
under both tension and bending conditions. Figure 2 shows results
for a 3-point bend specimen where the probe was moved along the
specimen axis to collect stress profile data under constant load
conditions. The actual stress contour (outer fiber bending stress)
is presented along with the eddy current measurements. In addition
to data collected with simple test specimens (Figs. 1 and 2),
results have also been generated with a residual stress specimen (a
disc with a pin shrunk into the bore) and the applicability to
residual stress measurements demonstrated.

As noted in Figures 1 and 2, an excellent correlation exists
between eddy current measurements and the magnitude of applied
stress. In addition, relatively little data scatter is encountered
for the low alloy steels investigated to date. In all cases
described above and the subsequent work with weldments, the eddy
current data were collected with a hand-held 0.25 in. (0.64 cm)
diameter, "pancake"-type probe driven at 400 kHz by a Zetek EM3300
nondestructive testing instrument. Prior to collecting data, the
test instrument was balanced to a zero reading with the probe in
contact with a stress-free portion of the test piece. The eddy
current response is a relative measure of the change in coil
impedance expressed in nondimensional units on the test instrument.

For the test conditions involved in this investigation, the
estimated depth of penetration of the eddy current field is about
0.005 in (0.013 mm). In addition, the probe is omnidirectional and
does not reflect the direction of stress.

4. APPLICABILITY TO WELDMENTS

An opportunity to establish the applicability of the eddy
current approach to the measurement of residual stresses in
weldments came about as the result of access to two large test

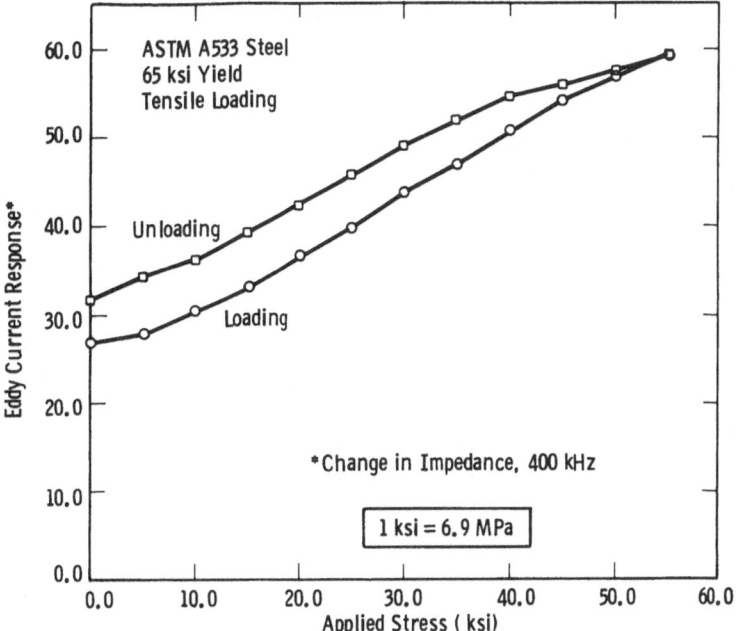

Figure 1. Eddy current response versus applied stress.

specimens fabricated several years ago to evaluate weld repair
procedures.[3] In this case, two ASTM A533 steel (65 ksi, 449 MPa
yield strength) specimens 18 in. long x 10 in. wide x 8-1/4 in.
thick (46 x 25 x 21.6 cm) were designed and fabricated to evaluate
the effect of peening on residual stresses in the weld repair of
stress relieved components. The test specimen design is shown in
Figure 3. In both specimens a "half bead" welding technique was
employed (per Section XI of the ASME Nuclear Pressure Vessel Code,
IWB-4421). One specimen was subjected to a hammer peening
operation intended to minimize residual stress development during
welding. The other specimen was not subjected to peening.
Following welding, residual stress measurements were conducted to
assess the impact of the peening operation on residual stress
development. In the original investigation, residual stress
measurements were made using a modified hole-drilling strain gage
technique. After four years in storage the specimens were
reexamined with the eddy current method described here.

4.1 Strain Gage Results

Residual stress measurements were made at four locations on
the top surface of each test block. Figure 4 shows the location of

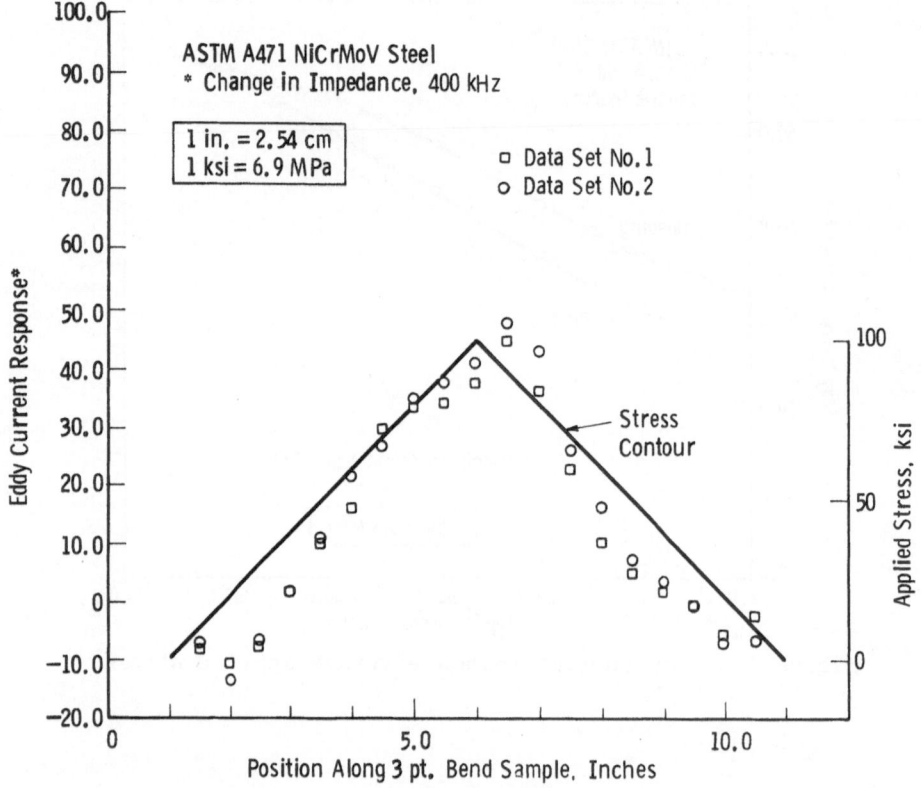

Figure 2.   Eddy current response for "As Loaded" Bend Specimen.

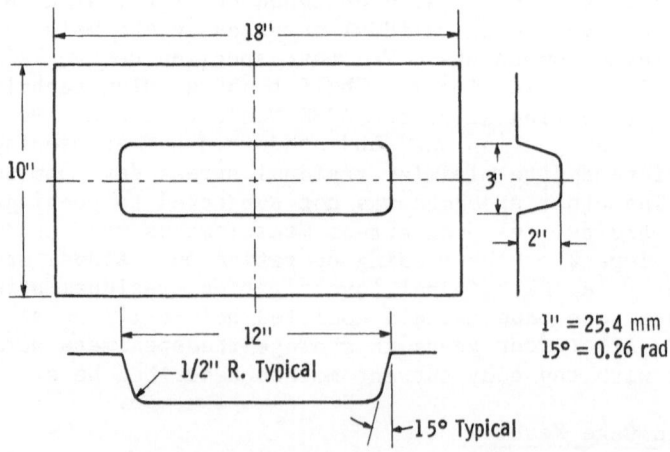

Figure 3.   Weld groove configuration.

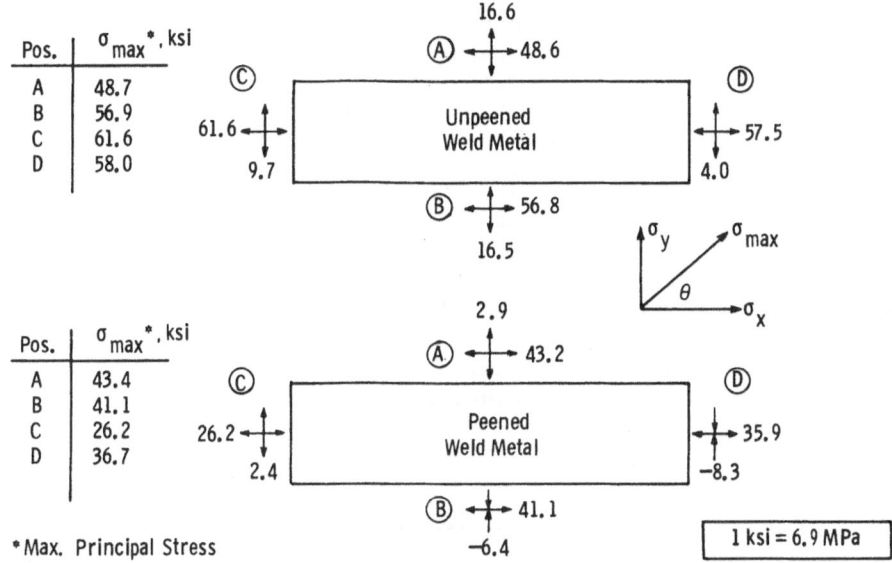

Figure 4.   Coordinate stresses from strain gage--metal removal
            analysis.

the hole drilling-strain gage measurements with respect to the weld
metal. With this method, local stress relaxation is obtained by
drilling a hole in the center of a small (0.35 in., 8.9 mm
diameter) three-gage rosette bonded to the specimen. An abrasive
air-jet machining technique was used to produce an essentially
stress-free hole 0.060 in. (1.5 mm) in depth and diameter. The
magnitude and direction of the coordinate stresses at the four test
locations are summarized in Figure 4. The maximum principal stress
values are also as expected from the specimen geometry and welding
process; the higher stresses align with the direction of welding.
Note that the maximum stress measured in the unpeened weldment,
61.6 ksi (425 MPa), σ_x at location C, is 95% of the base metal
yield strength. Note also that the maximum stress measured in the
test blocks is approximately 35% lower in the peened weldment than
in the unpeened weldment indicating a beneficial effect of peening.
Since the final layer of weld metal was not peended (per specification) and since the stress measurements were limited to relatively
shallow surface depths, it is reasonable to assume that somewhat
more stress reduction can be attributed to the peening operation.

4.2   Eddy Current Results

Eddy current measurements were made in the same regions of the test blocks as the hole drilling–strain gage measurements. However, because of the nondestructive nature of the eddy current approach it was possible to develop considerably more data and also to produce an estimated stress contour for each block. The results of a comparison of the eddy current measurements with the strain gage data are shown in Figure 5. In this case the eddy current data reflect measurements immediately adjacent to the rosette strain gages used in the hole drilling analysis. Note the excellent linear correlation between eddy current response and maximum residual stress value determined from the strain gage measurements. The correlation shown in Figure 5 was used as a calibration curve for all subsequent estimates of residual stress based on eddy current measurements. More specifically, these data were used to develop the estimated stress contours.

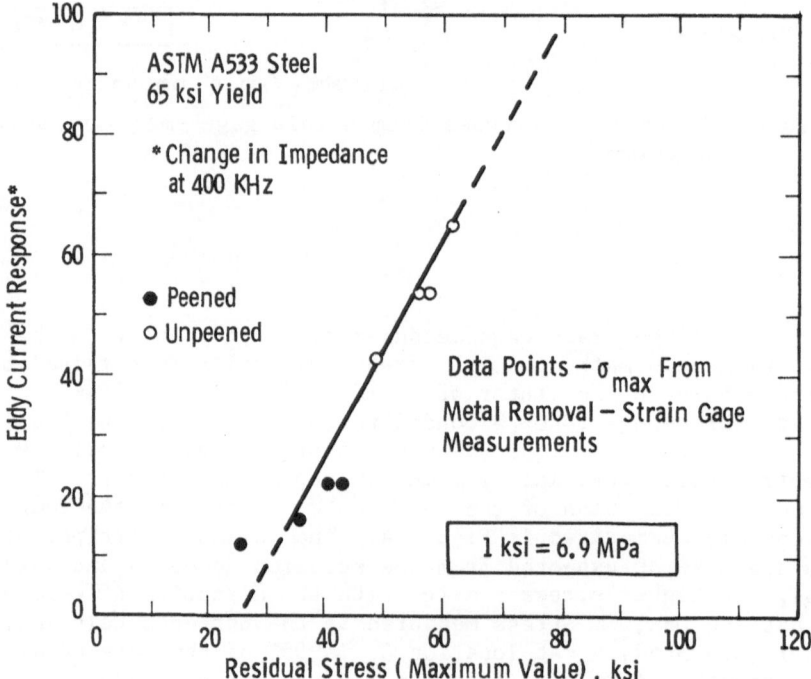

Figure 5.   E.C. response versus strain gage (hole drilling) measurements.

Figure 6 shows the eddy current measurement positions and identification used to establish stress contour data. Note that position No. 1 is immediately adjacent to the rosette gage and the

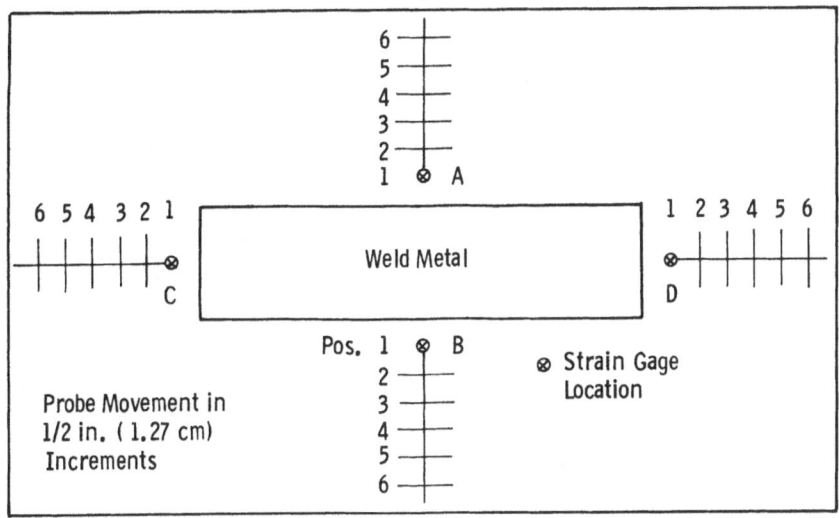

Figure 6.   Eddy current measurement positions on weldment.

other positions represent 0.5 in. (1.27 cm) increments along a line
perpendicular to the weld groove faces.

Figures 7 and 8 present the eddy current measurement profiles
for locations A and B and C and D, respectively. Note that in all
cases the eddy current measurements vary considerably upon moving
from position No. 1 to position No. 6. Note also that with few
exceptions the eddy current readings indicate significantly higher
stresses in the unpeened specimen compared to the peened specimen.
In general, the residual stresses appear to increase significantly
upon moving away from position No. 1. In the case of the unpeened
specimen, the estimated residual stresses measured beyond about
position No. 3 exceed the yield strength of the base metal (per the
calibration curve in Figure 5, an eddy current reading above 70
indicates yielding). Based upon these observations it appears that
the peening operation has a much larger impact on residual stresses
than the 35% reduction indicated from the strain gage data at
position No. 1.

5. DISCUSSION

The limited data developed in this investigation clearly
demonstrate the potential applicability of the eddy current
approach to the characterization of residual stresses in low alloy,
heavy section steel weldments. A good linear correlation between
eddy current response and strain gage measurements of residual
stress has been demonstrated. The eddy current approach is quick,
nondestructive in nature and can be used to develop surface

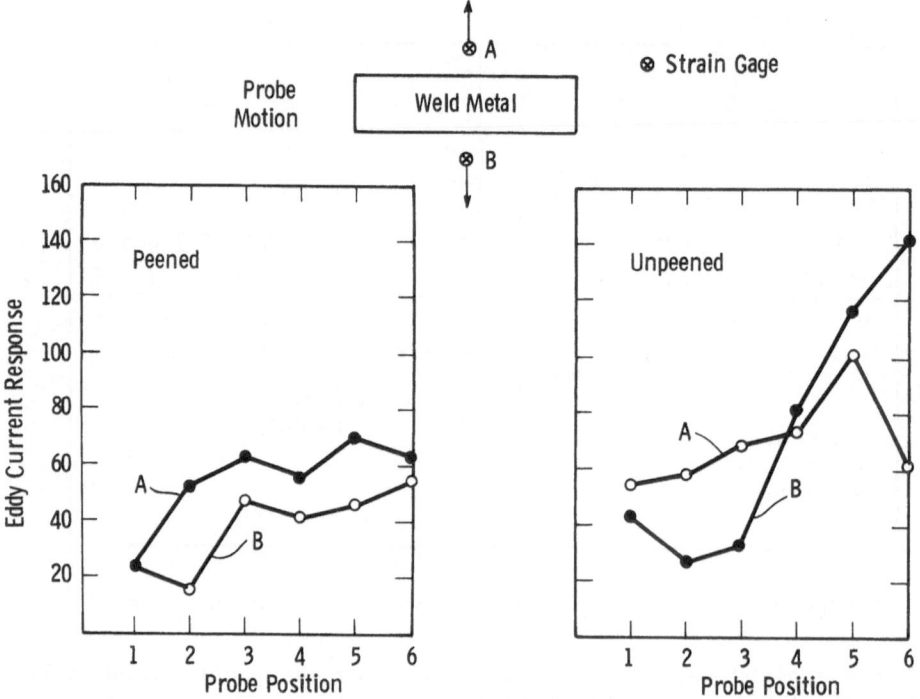

Figure 7. Eddy current measurement profile--locations A & B.

profiles of the residual stress field. Data developed with the
weld repair simulation samples clearly illustrate the advantages of
the eddy current method over the hole drilling technique.
Specifically, with the hole drilling technique, data can only be
developed at the strain gage locations. Thus, it is necessary to
estimate the position of maximum stress prior to strain gage
installation. If this estimate is not accurate, maximum stress
levels may be significantly under-estimated.

As noted previously, this paper represents a summary of
research in progress and a considerable amount of additional work
is required to adequately evaluate the use of eddy current
techniques to establish residual stresses. Some of the potential
problems include the effect of surface finish, material structure
and complex stress fields on eddy current measurements.

In this investigation it was noted that careful surface
preparation is required to yield reliable eddy current
measurements. Specifically, the variation in surface oxide
associated with the "as-welded" specimens examined in this
investigation created measurement problems and it was necessary to
remove the oxide to yield satisfactory results. Light hand sanding

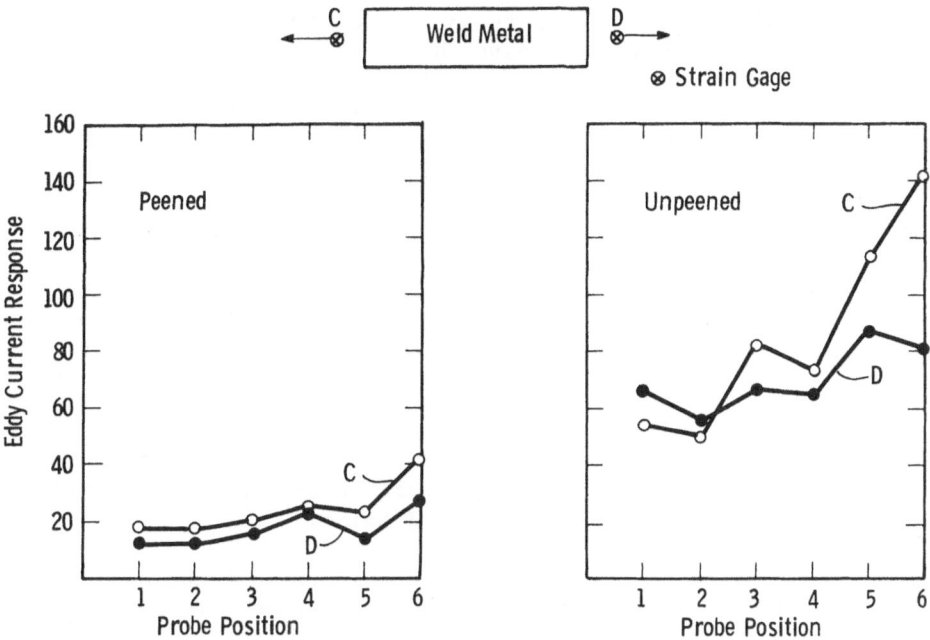

Figure 8.   Eddy current measurement profile--locations C & D.

with emery paper produced an adequate surface.   Surface preparation
experiments conducted on calibration specimens indicated that light
hand sanding does not affect applied stress measurements.

Comparison of the applied stress versus eddy current response
data in Figure 1 with the data in Figure 5 shows a significant
difference in results obtained with what appears to be similar
steels. Both sets of data were collected under identical eddy
current calibration conditions.   Note that neither the actual eddy
current versus stress values nor the slope of the data are consistent.   Comparison of calibration data with other heats of A533
steel are very similar to the data in Figure 1.   No explanation has
yet been found for this anomalous behavior. However, detailed
stress direction and microstructural evaluations are planned to
explore the impact of these variables.   Despite the inconsistencies
between calibration data, we believe that the data in Figure 5 are
accurate for the weldment in question since the calibration was
established on the same material.   However, the calibration problem
noted here serves to indicate the need to establish calibration
data on material representative of that to be examined for residual
stresses.

Presently, the work done to correlate eddy current response with applied and residual stresses has been conducted with conventional eddy current probes and simple test specimens. These coils are essentially omnidirectional (point source of electromagnetic energy) and do not have the capability to determine the direction of stresses. Consequently, for complex stress conditions which could develop in weldments, an omnidirectional coil could average the results and yield misleading data.

Despite problems such as the concerns noted here, the potential advantages of an eddy current nondestructive technique for the characterization of stresses justify further research and development effort in this important aspect of structural reliability.

6. CONCLUSIONS

The pertinent conclusions associated with the investigation described in this paper are summarized below.

1. An eddy current nondestructive inspection technique can be used to characterize the residual surface stresses in a low alloy, heavy section steel weldment.

2. A linear correlation exists between eddy current response and the magnitude of residual stresses measured by the hole drilling-strain gage technique.

3. The eddy current response to residual stress conditions is material dependent and a calibration must be developed for the material of concern.

4. Material surface conditions can influence the eddy current characterization of residual stresses. Heavy oxide layers must be removed to yield reliable results.

5. Conventional omnidirectional eddy current probes cannot measure the direction of stresses in a complex stress field.

7. ACKNOWLEDGEMENTS

The test specimens examined in this effort were supplied by G. E. Grotke of the Metallurgy Department. The hole drilling-strain gage residual stress analysis was conducted and reported by A. J. Bush of Materials Evaluation. David Chizmar of Materials Engineering conducted the experimental effort associated with all eddy current measurements. Financial support for this program was provided by the Corporate Research program.

8. REFERENCES

1. C. O. Ruud, "X-Ray Analysis and Advances in Portable Field
    Instrumentation", Journal of Metals, Vol. 31, No. 6, June 1979.

2. W. R. Junker and W. G. Clark, Jr., "Eddy Current
    Characterization of Applied and Residual Stresses", ARPA/AFML
    Review of Progress in Quantitative NDE, San Diego, CA, August
    1982.

3. Unpublished research work by G. E. Grotke and A. J. Bush,
    Westinghouse R&D Center.

MAGNETIC, ULTRASONIC, AND ANALYTICAL TECHNIQUES

FOR STRESS AND COLD WORK INDICATION

SOLVING INTERNAL STRESS MEASUREMENT

PROBLEMS BY A NEW MAGNETOELASTIC METHOD

Kirsti Tiitto

American Stress Technologies, Inc
Pittsburgh, Pa

ABSTRACT

Nondestructive measurement of internal stresses has
conventionally been restricted to the utilization of x-ray
diffraction method that yields absolute stress values under
certain physical conditions. This method is, however, accompanied
by a number of drawbacks, such as shallow depth of penetration and
high cost. It is also very difficult to use it to the continuous
monitoring of internal stresses in industrial processes.

In this paper, magnetoelastic interaction was used to
evaluate the internal stresses through analyzing the statistics
of magnetization transitions (Barkhausen noise). These transitions
were excited by a controlled magnetic field and detected electromagnetically. The average amplitude of the signal generated by
the transitions was used to characterize the transition statistics.
It has been shown that this amplitude is very sensitive to the
stresses in the material and can be applied to nondestructive
stress measurement in ferromagnetic materials.

This paper will present the results of stress measurements
using the above principle in several practical applications. The
results show the potential of this method both in the laboratory
for static measurements and in process control for dynamic,
continuous monitoring of the stress level. The advantages, as
well as the disadvantages, of this method over the conventional
ones are discussed.

INTRODUCTION

Due to the physical interaction between magnetostrictive and elastic lattice strains, the magnetic properties of ferromagnetic materials depend on their stress condition. This so-called magnetoelastic interaction appears as follows. If a piece of ferromagnetic steel is magnetized, it will elongate in the direction of applied magnetic field; conversely, if the same piece is stretched by an applied load, it will be slightly magnetized in the direction of the load. The same applies to compression, except that the resulting magnetization now occurs at 90 degrees to the direction of compressive load. This phenomenon can be observed by analyzing the Barkhausen noise,[1] as found by Gerlach and Lertes[2] in Germany as early as 1921. A year later, Zchiesche[3] published results showing a clear relation between the Barkhausen noise and material's stress condition. It took, however, more than half the century until this phenomenon received more attention and was exploited to determine the residual stress of ferromagnetic materials in practical applications.[4-8]

The Barkhausen noise results from the series of abrupt changes or jumps in the magnetization of a ferromagnetic substance when the magnetizing field is gradually altered. A qualitative indication of magnetoelastic interaction through Barkhausen noise results from the fact that increasing tensile stress is accompanied by growing Barkhausen noise level, and increasing compressive stress by decreasing noise level. Quantitative data may be obtained through calibration with known loads.

INSTRUMENTATION

Commercially available instrumentation was used in this work. It includes a sensor to excite and detect the magnetic noise, and a central unit to control the sensor and process the signal. The central unit automatically evaluates and displays in a numerical form the measurement result, called magnetoelastic parameter here.

The instrumentation available allows for both static and dynamic measurements. Most of the results reported in this work were obtained by using the static laboratory unit shown in Fig. 1. This unit can measure point by point in either single or continuous mode of operation. In single mode, the result is obtained within 4 seconds. For some measurements, the dynamic model equipped with a sensor that allows the measurements on a moving surface up to speeds 5 m/s was used.

The measurements are conducted nondestructively on the sample surface without any special surface preparation. With the sensors mentioned above, the magnetoelastic parameter can be measured in

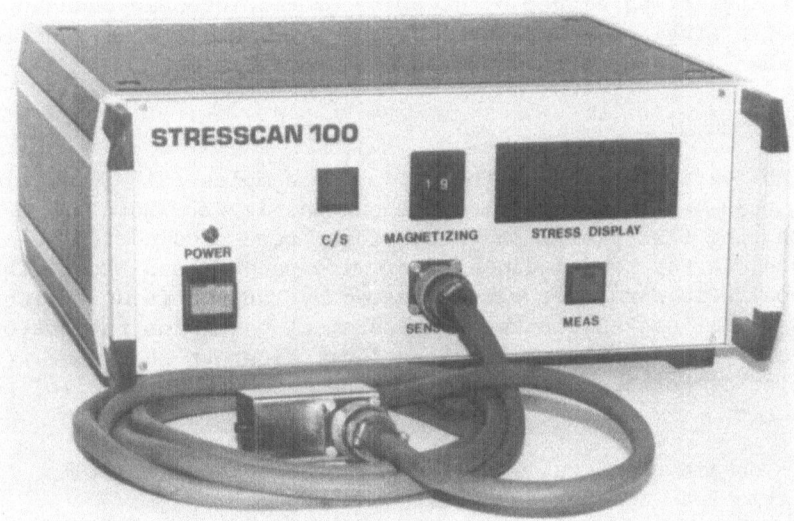

Fig. 1. The instrument used for static measurements.

any direction parallel to the sample surface. Depending on the
sensor configuration, the area of measurement varies from 0.0016
in^2 (1 mm^2) up to 0.4 in^2 (250 mm^2). The depth of measurement
depends on the frequency range of noise signal and can hence be
varied. For the range of 500-10000 Hz used in this work, it is
0.008 - 0.012 in (0.2 - 0.3 mm).

EXPERIMENTAL PROCEDURE

Calibration Tests

     The magnetoelastic parameter as measured in this work
represents the stress level of the material in relative units.
Since an absolute stress value is desirable in some applications,
calibration tests were conducted to study the relation between
this parameter and applied stresses in different materials. Four
different steels with yield strength ranging from 34 to 103 ksi
(234 to 710 N/mm^2) were used in these tests. The calibration
samples were typically 0.5 in (12 mm) wide, 4 in (102 mm) long, and
0.25 in (6 mm) thick. They were stress relieved at 1200 F (650 C)
for 2 to 3 h and cooled slowly to room temperature before
calibration. These test pieces were then strained in tension
and compression within the elastic limit by using a tensile
machine together with strain gauges. Simultaneously, the
magnetoelastic parameter was recorded in the direction of loading.
Depending on the operation mode of the measuring unit, the tests

could be performed either by stepwise or continuously changing loads. In single type of operation, approximately 12 measurements were made in the tension and 4 to 6 in compression.

Weld Seam Samples

The configurations of the weld seam samples (SAE 1020, yield strength 36 ksi (248 N/mm^2)) are shown in Fig. 2. Both types of samples, the flat plates and T-sections, were studied in the as-received, stress relief annealed or wire peened condition. The magnetoelastic parameter was evaluated by the static unit with 1/8 in (3 mm) distance between measurement points on the bottom surface of the samples at two locations, as shown in Fig. 2. The direction of measurement was both perpendicular and parallel to the weld seams.

To obtain the actual stress values of these weld seams, calibration was conducted as discussed above.

Tube Samples

In order to study the effect of stress relief annealing, AISI 52100 bearing tubes with wall thickness of 0.24 in (6 mm) were measured both statically and dynamically in the as-received and stress relief annealed (1250 F (677 C) / $\frac{1}{2}$h) condition. With the static unit, the measurements were conducted at spacings of 0.2 in (5 mm) around the tubes. In the dynamic measurements, the tube sample was rotated under the sensor, and the magnetoelastic parameter was simultaneously recorded continuously all around the tube.

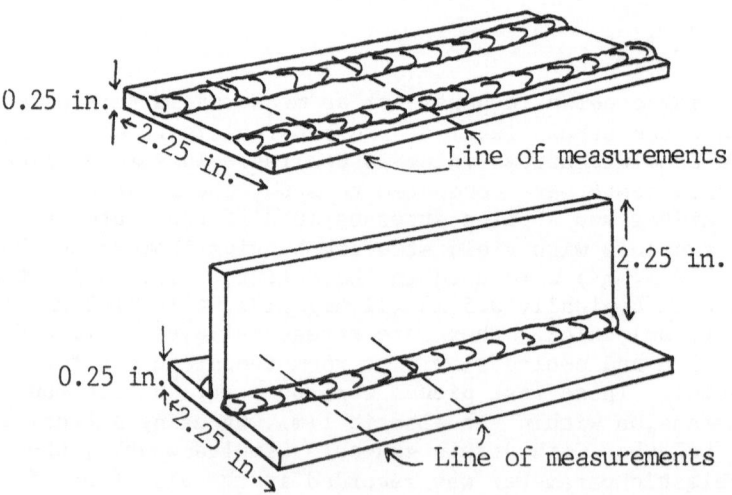

Fig. 2. Welded flat plate and T-section samples.

RESULTS AND DISCUSSION

Calibration Curves

The calibration curves obtained for four different steels with yield strength from 34 to 103 ksi (234 to 710 N/mm^2) are shown in Fig. 3. It can be seen that the curves are close to linear from approximately one third of yield stress in compression to approximately two thirds of yield stress in tension. Close to the yield stress, each of the calibration curves saturates. This phenomenon is especially pronounced for the steel with highest yield strength.

It is further observed in Fig. 3 that for Steel 4, the magnetoelastic parameter changes from 8 to 106 (106/8=13) with changing stress from close to yield stress in compression to close to yield stress in tension. For Steel 2, this change (190/8=24) is even higher. It is thus evident that the magnetoelastic parameter is very sensitive to a change in stress. The experimental curves of Fig. 3 can be used to convert the magnetoelastic parameter into actual stress values in materials under inspection.

Residual Stresses of Weld Seams

The results of stress measurements on the welded flat plate and T-section samples depicted in Fig. 2 are given in Fig. 4. The most interesting results can be listed as follows:

1.   The as-welded T-section and flat plate samples have high tensile stresses close to the yield stress parallel to one of the weld seams, see Fig. 4a and d. The other seam has lower parallel stresses; it was most probably welded first and "stress relieved" during the welding of the second seam.

2.   In the welded T-section samples, stress relief annealing has decreased the stress levels close to zero in both directions (Fig. 4b), whereas wire peening has generated beneficial compressive stresses (Fig. 4c).

3.   Stress relief annealing has not been able to eliminate the high tensile stresses parallel to the weld seams in the flat plate sample, as can be observed in Fig. 4e.

The present results suggest that the stresses in weld seams can indeed be close to the yield stress of the material and are not always eliminated by the stress relief annealing.

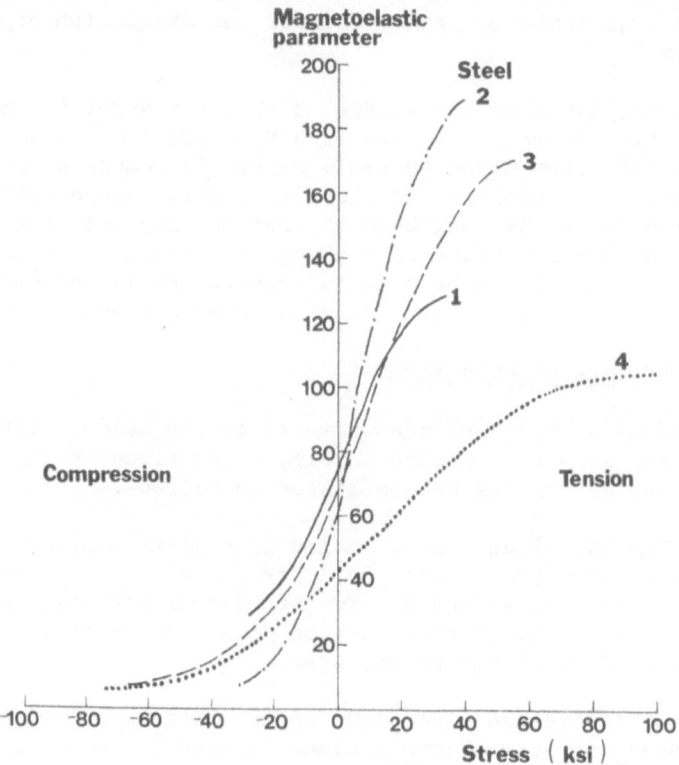

Fig. 3.   Calibration curves for four different steels with yield
          strength of 34 ksi (Steel 1), 36 ksi (Steel 2), 58 ksi
          (Steel 3) and 103 ksi (Steel 4).

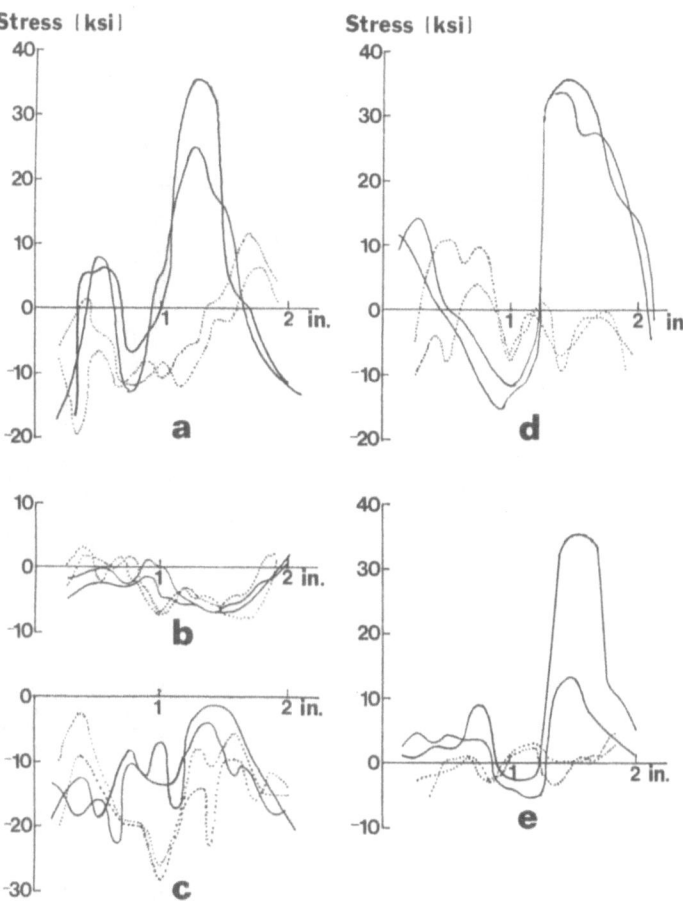

Fig. 4.   Residual stresses of weld seams in the (a) as-welded, (b)
          stress-relieved and (c) wire-peened T-section samples; (d)
          as-welded and (e) stress-relieved flat plate samples of
          SAE 1020.
          For explanation, see Text and Fig. 2.
          Solid: parallel stresses; dotted: perpendicular stresses.

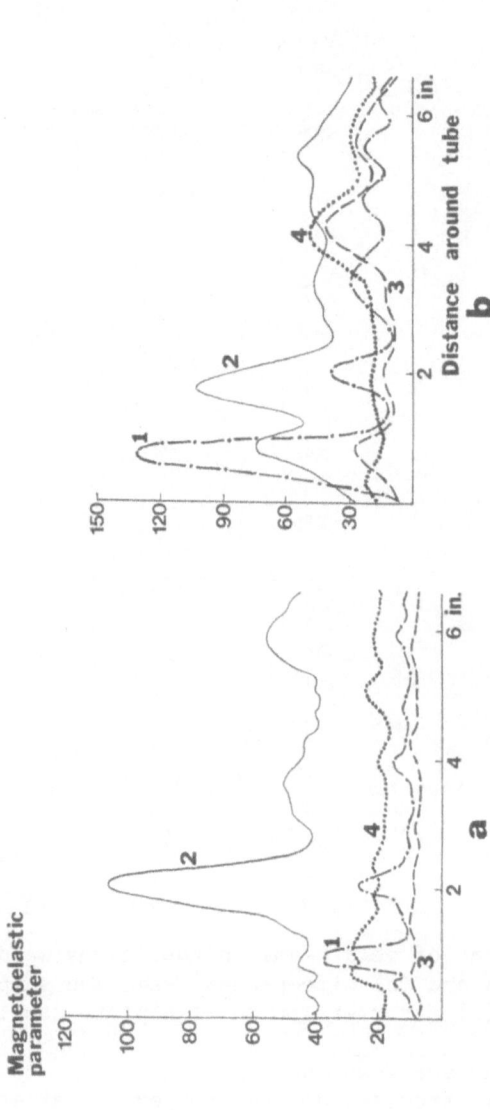

Fig. 5. The magnetoelastic parameter as a function of distance around AISI 52100 bearing tube, as measured with (a) the static unit and (b) the dynamic unit.

Curve 1: as-received tube, parallel stresses
Curve 2: as-received tube, perpendicular stresses
Curve 3: stress relief annealed tube, parallel stresses
Curve 4: stress relief annealed tube, perpendicular stresses

Effect of Stress Relief Annealing

Figure 5 shows the measurement results on AISI 52100 bearing tubes as obtained in the static and dynamic tests. It appears that both units give high values of the magnetoelastic parameter in the as-received tube in parallel and perpendicular direction to the tube axis in some areas. Since calibration test pieces in this specific material were not available, only estimates of the stresses present can be made. It is most probable that the high values mentioned above represent high tensile stresses. During the stress relief annealing the level of the curves goes down and simultaneously the high tensile stresses in some areas are eliminated to a great extent. Comparison of the values measured in the tubes to the calibration curves of Fig. 3 suggests that the stress relief annealing has introduced slight compressive stresses in the surface of the tubes.

CONCLUSIONS

The results presented above have shown that the magnetoelastic method can give a qualitative, and in many cases a quantitative indication of the stress condition in ferromagnetic materials. The main advantage of this method is that the measurements can be conducted nondestructively on both static and moving samples. Also, no special surface preparation is needed. Lightweight, portable battery-driven units can easily be taken to the field. Special sensor constructions already available also enable the measurements on complicated structures and parts like gears, camshafts, crankshafts, bearings, etc.

The disadvantage of this method is that it does not directly yield the absolute stress values. In cases where it is desirable to know the stresses accurately, calibration must be conducted. The magnetoelastic method is best suited to comparative stress measurements where high accuracy is not necessary and has already found a number of promising applications in the industry.

REFERENCES

1. H. Barkhausen, Phys. Zeitschrift 20:401 (1919).
2. W. Gerlach and P. Lertes, Phys. Zeitschrift 22:568 (1921).
3. K. Zschiesche, Phys. Zeitschrift 23:201 (1922).
4. J. Barton and F. Kusenberg, Trans. ASME, J. Eng. Power 10:349 (1974).
5. S. Tiitto, Acta Pol. Scand., Ph 119:3 (1977).

6.   M. Zschau and F. Eichhorn, Anwendbarkeit der Analyse des
     Barkhausen-Rauschens zur Bestimmung von
     Eigenspannungen, Inst. für Schweisstechn.
     Fertigungsverfahren der RWTH Aachen, Aachen,
     Dec. 1978.

7.   L. P. Karjalainen, M. Moilanen, and R. Rautioaho,
     Evaluating the residual stresses in welding from
     Barkhausen noise measurements, Materialprüf. 22:196
     (1980).

8.   S. Tiitto, Ein neues Verfahren für die zerstörungsfreie
     Ermittlung der Eigenspannungen in ferromagnetischen
     Stählen, Schweissen und Schneiden 32:449 (1980).

NONDESTRUCTIVE DETERMINATION OF RESIDUAL AND APPLIED STRESS

BY MICRO-MAGNETIC AND ULTRASONIC METHODS

Eckhardt Schneider, Iris Altpeter, and Werner Theiner

Fraunhofer-Institute
für zerstörungsfreie Prüfverfahren (Izfp)
D-6600 Saarbrücken, FRG

INTRODUCTION

Ultrasonic techniques can detect volume and surface stresses and they are not limited to the investigation of ferromagnetic materials as the micromagnetic methods. The magnetic nondestructive techniques are restricted by eddy current damping to surface near regions. All ferromagnetic and magnetoelastic quantities are sensitive to the mechanical stress and micro-structural state of the test specimen, whereas the ultrasonic velocity is mainly influenced by changes of the stress level and the texture. The main problem is: how to separate the different influences in the measuring quantities which is required to realize a quantitative nondestructive stress-measuring device.

MAGNETIC TECHNIQUES

Magnetic and magnetoelastic measuring quantities show very different reactions in the tensile and compressive stress region.[1-3] A quantitative description of these interactions between the ferromagnetic state represented especially by Bloch walls and micro- and macro-stresses is not given up to now. Therefore, in the Fraunhofer-Institut für zerstörungsfreie Prüfverfahren, different stress-sensitive quantities are measured in combination. By this, a possibly small sensitivity of one measuring quantity can be overcome.[1] Furthermore, one can decide in many cases between ambiguous values, and can detect changes in microstructure.[2,3] The various micromagnetic quantities which can be used for stress measurements on components are: the coercivity (H_c), the magnetic (M) and acoustic (A) Barkhausen noise, the

115

incremental permeability (μ_Δ), and the dynamic magnetostriction (EMUS$_\lambda$). Different combinations of these quantities are possible (see also the contribution of W. Theiner and H. Willems in this volume) which can be used under practical test conditions.

The principle and the results of one application, namely the hardness and residual stress test on low pressure turbine blades using the magnetic Barkhausen noise, are shown in Figs. 1-3. Because there are two unknown quantities--the hardness and the stress--one has to measure at least two nondestructive quantities. In our case, the corecivity (H_{cM}) and the maximum (M_{MAX}) deduced from the rectified Barkhausen noise signal M(H) are measured. The microstructural state can be determined by the correlation HV10$\longleftrightarrow$ H_{cM} (Fig. 1). Both quantities, H_{cM} and M_{MAX}, are then used to calibrate the $M_{MAX,H_{cM},i}$ values (Fig. 1) in stress units in conjunction with the x-ray method. Because the magnetic transducer detects signals out of a depth $\leq$300 µm, whereas the x-ray method probes only a surface layer less than 20 µm, a direct correlation will be observed only if there are no sharp stress or structural gradients within the given interacting volumes.

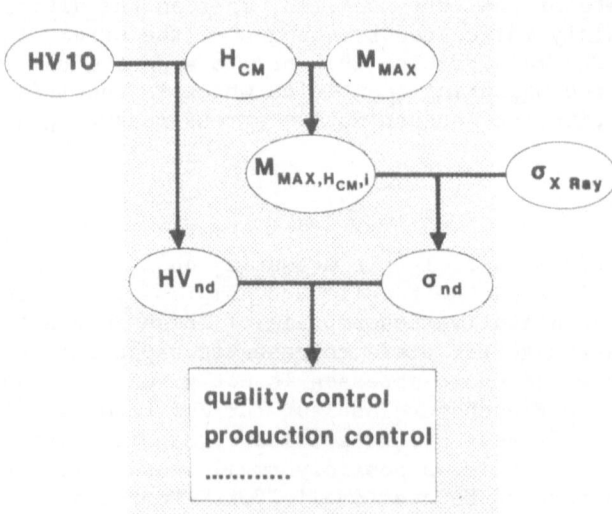

Fig. 1.   Hardness and residual stress determination by magnetic
          nd techniques.

Tests on saw blades, welds and low-pressure turbine blades are carried out by the EMAG analyzer developed in the Fraunhofer-Institut für zerstörungsfreie Prüfverfahren. This unit generates and controls the magnetic field excitation of the electromagnet. The exciting amplitude and the frequency (20-100 Hz) can be selected. The tangential field strength H is measured by a Hall probe; the Barkhausen noise signals by air coils or tape recorder heads. After further signal conditioning the quantities H_{CM} and M_{MAX} are deduced from the M(H) curves, and hardness and stress values are calculated as outlined above.

Hardness and residual test results are presented in Figs. 2 and 3. In Fig. 2 the hardness (HV10), the coercivity (H_{cM}) and the calculated EMAG-hardness values HV_{zf} are compared for the same testing trace. One can see that the nd quantity is strongly correlated to the conventional hardness test results. The surface resolution of the magnetinductive probe is ∿4 mm. The further testing procedure is outlined in Fig. 3. From the stress-sensitive magnetic Barkhausen noise signal M_{MAX} and the structural-sensitive quantity H_{cM}, the EMAG-analyzer calculates the stress value σ_{zf}. These σ_{zf} values are compared in Fig. 3 with the σ_{x-ray} results. Since the error bar in the case of the x-ray results is ∿±60 N/mm^2, the agreement with the σ_{zf} values is quite good. The measuring time for the hardness (HV_{zf}) and stress (σ_{zf}) test was ∿0.5 seconds.

ULTRASONIC TECHNIQUES

The birefringence of plane-polarized shear waves enables a method for the determination of stress which only exploits the times-of-flight of two linear-polarized shear waves.[4] By using the shear modulus μ and the third-order elastic constant n, the difference of the two principal stresses in the plane perpendicular to the propagation direction are determined.

In order to separate the influence of texture against the influence of stress on the sound velocity, additional information is necessary. To obtain this information, ultrasonic absorption and dispersion measurements have been investigated.

According to the vibrating-string model,[5] the dislocation loop length between the pinning points increases with increasing static stress. These dislocations interact with the superimposed alternating stress field of an elastic wave, and therefore affect the change of the absorption as a function of stress.

As an experimental result, Figs. 4 and 5 show the residual stress and the absorption in a sample with a bi-axial stress state. The residual stress state was determined by x-ray diffraction (100 μm surface etched), and the absorption

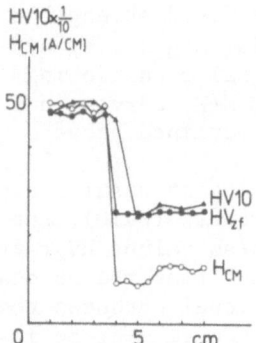

Fig. 2.   Hardness tests on low-pressure turbine blades using the
          magnetic Barkhausen noise.

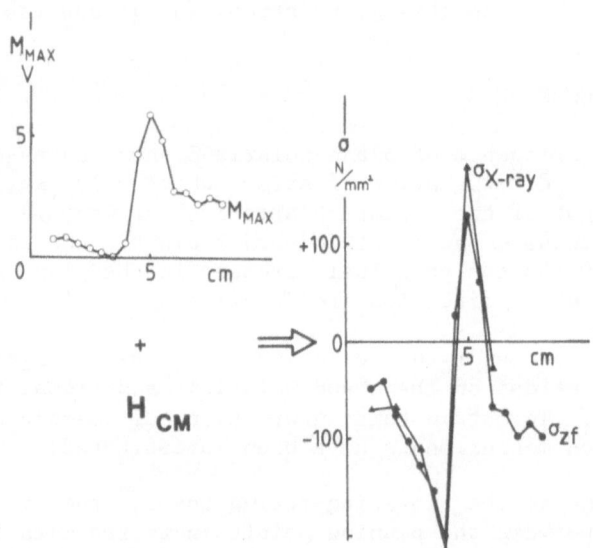

Fig. 3.   Residual Stress determination on low-pressure turbine
          blades using the magnetic Barkhausen noise.

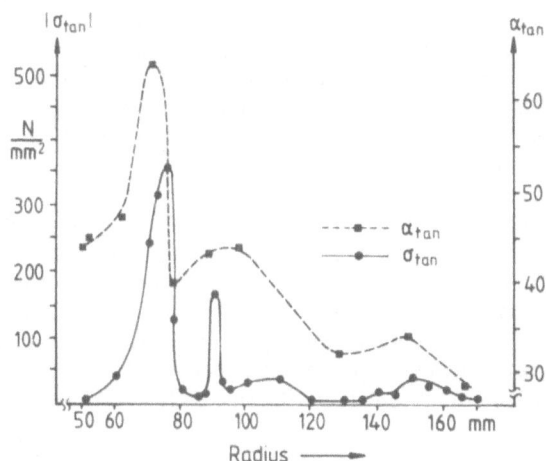

Fig. 4. Residual stress ($\sigma_{tangential}$) and attenuation coefficient
($\alpha_{tangential}$) of a shear wave polarized parallel to the
principal stress direction vs the radius of a saw blade.

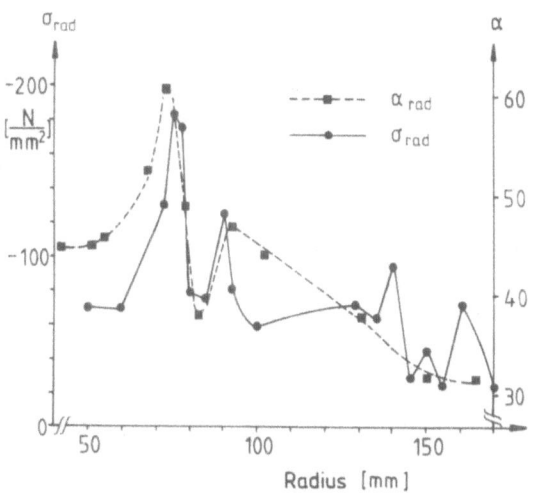

Fig. 5. Residual stress (σ_{radial}) and attenuation coefficient
(α_{radial}) of a shear wave polarized parallel to the
principal stress direction vs the radius of a saw blade.

measurements were carried out with a 3.3 MHz EMAT. This stress
dependence of the absorption cannot be used for an absolute stress
determination, because the absorption coefficient depends on
micromagnetic parameters which usually cannot be determined in
technical materials. Additionally, the measurement of the
absolute ultrasonic absorption turns out to be very difficult.
But the relative change of the absorption determined by changing
the polarization direction enables one to determine the residual
stress in textured materials.

    Combining the results of the time-of-flight and the absorption
measurements, two-dimensional stress states in slightly textured
samples can be characterized quantitatively. As an example, Fig. 6
shows the difference of the principal stresses together with the
results found by the x-ray method vs the distance from the center
of a textured saw blade. In samples with a stronger texture,
mainly caused by plastic deformations like rolling, the

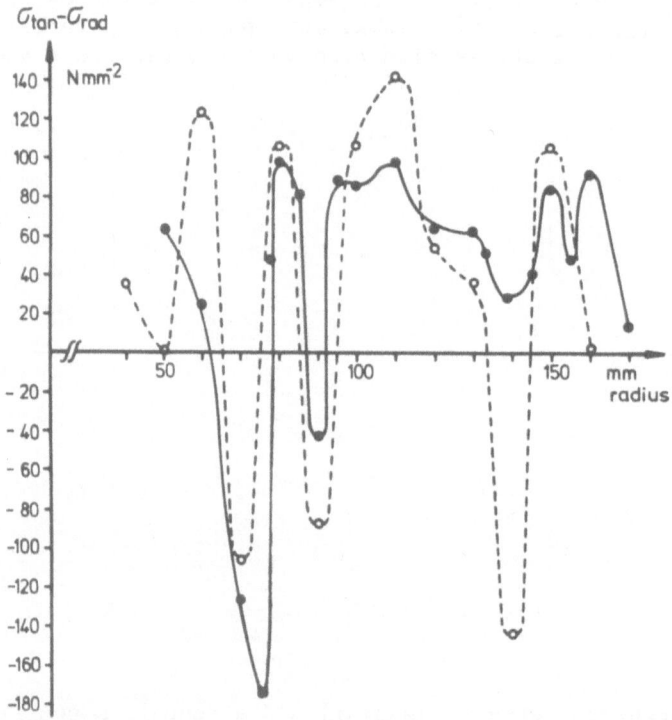

Fig. 6. Difference of the two principal stresses $\sigma_{tangential}$ -
         σ_{radial} vs the radius of a textured saw blade.
         ---o--- ultrasonic method
         ---●--- x-ray diffraction

configuration of the dislocations is changed and depends on the direction of the deformation. In this kind of sample the combination of time-of-flight and absorption measurements in order to separate the texture influence against the influence of stress does not work. Here, the second possibility to obtain additional information in order to separate the different influences might be given by the frequency dependence of the sound velocity in textured samples.

An ultrasonic scattering theory has been worked out[6] which allows one to calculate the scattering coefficients and velocities of longitudinal and transverse waves in polycrystals as as function of the wave number k times grain radius a without limitation to the Rayleigh region. The theory includes mode conversion and multiple scattering and can be used to describe ultrasonic propagation in polycrystals with randomly oriented grains, as well as in those with preferred grain orientation. The calculations until now were carried out for compressional and shear waves in polycrystals of cubic symmetry with randomly oriented grains in second-order perturbation theory using the assumption that the changes in the elastic constants and in the density of the materials from grain to grain are small. The asymptotic values at low ka (Rayleigh scattering) are exactly the same as the well-known results from Bhatia and Moore. Numerical calculations yield a dispersion effect in the range of order 10^{-3} to 10^{-2}, which mainly is affected by the anisotropy factor. At present, the calculations are extended for waves in polycrystals with preferred grain orientation. In first experiments the time-of-flight of shear waves polarized parallel and perpendicular to the rolling direction in steel plates are measured in a frequency range of 2 to about 8 MHz. The results indicate that the 'relative time-of-flight difference versus frequency' is indeed influenced by the degree of the texture and are thus encouraging their theoretical analysis as indicated above.

ACKNOWLEDGEMENT

This contribution is based on work performed with the support of the German Ministry for Research and Technology and the European Community for Carbon and Steel.

REFERENCES

1. W. Theiner and P. Höllerm in: "Eigenspannungen und Last-spannungen," V. Hauk and W. Macherauch, eds., Carl Hanser Verlag, München, Wien (1982), pp. 156-164.

2. W. Theiner and I. Altpeter, in: "Germany/United States Workshop on Research and Development to New Procedures in NDT," Springer Verlag, Berlin (1983).
3. I. Altpeter, W. Theiner, and B. Reimringer, in: Eigen-spannungen," Deutsche Gesellschaft für Metallkinde (1983).
4. E. Schneider and K. Goebbels, Determination of residual stress by time-of-flight measurements with linear-polarized shear waves, in: "Ultrasonics Symposium Proceedings," IEEE (1981).
5. A. Granato and K. Lücke, Theory of mechanical damping due to dislocation, J. Appl. Phys. 27:583, 789 (1956).
6. S. Hirsekorn, The scattering of ultrasonic waves by polycrystals, J. Acoust. Soc. Am. 72:1021 (1982); S. Hirsekorn, The scattering of ultrasonic waves by polycrystals II. shear waves, to be published in J. Acoust. Soc. Am. (1983).

THE NEED FOR ANALYTICAL/EXPERIMENTAL ORCHESTRATED APPROACHES

TO SOLVE RESIDUAL STRESS PROBLEMS IN REAL STRUCTURES

Koichi Masubuchi

Department of Ocean Engineering
Massachusetts Institute of Technology
Cambridge, Massachusetts 02139

ABSTRACT

This paper discusses the need for analytical/experimental
hybrid approaches to solve residual stress problems in real structures. Analytical models are used as the basis, and experimental
data obtained on selected locations are used to calibrate the analytical models and to improve their accuracy. By properly combining
recent analytical and experimental techniques, it is possible to
significantly improve the state-of-the-art of studying residual
stresses and distortion and their effects on service behaviors of
welded structures.

INTRODUCTION - PROBLEMS AND POSSIBLE SOLUTIONS

This paper discusses the need for analytical/experimental hybrid
approaches to solve residual stress problems in real structures.
Today, welding is widely used for fabricating many structures including aerospace structures, bridges, pipelines, pressure vessels, and
ships. However, welded structures are by no means free from problems.
One of the major problems associated with welding is related to residual stresses and distortion. Due to localized heating by the welding
arc, a weldment always contains residual stresses. The maximum value
of the stresses in regions near the weld is as high as the yield
stress of the material. Various types of distortions also occur.
The residual stresses and distortion cause complex effects, most of
which are harmful to the integrity of a welded structure. For example, transient thermal stresses and metal movement during welding
may cause weld cracking and joint mismatching. High tensile residual
stresses in areas near the weld may cause premature fractures of the

123

welded structures. Distortion and compressive residual stresses in
the base metal may reduce the buckling strength of a structural
member subjected to compressive loading.

Because of their significant, mostly adverse, consequences,
considerable research has been carried out on residual stresses and
distortion since the 1930's. Also studied have been their effects
on the service behaviors of welded structures. Results have been
reported in numerous publications. One of the most comprehensive
sources of information regarding these problems is a recently published book by Masubuchi[1]. This book, which can be useful to design-
ers and fabricators of welded structures, presents a systematic
summary of knowledge on residual stresses and distortion in welded
structures. The information presented in this book can be regarded
as a "first generation" or "pre-computer age" system for analyzing
these problems.[2]

Figure 1 shows schematically a system for analyzing residual
stresses and distortion in welded structures, and how it may be used
by practicing engineers. The system is composed of two parts,
Analysis 1 and Analysis 2. Analysis 1 is to predict residual stresses

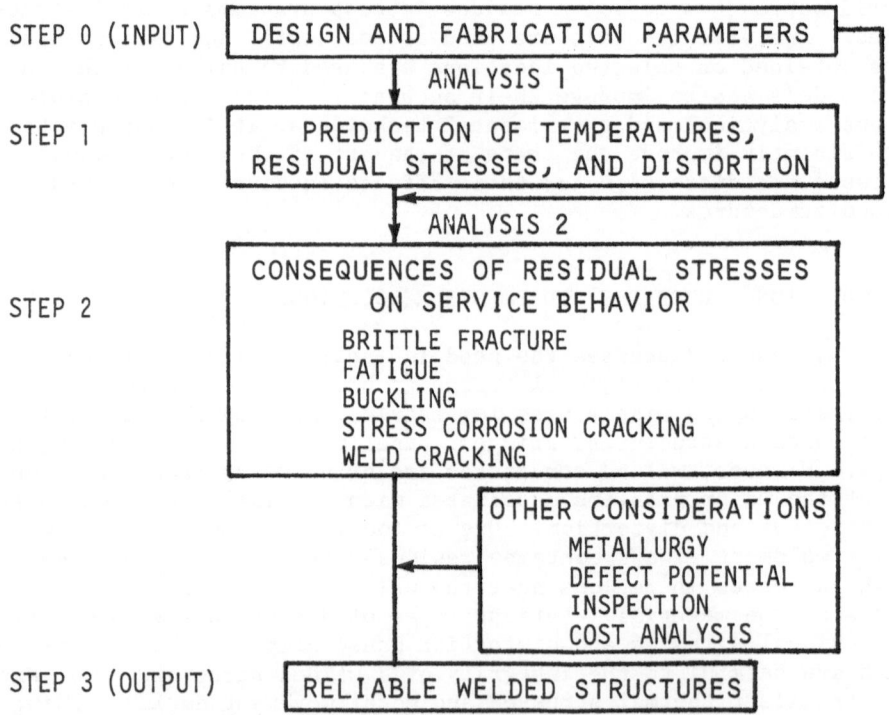

Fig. 1. Systematic approach for predicting residual stresses
         due to welding and their consequences.

and distortion due to welding, while Analysis 2 is to determine their effects on service behaviors of welded structures. What a practicing engineer wants to do is to select design and fabrication parameters (such as plate thickness, joint design, welding processes and condilions, etc.) in such a way that adverse effects of residual stresses and distortion can be reduced to acceptable levels. It is much better to achieve this goal in the early stages of design and fabrication rather than have to confront the problem in later stages of fabrication. In actual practice, the engineer also must consider other subjects including metallurgical properties of weldments, cracking and other defects, inspection, fabrication cost, etc.

In order for such a system to be useful, the system must be able to handle weldments in complex geometries, not just simple weldments. In fact, the system is most useful if it can handle the weld geometries used in the actual structure being designed and constructed. Also, the system must be accurate, and the necessary computations must be done in a short time with a reasonable cost. Unfortunately, analytical systems available today are far from complete.

Despite numerous studies that have been made on residual stresses and distortion due to welding, almost all of the information generated so far has been on test specimens mainly in simple geometries such as simple butt welds, fillet welds, and circular patch welds. There have been some studies on weldments in more complex geometries such as pipes and spherical shells; however, even in these cases, studies are generally made on specimens simulating actual welds and very little information has been generated on residual stresses and distortion in actual welded structures such as pressure vessels and ships. This is primarily due to the following:

(1) Analytical Limitations. Without use of high-speed computers, it is not possible to analyze residual stresses and distortion in practical weldments. In fact, even in such a simple case as welding along the longitudinal edge of a rectangular plate, it is almost impossible to manually calculate changes of thermal stresses during welding, as discussed in a later part of this paper; and

(2) Experimental Limitations. Many experimental techniques for measuring residual stresses, such as those using electrical and mechanical strain gages, are destructive; therefore, they cannot easily be used for actual structures. There are some non-destructive residual stress measuring techniques such as the X-ray diffraction technique. However, these techniques tend to be rather slow and they are not easily used for large welded structures.

The author believes that this is changing now. With the development of computer technology and such computational methods as the finite element method, it is now possible to analyze residual stresses nd distortion in a fairly complex weldment in a short time and with a reasonable cost. Researchers at M.I.T. and other laboratories in several countries have developed a series of computer programs for calculating transient thermal stresses, residual stresses, and distortions in weldments in a variety of geometries. It should be noted that these analytical developments have been made mainly by specialists in applied mechanics, not by welding engineers. One of the problems associated with the analytical studies is how to select a proper combination of welding characteristics (such as the arc efficiency), and physical and mechanical properties of the material (such as heat conductivity, transformation characteristics, yield stress, etc.), especially at elevated temperatures. By feeding different sets of values to a computer program, different computational results are obtained.

Advances also have been made of experimental techniques for measuring residual stresses and distortion allowing measurements to be made both more accurately and faster. A notable example is the advanced high-speed X-ray diffraction technique developed by Ruud.[3] Other examples which show promise for possible future applications include the ultrasonic technique for measuring residual stresses and the laser interferometry for measuring out-of-plane distortion.[4] In fact, several papers presented at this symposium cover recent developments in some experimental techniques which may be used for welding applications. Another important development is the application of computers in processing experimental data. With use of modern computers, processing of data can be handled not only in a short time but also automatically (without having technicians attending at the location) and remotely (data can be stored in a location away from where the welding is being performed).[5] It should also be noted that these developments have been made by specialists in applied physics and electronics, not by welding engineers.

The author believes that it is now possible to extend the analysis of residual stresses and distortion to cover real structures. A most promising way is to develop a hybrid method combining analytical and experimental techniques. The fundamental principle is to use analytical models as the basis and use experimental techniques for calibrating the analytical models. Through this method the accuracies of analytical models can be improved significantly. The major part of the remainder of this short paper is used to present a brief summary of the current state-of-the-art of the analysis of residual stresses and distortion in weldments, based primarily on results obtained at M.I.T., as well as how the analytical technique could be improved.

RESIDUAL STRESSES AND DISTORTION DUE TO WELDING

Figure 2 shows schematically changes of temperature and thermal stresses during welding. A weld bead is being laid along the x-axis of a plate. The welding arc, which is presently located at the origin, 0, is moving at speed, v, (see Fig. 2a).

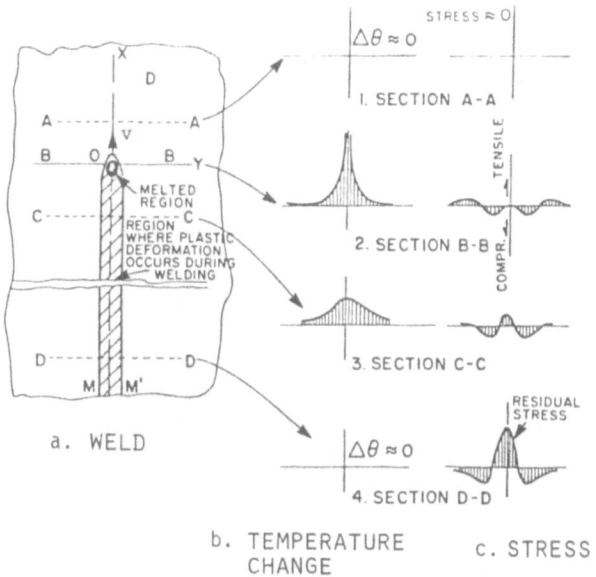

Fig. 2.   Schematic representation of changes in temperature
and stresses during welding.

Figure 2b shows temperature distributions along several cross
sections. Along Section A-A, which is ahead of the welding arc, the
temperature changes due to welding, ΔT, are almost zero. Along
Section B-B, which crosses the origin, the temperature distribution
is very steep. Along Section C-C, which is some distance behind the
arc, the distribution of temperature change is as shown in Figure
2b-3. Along Section D-D, which is very far behind the arc, the temperature changes due to welding again diminish.

Figure 2c shows distributions of thermal stresses in the x-
direction, σ_x, along these sections. Stresses in the y-direction,
σ_y, and shearing stress, τ_{xy}, also exist in a two-dimensional stress
field. Along Section A-A, thermal stresses caused by welding are
almost zero. The stress distribution along Section B-B is shown in
Figure 2c-2. Because molten metal does not support a load, stress
underneath the welding arc is close to zero. Stresses in areas near
the arc are compressive, because thermal expansions in these areas
are restrained by the surrounding metal that is at lower temperatures.
Since the temperatures of these areas are quite high and the yield
stress of the material is low, stresses in these areas are as high
as the yield stresses of the material at corresponding temperatures.

Stresses in areas away from the weld are tensile, to balance the
compressive stresses in areas near the weld.

Stresses are distributed along Section C-C as shown in Figure
2c-3. Since the weld metal and base metal regions near the weld
have cooled, they try to shrink causing tensile stresses in areas
near the weld. Figure 2c-4 shows the stress distribution along
Section D-D. High tensile stresses are produced in areas near the
weld, while compressive stresses are produced in areas away from the
weld. The distribution of residual stresses that remain after
welding is completed is shown in Figure 2c-4.

The cross-hatched area, MM', in Figure 2a shows the region where
plastic deformation has occurred due to welding. The small, egg-
shaped region near the origin indicates the region where the metal
is melted. The region outside the cross-hatched area has remained
elastic during the entire welding thermal cycle.

Figure 3 shows schematically how a rectangular plate deforms
when arc welding is performed along its upper longitudinal edge.
Since temperatures are higher in areas near the upper edge, these
areas expand more than areas near the lower edge causing upward
movement of the center of the plate, δ, as shown by Curve OA in
Figure 3. When welding is completed and the metal starts to cool,
the plate deforms in the opposite direction. If the material were
completely elastic in the entire region of the plate during the

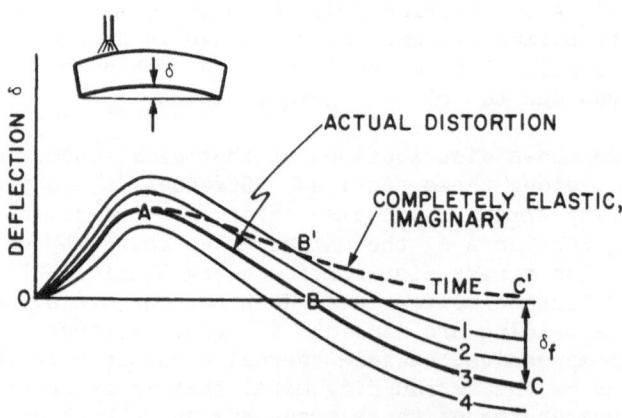

Fig. 3.   Deformation of a rectangular plate during arc welding
          along the upper longitudinal edge.

entire period of the heating and cooling cycle, the plate would
deform as shown by Curve AB'C' returning to its original shape with
no residual distortion. However, this does not happen in welding a
real material, be it steel, aluminum, or titanium. In case of welding
a real material, plastic deformations occur in areas near the upper
edge while the temperatures are high and yield stresses of the
material are low. As a result of the compressive plastic strains
(shrinkage) that are produced in areas near the upper edge, the plate
continues to deform after passing its original shape resulting in the
negative final distortion, δ_f, when the plate cools down to its
initial temperature, (see Curve OABC).

The above discussed phenomenon of having thermal stresses and
distortion in one direction during the process and resulting in residual stresses and distortion in the opposite direction after the completion of the process is typical for many manufacturing processes,
such as welding, flame cutting, casting, and heat treatment, which
are known to produce residual stresses..

DEVELOPMENT OF METHODOLOGIES FOR ANALYZING RESIDUAL STRESSES AND
DISTORTION IN WELDMENTS

Figure 4 shows schematically various methods for analyzing
residual stresses and distortion in weldments. The most orthodox
method is to conduct an analytical simulation. This method involves
the following steps:
Step 1: Analysis of heat flow
Step 2: Analysis of transient thermal stresses during welding
Step 3: Determination of incompatible strains after welding is
         completed
Step 4: Determination of residual stresses and distortion due
         to the incompatible strains.

In the completely analytical simulation, all the steps are performed analytically. In analyzing residual stresses and distortion,
Step 3 is the most important. If transient thermal stresses were
completly elastic in all portions of the weldment during the entire
process of welding, no residual stress would remain and the weldment
would return to its original shape when it cooled to the original
temperature. In welding real materials, however, incompatible
strains are produced as a combined effect of strains due to plastic
deformation, solidification of the weld metal, and strain changes
caused by phase transformations which occur in some materials including steels. Once the distribution of incompatible strains is determined, it is then possible to analytically determine residual stresses
and distortion by an elastic analysis.

If one is mainly interested in determining residual stresses and
distortion which remain after welding is completed, it is possible to

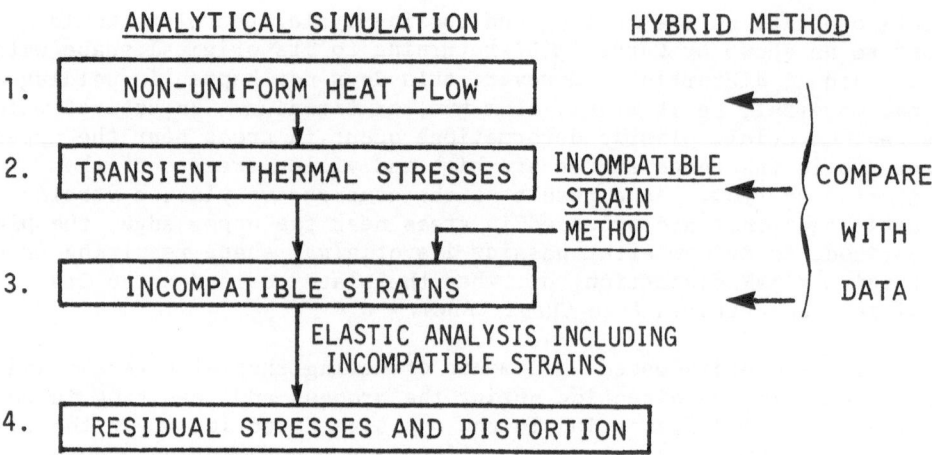

Fig. 4.   Methodologies for analyzing residual stresses
and distortion in weldments.

develop a second method which by-passes Steps 1 and 2. The distribution of incompatible strains may be determined experimentally or
even by assumption as well as analytically. This method may be called
the "incompatible strain" method or the "inherent strain" method. A
unique characteristic of residual stresses in weldments is that in
most cases, incompatible strains are confined to small areas near the
weld. Therefore, analyses based on relatively simple distributions
of incompatible strains often provide results which are reasonably
accurate. The major advantage of this method is that it can be used
for analyzing residual stresses and distortion in complex weldments
on which the complete analytical simulation is either very difficult
or expensive.

Studies Before Around 1960 (Without Computers)

Obviously, people have been interested in analytical simulation
for a long time since they began to study residual stresses and distortion in weldments. For example, in a paper published in 1936,
Boulton and Lance-Martin[6] discussed the transient thermal stresses
that occur in a rectangular plate during welding along its longitudinal edge, as shown in Figure 3. Although they had to make a number
of assumptions to simplify the analysis, they managed to study general
tendencies of residual stresses in this case. However, because of
limitations of computational capabilities available with manual calculations, only limited studies were done on transient thermal stresses
and metal movement during welding. Before the use of modern computers,
studies on residual stresses based on the analytical simulation method
were limited to simple cases such as:
      (1) Spot welding in which temperatures and stress changes are
           axially symmetric, and

(2) Instantaneous heating along the edge of a rectangular plate
     in which temperature and stress changes are functions of
     only one axis (the lateral distance from the heated line).
However, these cases are far too simple to represent weldments used
in actual structures.

Therefore, many studies discussed the residual stresses and distortion that remain after welding is completed, such as the final
distortion, δ_f, shown in Figure 3. In fact, these studies have provided useful information to practicing engineers. What was (and still
is) important is the study of how residual stresses and distortion
are affected by various design and fabrication parameters, such as
plate thickness, joint design, welding conditions, etc. (see Figure
1). Many studies, which were mainly empirical, were made to generate
much of the practical information needed. These studies may be
called "parametric studies", because their main objectives have been
to determine relationships between various design and fabrication
parameters, and residual stresses and distortion. Analytical studies
based on the incompatible strain method were often used to develop
analytical relationships among some of the parameters and residual
stresses and distortion. Masubuchi's book[1] presents a most comprehensive summary of results obtained in these studies.

Computer-Aided Studies Since Around 1960

Since around 1960, there has been a surge of interest in research on residual stresses and distortion in weldments, by use of
computers. Tall[7], in his Ph.D. thesis published in 1961, developed
a simple computer program on transient thermal stresses that occur
during bead-on-plate welding along the center line of a strip (see
Figure 2). The temperature distribution was treated as two-
dimensional; however, in analyzing stresses it was assumed that (1)
longitudinal stress, σ_x, is a function of the lateral distance from
the weld line y only, and that (2) σ_y and τ_{xy} are zero. In this
paper such an analysis is called "one-dimensional". In 1968, based
upon Tall's analysis, Masubuchi et al[8] developed a FORTRAN program
on one-dimensional analysis of thermal stresses during welding.

Since around 1970 a number of investigators in various parts of
the world have developed computer programs calculating transient
thermal stresses, residual stresses, and distortion in weldments in
different geometries.[9,10] At M.I.T., for example, the following
programs have been developed:
  1. Basic one-dimensional program
  2. Two-dimensional finite element program for plane stress
      analysis
  3. Two-dimensional finite element program for plane strain
      analysis
  4. Programs for analyzing some practical problems
      a. Girth welding of a cylindrical shell

        b.   Built-up beams.
The basic one-dimensional program is useful for calculating stresses
during welding along a longitudinal edge of a rectangular plate, as
shown in Figure 3.   The 1-D program, which is very inexpensive to run,
is useful in calculating residual stresses and distortion, of a weldment which is very long in one direction.   The 1-D program is similar
to the beam theory in stress analysis.   The 2-D plane stress program
is suited for calculating stresses during butt welding thin plates.
It should be noted that the stress analysis of a simple butt weld is
complicated enough to require a finite element method.   The 2-D plane
strain program is suited for calculating stresses in a cross section
of a butt weld between thick plates.   A finite element program for
the three dimensional analysis could be developed; however, the program is extremely expensive to run.

        In addition, investigations have been made to develop finite
element programs for calculating residual stresses in complex weldments such as butt welding between cylindrical and hemispherical
shells.[11] However, because of the complexities of analyses required
for analytical simulations, the incompatibility method has frequently
been used in dealing with stresses in weldments in complex geometries.

FUTURE DEVELOPMENTS

        With the increased used of computers, our capability for analyz-
ing residual stresses and distortion has improved significantly
during the last 12 years or so.   However, there is still considerable
room for further improvement. Among various possibilities of future
improvement, the author would like to stress the need for combining
recent developments in analytical and experimental techniques, especially to improve the accuracy of computer analyses.

        As stated earlier, one of the problems associated with computer
analysis is how to select a proper combination of welding characteristics (such as the arc efficiency), and physical and mechanical
properties of the material (such as heat conductivity, transformation
characteristics, yield stress, etc.), especially at elevated temperatures.   When different values of welding characteristics and
material properties are selected and fed to a computer program,
different computational results may be obtained as shown by Curves 1,
2, 3, and 4, in Figure 4.   Then, the engineer may have difficulties
in selecting the right result.   By monitoring data (either temperatures, strains, or deformation) on a few locations, it becomes possible to calibrate the analysis and pick the right combination of
values to be fed to the computer program.   Thus the accuracy of the
analysis can be improved significantly.   This method, which uses
analytical solutions as the basis, and then experimental data on
selected locations for calibrating the analysis, may be called the
analytical/experimental hybrid method.

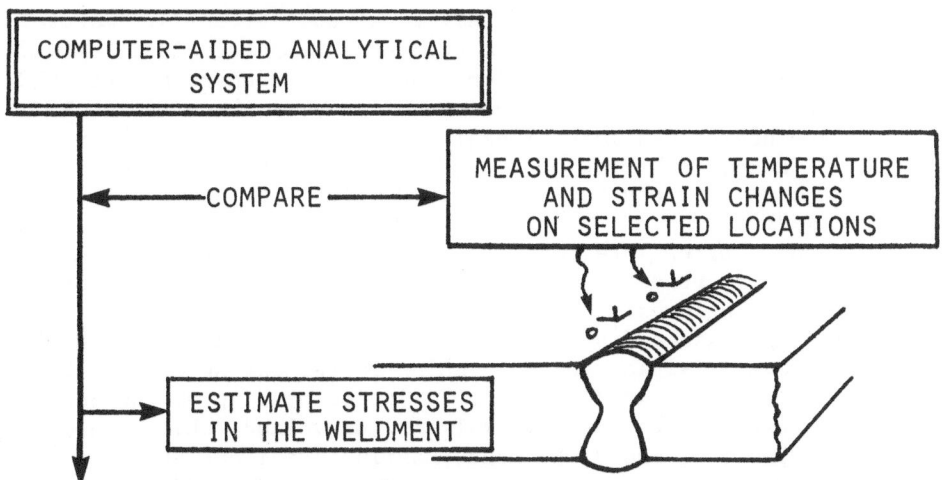

Fig. 5. Use of the computer-aided analytical system for studying
residual stresses in a heavy weldment and their consequences.

An example in which the hybrid method is very useful in the
determination of residual stresses in a heavy weldment, is shown in
Figure 5. In all of the stress measurement techniques in use today,
including strain gages and X-ray diffraction techniques, the stresses
are determined by measuring strains on the surface of the specimen.
In order to measure residual stresses in a heavy weldment, the surface on which the measurements are to be made must be exposed by
drilling or sectioning. These methods are not only time consuming
and costly, but they also provide stress data only at the drilled
hole or along the sectioned block. On the other hand, what we are
really looking for is the distribution of residual stresses in the
entire weldment. Fortunately, we now have the advanced high-speed
X-ray diffraction technique, which is capable of measuring residual
stresses non-destructively in a short time. Obtaining residual stress
data on a few selected locations does not provide important information in itself; however, the results can be of a great value if they
are used to calibrate analytical models. Through calibration, the
accuracy of the analytical models can be improved significantly. The
analytical model can provide information about the distribution of
residual stresses in the entire weldment. It is also possible to
further extend the analysis to study effects of residual stresses on
the service behaviors of the weldment.

REFERENCES

1.   K. Masubuchi, "Analysis of Welded Structures - Residual Stresses,
     Distortion, and Their Consequences", Pergamon Press, Oxford/
     New York (1980).

2.   V. J. Papazoglou, K. Masubuchi, E. Goncalves, and A. Imakita,
     "Residual Stresses Due to Welding; Computer-Aided Analysis
     of Their Formation and Consequences" presented at the 1982
     Annual Meeting of the Society of Naval Architects and Marine
     Engineers, New York (1982).
3.   C. O. Ruud, "A Review of Selected Non-Destructive Methods for
     Residual Stress Measurement," NDT International, pp. 15-23
     (1981).
4.   W. Cook, "Laser Interferometer for Weld Deformation Measurement,"
     a report prepared under Subject 13.713 Special Problems in
     Ocean Engineering, M.I.T. (1982).
5.   J. E. Agapakis and K. Masubuchi, "A Flexible Microcomputer-Based
     data Acquisition and Processing System for Welding Experi-
     ments," presented at the 1983 Spring Meeting of the Society
     of Experimental Stress Analysis, Cleveland, Ohio (1983).
6.   N. S. Boulton, and H. E. Lance-Martin, "Residual Stresses in Arc
     Welded Plates," Proc. Inst. Mech. Eng., Vol 133, pp. 295-339
     (1936).
7.   L. Tall, "The Strength of Welded Built-Up Columns," Ph.D. Thesis,
     Lehigh University (1961).
8.   K. Masubuchi, B. Simons, and R. E. Monroe, "Analysis of Thermal
     Stresses and Metal Movement During Welding," RSIC-820,
     Redstone Scientific Information Center, Redstone Arsenal,
     Alabama (1968).
9.   K. Masubuchi, "Applications of Numerical Analysis in Welding, in
     Welding in The World, Vol. 17, No. 11/12, 268-291 (1979).
10. V. J. Papazoglou, A. Imakita, and K. Masubuchi, "Annotated
     Bibliography on Numerical Analysis of Stresses, Strains, and
     Other Effects Produced by Welding," Document X-996-81,
     Commission X of the International Institute of Welding (1981).
11. Y. Fujita and T. Nomoto, "Welding Deformations and Residual
     Stresses Due to Circumferential Welds at the Joints Between
     Cylindrical Drum and Hemispherical Head Plate," Document
     X-985, Commission X of the International Institute of
     Welding (1981).

INTERNAL FRICTION OF Fe-BASED BINARY ALLOYS AT HIGH FREQUENCY

M. Deka and N. Eberhardt*

Department of Electrical and Computer Engineering
Lehigh University
Bethlehem, Pennsylvania

ABSTRACT

Ultrasonic absorption data in the range 50 KHz - 1.2 MHz are presented for the binary alloys Fe-7.8% Cr, Fe-3.8% Cr, Fe, Fe-1.5% Si, Fe-3.6% Mo and Fe-0.035% C. The magnetoelastic part of absorption is obtained by subtraction of the absorption obtained under magnetic saturation. It is found that the predominant part of absorption in all alloys is magnetoelastic in nature. It is predominantly the shear-viscosity coefficient that causes the damping. The bulk-viscosity coefficient is zero or near zero. Small coldwork causes a strong decrease of absorption via a decrease in permeability. Occasionally, in accordance with the theory by Becker and Döring, a broad absorption maximum appears in the investigated frequency range.

INTRODUCTION

Internal friction results in absorption of ultrasonic waves. This effect is distinct from the Rayleigh scattering from grain boundaries, which has been demonstrated to strongly depend on grain size and thus, via the Hall-Petch relation, can be used to determine yield strength.[1] Absorption, on the other hand, may or may not have any relation to grain boundaries. The knowledge of absorption is a useful complement to the mentioned Rayleigh scattering methods. Particularly at lower frequencies and large grains an absorption "background" enters into the picture and is

---

*This research was performed under a Grant from Bethlehem Steel Corporation.

not distinguishable from the scattering loss. Therefore, it needs
a separate study.

Many theories and hypotheses exist on mechanisms of ultrasonic
absorption.[2] Even though much experimental work has been reported
on scattering, not many data are available on absorption in ferromagnetic polycrystalline material, Comp (2), p. 536.[9] Especially
the peak, as predicted by Döring,[7] has not been observed very
often. In 1981 our group has reported a successful extension of
the resonant method to well above 1 MHz, which also can distinguish
between longitudinal and transverse vibrations,[3,4] In the present
paper we report results of absorption measurements in several
Fe-based binary alloys: Fe, Fe-7.8% Cr, Fe-3.8% Cr, Fe-1.5% Si,
Fe-3.6% Mo, Fe-0.035% C. We try to determine how much of the
absorption is magnetoelastic in nature and whether our results are
compatible with Döring's theory of magnetoelastic absorption. We
also find residual strain and small coldwork to be factors of
strong influence on absorption, something that has often been
neglected in previous measurements.

METHOD OF MEASUREMENT

This brief review will include certain simplifications adapted
after the first publication.[3] In principle we measure the
transmission spectrum through loosely coupled cylindrical
resonators made from the material to be investigated. Generally,
commercial broadband shear-wave transducers are used. One
successful way of loosely coupling is as follows (Fig. 1): steel
spheres of 0.5 mm diameter (from ball-point pens) are honed down
to half-spheres and epoxied to the surface of the transducers.
Now the cylinder is loosely clamped between two such transducers,
being contacted only by the half-spheres. The contact force is
of the order of one gram. The choice of contact point location is
essential. Often it is possible to verify vibrational patterns
of a given mode of vibration by contacting at various points along
the cylinders, use various orientations (with respect to polarization) and monitor the transmitted signal at resonance. A combination of small pressure and choice of contact points near a
vibrational node leads to sufficiently loose coupling and thus
insures a large external Q-factor, Q_e. The full transmission
bandwidth is equal to Q_L^{-1}, the inverse loaded Q. The general
relation $Q_L^{-1} = Q_e^{-1} + Q_o^{-1}$ exists. Q_o, the internal Q, is the
quantity to be measured. Hence the need to minimize Q_e^{-1}. Using
resonators of extremely low absorption, such as quartz, we find
Q_e^{-1} to lay between 1.5×10^{-5} and 6.6×10^{-5}. The transmission
test set is computer controlled with automatic curve fitting and
determination of $Q_L^{-1} \simeq Q_o^{-1}$. An internal Q will further on be
called Q_t or Q_ℓ, depending on whether standing transverse or
longitudinal waves are involved.

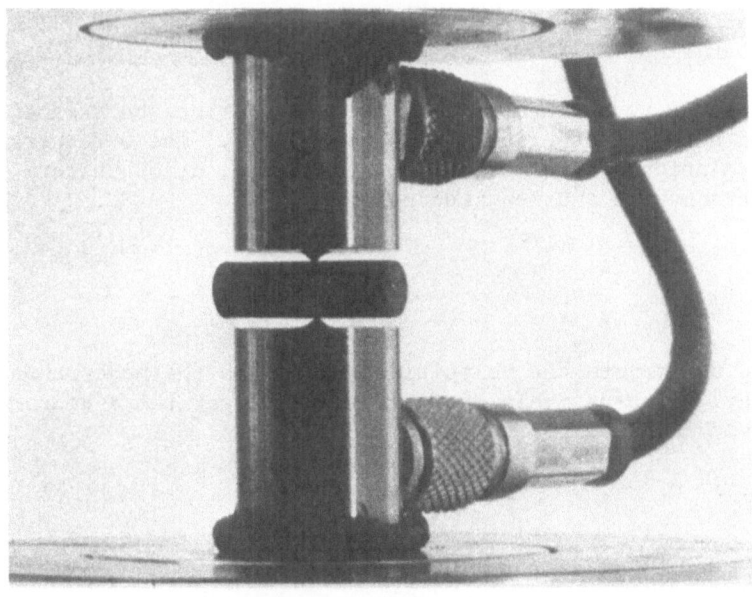

Fig. 1. Cylindrical resonator coupled to shear wave transducers.

        In Reference 3 we have reported that any (mixed) resonant mode
whose frequency can be accurately computed can be used to
separately determine absorption associated with transverse and
longitudinal movement. Meanwhile, we have reduced our method to a
simpler and more accurate procedure: to measure the absorption
and velocity of purely transverse waves (vibrations), we select
only torsional modes. Their resonant frequencies are given by the
simple condition that the cylinder length be an integer multiple of
$\lambda_t/2$, where λ_t is the wavelength of shear-waves. Hence, only the
shear-modulus is involved. The inverse full bandwidth of such
modes gives the transverse Q_t directly.

        As far as longitudinal vibrations are considered, we found in
cylinders of aspect ratio one (diameter = height), one resonance
that almost exclusively depends on longitudinal velocity alone.
Hence, the Q-factor of this resonance is directly the longitudinal
Q_ℓ. This is an almost compressional mode, a cylindrical equivalent
of the "breathing sphere" mode. Use of this resonant mode should
give good accuracy in the measurement of bulk viscosity, which can
not be directly obtained by wave methods, where bulk and shear
viscosity simultaneously contribute to absorption.

PHENOMENOLOGY OF ULTRASONIC ABSORPTION IN ISOTROPIC SOLIDS,
SINUSOIDAL TIME VARIATION

It seems to be necessary to properly define the parameters
measured here and other derivable parameters. The Q-factors have a
simple relation to the attenuation constants, α, of uniform plane
waves, transverse and longitudinal.

$$\alpha_t = \frac{\omega_o}{2V_t} Q_t^{-1} \qquad (1)$$

and

$$\alpha_\ell = \frac{\omega_o}{2V_\ell} Q_\ell^{-1} \; , \qquad (2)$$

where V_t and V_ℓ are the velocities, and ω_o is the mode resonant
frequency. Alternatively, sometimes the "logarithmic decrement,"
δ, is used in the literature:

$$\delta_t = Q_t^{-1} \pi \qquad (3)$$

and

$$\delta_\ell = Q_\ell^{-1} \pi \; . \qquad (4)$$

An equivalent way to describe absorption properties with
sinusoidal time variation is to introduce imaginary parts to the
elastic constants. For instance, according to Reference 5,
Chapter 4, the imaginary part of the shear modulus is

$$\zeta = \rho \, V_t^2 \, Q_t^{-1} \; . \qquad (5)$$

This is known as shear viscosity coefficient. ρ is the density.
The imaginary part of the bulk modulus is

$$\xi = \rho \, V_\ell^2 \, Q_\ell^{-1} \; . \qquad (6)$$

This is known as bulk or volume viscosity coefficient. From the
last two equations, it follows that:

$$\frac{Q_t^{-1}}{Q_\ell^{-1}} = \left(\frac{V_\ell}{V_t}\right)^2 \left(\frac{\xi}{\zeta} + \frac{4}{3}\right)^{-1} \; . \qquad (7)$$

According to this, if the bulk viscosity approaches zero, for steel
the ratio between the Q's has an upper limit

$$\frac{Q_t^{-1}}{Q_\ell^{-1}} \leq \frac{3}{4} \left(\frac{V_\ell}{V_t}\right)^2 \approx 2.5 \; . \qquad (8)$$

Interestingly, as will be seen later on, this means that in
most of our measured specimen the bulk viscosity is found to be
small, as compared to the shear viscosity.

Actually a more appropriate way of comparing the absorption as

caused by shear strain with the one caused by compression would be
to use the loss tangents of both moduli. With the shear modulus
$G = \rho V_t^2$ and the bulk modulus $K = (V_\ell^2 - 4V_t^2/3)$, one obtains for
the ratio between the loss tangents:

$$\frac{\tan\theta_B}{\tan\theta_S} = \frac{\xi/K}{\zeta/G} = \frac{(V_\ell/V_t)^2 Q_\ell^{-1}/Q_t^{-1} - 4/3}{(V_\ell/V_t)^2 - 4/3} . \tag{9}$$

Those two loss tangents have fundamental physical significance
beyond phenomenology.

RESULTS FOR SOME Fe-BASED BINARY ALLOYS

The strongest absorption was found in the Fe-7.8% Cr alloy.
Figure 2 shows the results in specimen of various grain sizes. No
monotonic grain size dependence is perceivable. In fact, as will
be demonstrated later on, the dominating effect must be from
residual strain, rather than grain size, possibly even from some
other properties of the microstructure. Thus, there are variations
from one specimen to another which cannot be removed by annealing
below the recrystallization point. Figure 3 shows the identical
material, identically annealed (800°C), but machined into cylinders
by turning on a lathe as against grinding in Fig. 2.

What seems to be significant, though, is that all the curves
either show a flat maximum or suggest the appearance of a maximum
outside the measured frequency interval. This is repeated in the
other alloys and is well consistent with Döring's theory of micro-
eddycurrent loss. According to this theory,[6] the Q-factor would be
given by:

$$Q = \frac{\chi_0 E_s \eta^2 C}{\pi I_s} \cdot \frac{f/f_0}{1 + (f/f_0)^2} \tag{10}$$

with

$$f_0 = \frac{R}{96\chi_0 \ell^2} ,$$

where χ_0 = initial magnetic susceptibility, λ = magnetostriction,
R = resistivity, ℓ = magnetic domain size, I_s = saturation
magnetization, C = an elastic constant, depending on material and
mode of excitation, and E_s = Young's modulus in magnetic saturation.
The last factor in eqution (10) determines a peak at the frequency
f_0. With reasonable values of magnetic domain size, this frequency
is between 10 KHz and 1 MHz.

Figure 4 shows the results for two different grain sizes of
an Fe-3.8% Cr alloy. Even though the peak seems to shift towards
lower frequencies with increasing grain size, one should be
careful with any interpretation at this point.

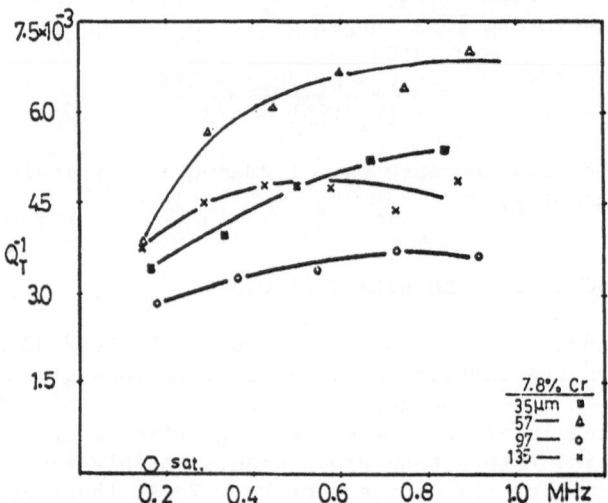

Fig. 2. Absorption of transverse waves in Fe-7.8% Cr alloy at various grain sizes.

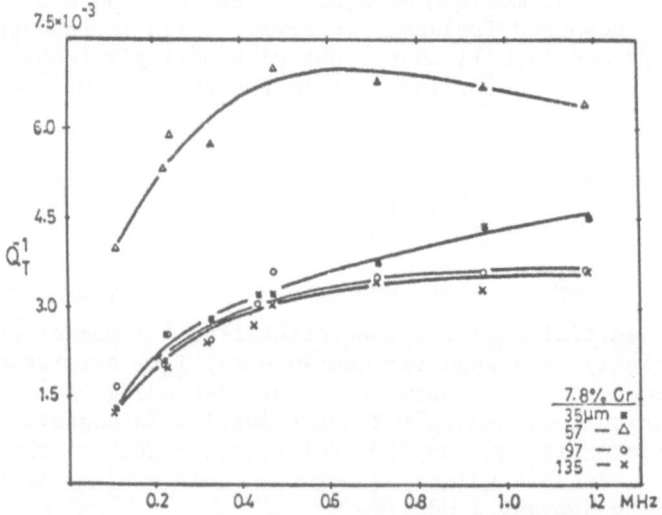

Fig. 3. Absorption of transverse waves in Fe-7.8% Cr alloy at various grain sizes, differently machined.

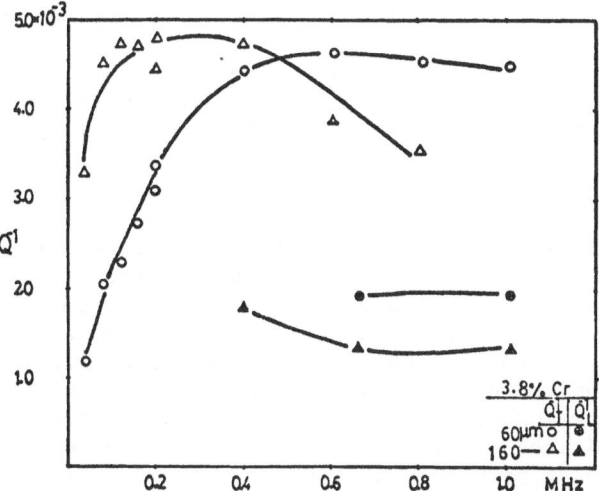

Fig. 4. Absorption of transverse and longitudinal waves in
         Fe-3.8% Cr alloy at two different grain sizes.

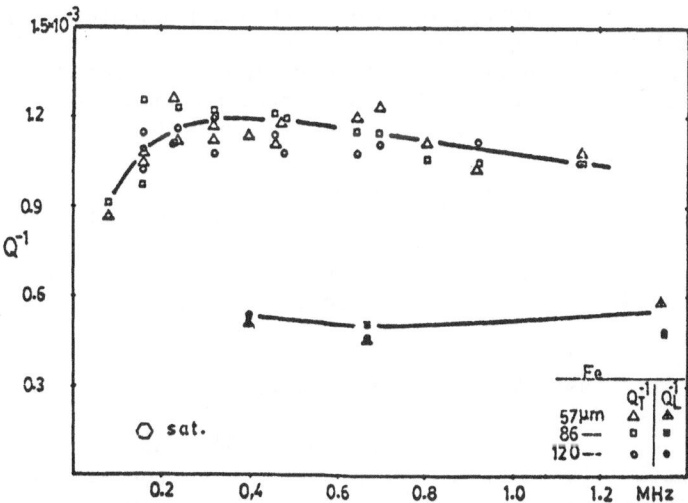

Fig. 5. Absorption of transverse and longitudinal waves in
         electrolytic iron at various grain sizes.

Included are also some measured points of Q_ℓ^{-1}. These data are more scarce, since each cylinder only yields one point. Also, there is more scattering of data because of the mentioned nonuniformity between different cylinders of the same material.

Figures 5 and 6 contain similar data for pure Iron (Electrolytic Iron) and Fe–0.035% C. Here, any grain size dependence seems to disappear. Also, we generally observe less scattering of data. The Cr-alloys, with their large absorption, seem to be most sensitive to residual strain.

In the last three alloys Fe, Fe–0.035% C, and Fe–3.8% Cr, an interesting observation on shear and bulk viscosity can be made (Fe–7.8% Cr has been excluded because of uncertain results in Q_ℓ^{-1}, due to residual strain): the ratio between the loss tangents, equation (9), is small. Considering the large error limits resulting from the need to use a different cylinder for each longitudinal resonance, one might very well conclude that the data are scattered around zero. Negative ratios have no physical reality and must be merely a result of measurement error. The average of 24 measured points was 0.058.

The question arises to what extent is this a general feature? It is quite conceivable that compression does not result in a significant damping, except for the small adiabatic-thermal conductivity effect. Shear strain, however, seems to be a natural candidate to cause "internal friction." For that, at least for micro-eddycurrent damping, our measurements present strong evidence. So far, we have not been able to corroborate much information on this from existing theoretical or experimental literature. We find contradicting opinions even on the proper phenomenology of damping. Döring's theory is based on axial strain only.[7]

Finally, we present some data on the absorption in Fe–1.5% Si (Fig. 7) and Fe–3.6% Mo (Fig. 8). Both have been annealed at 800°C for 1 hour. Here only Q_t^{-1} has been measured. Absorption in both alloys seems to be extremely sensitive to residual strain. Annealing at 700°C results in up to 50% lower Q_t^{-1} in the Mo alloy.

MAGNETOELASTIC ABSORPTION VERSUS BACKGROUND

There seems to prevail a general assumption that damping in ferromagnetic materials is predominantly magnetoelastic. We have tried to check this at the higher frequency range and determine how much of other mechanisms might be involved also.

For that reason we have modified our experimental set up to include the possibility to magnetically saturate the cylinders during measurement of ultrasonic absorption. Figure 9 shows the

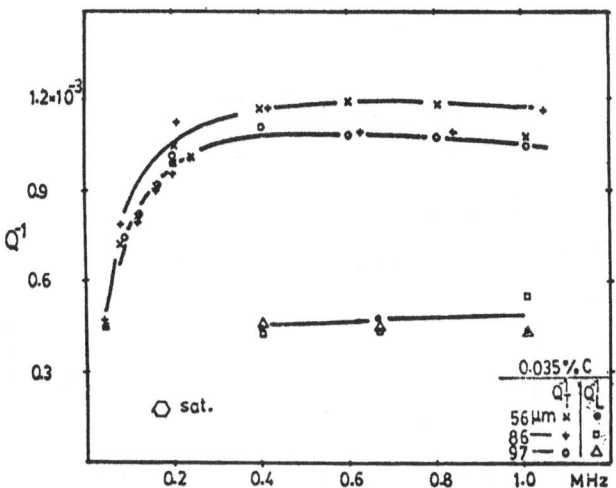

Fig. 6.   Absorption of transverse and longitudinal waves in 0.035%
          carbon steel at various grain sizes.

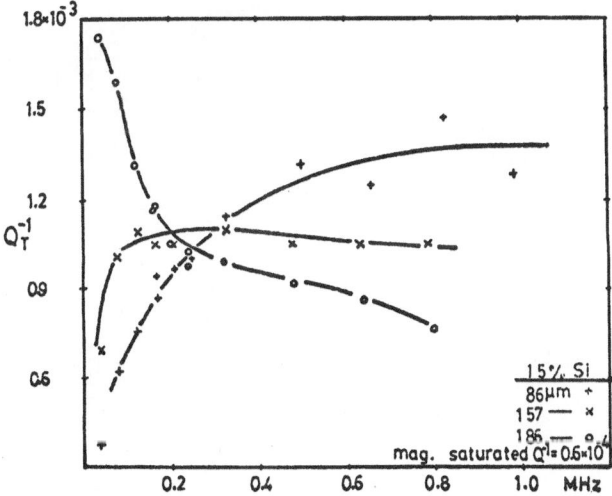

Fig. 7.   Absorption of transverse waves in Fe-1.5% Si at various
          grain sizes.

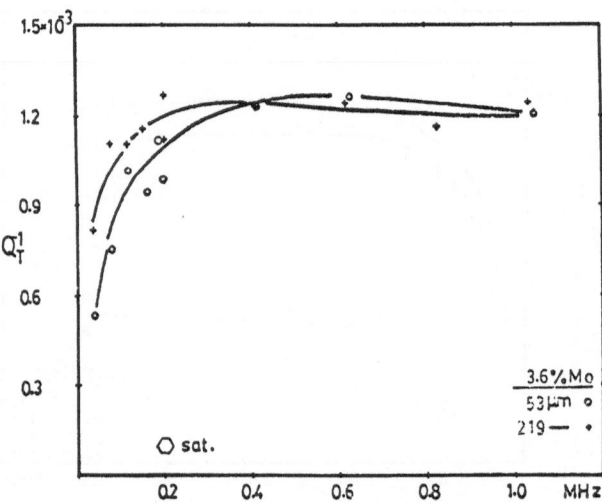

Fig. 8. Absorption of transverse waves in Fe-3.6% Mo at various
         grain sizes.

Fig. 9. Configuration for measurement under saturation with
         elongated cylinder between magnetic polepieces and
         ultrasonic transducers.

configuration. It becomes necessary to introduce additional
support for the cylinder so that a gap is maintained to both
polepieces. We have used small pads of styrofoam--or sharp brass
points in the axis of each polepiece. For torsional vibrations,
this works if the points are exactly contacting in the axis of the
cylinder. Still the difficulty is external loading of the
resonator by such support. For that reason our results, presently,
can only give an upper limit of any residual non-magnetoelastic
absorption. This limit was determined to be around $Q^{-1} = 10^{-4}$.

Figure 10 shows an example of how Q_t^{-1} declines with increasing
magnetic flux and comes to a constant level in saturation.

Figure 11 presents Q_t^{-1} of Fe and Fe-7.8% Cr. One sees that
both drop to a similar level under magnetic saturation. The
Cr-alloy can drop by up to a factor of 25 and Fe by a factor of 6.
Even though the actual "background" probably is lower, it is
interesting to compare this level with absorption in a nonmagnetic
metal. In pure copper we find $Q_t^{-1} \simeq 0.7 \times 10^{-3}$, which is
significantly larger. Cu, as an fcc material, has a large
dislocation density, causing dislocation damping. Fe has few
dislocations, and those have the tendency to be pinned by
impurities. We believe that the background in our Fe-alloys is
mainly damping by intercrystalline thermal currents. But at
present our experimental method does not accurately separate such
a small background from the large magnetoelastic damping.

EFFECT OF RESIDUAL STRAIN AND COLDWORK

Early in our measurements we noted the strong influence of
small coldwork on ultrasonic absorption. Clamping a cylinder in a

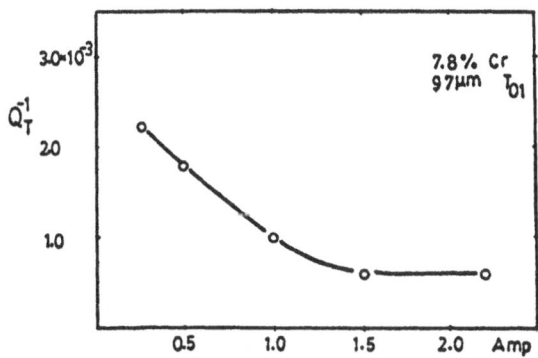

Fig. 10.   Decline of Q_t^{-1} in Fe-7.8 Cr with increasing magnet
           current.

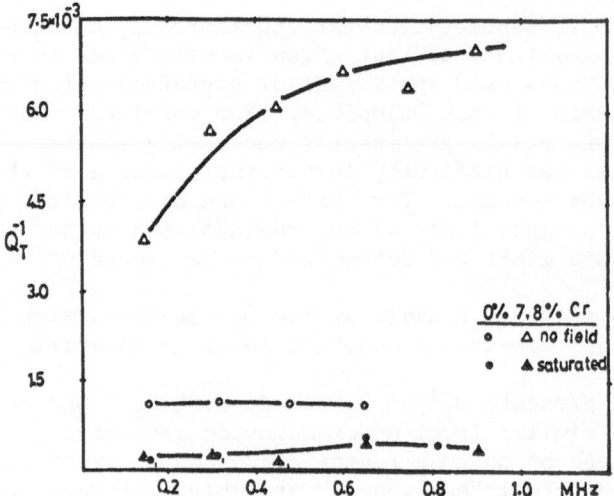

Fig. 11. Q_t^{-1} in Fe and Fe-7.8 Cr without and with magnetic
          saturation.

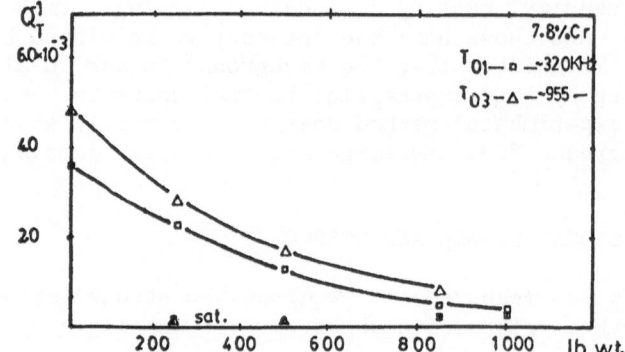

Fig. 12. Drop of Q_t^{-1} after successive compression of cylinder by
          increasing force at two frequencies and with magnetic
          saturation.

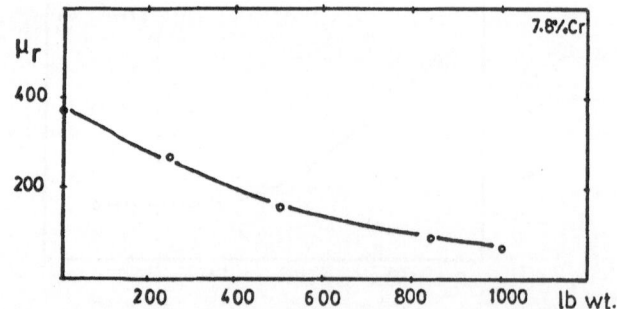

Fig. 13. Drop of initial permeability after successive compression
          of cylinder by increasing force.

lathe-chuck, without hardly any visible clamp marks left, could reduce Q_t^{-1} by a factor of 5. Also, such change sometimes was not completely reversible by strain release annealing. Accordingly, careful runs of annealing and absorption measurements had to be conducted to insure meaningful results.

Among other attempts, we have studied Q_t^{-1} as a function of compression type coldwork. The cylinders were axially compressed between two flat surfaces, in a hydraulic rig. Figure 12 shows how the Q_t^{-1} drops in Fe-7.8% Cr after having been successively compressed with increasing force. After the 500 lb compression, the length was diminished by 0.4%. We also include measurements under magnetic saturation. One notices that the background is virtually unaffected by the coldwork. Obviously, the coldwork effect is restricted to the magnetoelastic part of the absorption.

One might suspect that the reduction in Q_t^{-1} is due to a drop in initial magnetic permeability. That is verified by parallel measurements of permeability, Fig. 13. Indeed, Q_t^{-1} and $\mu_{init.}$ drop roughly proportionally as Döring's theory, equation (10), requires. The permeability measurements were performed on a second cylinder of the same size and material, treated identically. It has an axial bore, through which a few turns of thin copper wire were threaded, thus producing a toroidal transformer. By sending an AC-current of a few Hz through one winding and measuring the induced voltage in the other, $\mu_{init.}$ can be determined.

The decrease in permeability due to residual strain is explained by magneto-mechanical coupling and is well documented.

CONCLUSIONS

We have presented ultrasonic absorption data on several Fe-based binary alloys between 50 KHz and 1.2 MHz. It is verified that the absorption in this frequency range is predominantly magnetoelastic in nature. Occasionally, a broad maximum appears in accordance with the theory by Becker and Döring.

The absorption is produced predominantly by shear strain via the shear viscosity coefficient. This seems to be a fact not predicted by current theoretical understanding. Compare the results in Reference 10 on single crystals.

Small coldwork strongly affects the absorption. The cause is a change in initial permeability due to internal strain.

ACKNOWLEDGEMENT

    The authors are much indebted to our metallurgical coworkers,
Prof. Y.T. Chou and S. Tantichareonkiat, for preparing the specimen
and to the Bethlehem Steel Corporation for casting the special
alloys. We also thank Mr. J.A. Bohar for working out the
permeability measurement and doing most of the testing.

REFERENCES

1.   R. Klinman, G. R. Webster, F. J. Marsh, E. T. Stephenson,
     "Ultrasonic Prediction of Grain Size, Strength, and
     Toughness in Plain Carbon Steel," Material Evaluation 38
     (1980), pp. 26–32.
2.   A. S. Nowick, B. S. Berry, "Anelastic Relaxation in
     Crystalline Solids," Academic Press, 1972.
3.   N. Eberhardt, A. Tverdikhlebov, "A Resonance Method for
     Measurement of Longitudinal and Transverse Ultrasonic Wave
     Velocities and Their Attenuations," in: Review of Progress
     in Quantitative Nondestructive Evaluation, pp. 639–646,
     Vol. 1, Plenum Press, New York, 1982.
4.   A. Tverdokhlebov, N. Eberhardt, "An Improved Approach to the
     Computation of Sonic Resonant Frequencies of Cylinders,"
     Ibid., pp. 647–651.
5.   A. B. Bhatia, "Ultrasonic Absorption," Oxford, 1967.
6.   W. P. Mason, Phys. Rev. 83, p. 683 (1951).
7.   R. Becker, W. Doring, "Ferromagnetismus," p. 379, Springer,
     1939.
8.   R. M. Bozorth, "Ferromagnetism," Van Nostrand, 1951, p. 610.
9.   W. P. Mason, Rev. Mod. Physics 25, p. 136 (1953).
10. S. Levy, R. Truell, "Ultrasonic Attenuation in Magnetic Single
     Crystals," Rev. Mod. Phys. 25(1), 1953, pp. 140–145.

GENERAL ULTRASONIC CONSIDERATIONS AND APPLICATIONS

THE INVERSE PROBLEM IN MATERIALS CHARACTERIZATION THROUGH

ULTRASONIC ATTENUATION AND VELOCITY MEASUREMENTS

Emmanuel P. Papadakis

Ford Motor Company
Manufacturing Processes Laboratory
24500 Glendale Avenue
Redford, Michigan 48239

ABSTRACT

The inverse problem in materials characterization is most often skipped over in favor of a correlation because the inverse problem is so difficult and the correlation is only tedious, not hard. In this talk the basic difficulty in the inverse problem field will be illustrated by two examples: finding grain size in metals when ultrasonic attenuation is measured, and finding graphite shape in cast iron when ultrasonic velocity is measured.

The basic difficulty arises because the measured quantity, attenuation or velocity, is a function of several variables. Thus, any single variable among the latter cannot be written as a single-valued function of the measured quantity.

For instance, attenuation α, caused by grain scattering, is a function of frequency f, grain diameter D, grain substructure μ, grain size distribution "GSD", and the ratio of the grain diameter to the ultrasonic wavelength D/λ. The total attenuation "ATT" is also a function of geometrical beam spreading "BS" (which depends on sample anisotropy) as well as on physical absorption mechanisms "ABS" which in turn are functions of frequency and other parameters.

We have

$$ATT = \alpha + ABS + BS \qquad (1)$$

The portion BS can be calculated and subtracted out immediately as long as the macroscopic sample anisotropy is known. The portion ABS may then be separable by multi-frequency measurements or by changing other environmental parameters (temperature, magnetic field) which may influence it, or by calculating it in the simplest cases like thermoelastic losses. The remainder, α, is frequently numerically the largest of the three terms. The term α is written symbolically as

$$\alpha = \alpha \ (D, \ GSD, \ \mu, \ D/\lambda, \ f) \tag{2}$$

The variables combine in such a way that there is a function F of D, GSD, and D/λ; another function n of D/λ; a coefficient A which is a function of n; and the term μ^2. The result is

$$\alpha = AF \ \mu^2 \ f^n \tag{3}$$

The multiple factors discussed above make it obvious that the simplistic hope of finding

$$\overline{D} = \overline{D} \ (\alpha) \tag{4}$$

will not be fulfilled in engineering materials, and that many factors must be taken into account.

The talk will present further clarifications of these challenging research opportunities using ultrasonic attenuation and velocity.

I. GENERAL

The entire field of nondestructive inspection technology deals with the "inverse problem." Yet, until roughly the last decade[1], hardly any progress was made in dealing with it rigorously.

Two broad areas were studied. One was the limited class of the "inverse problem" under the conditions where "everything else was held constant." This class is actually the "forward problem" in a reversible form. If nothing else changes, the dependent variable in the forward problem is usually monotonically related to the independent variable, and either can be found from the other. Extensive research has been done on the "forward problem," and forms the basis for beginning to understand the general "inverse problem."

The second broad area studied was the "correlation problem" where it was accepted that the "inverse problem" could not be solved with the limited number of variables which were measurable. Approximate predictive curves are produced by correlation, and the inverse quantity is accepted with errors. The method has great utility in practice. The "correlation problem" can be handled without physical theory by using only statistical rigor.

The "inverse problem" in full generality deals with the situation in which "everything else is not held constant." One begins with the "forward problem" which yields an equation for a measurable quantity Q. This could be a sum of two or more functions, say R and S, which are each functions of several variables X, Y, Z, U, V, W, etc. R and S may have some variables in common; R may contain some variables which S does not, and vice versa. The inverse problem requires the determination of the value of one variable, say V, from the set.

For materials characterization, this general problem is not trivial and has not been solved, except in a few specific cases.

II. ULTRASONIC ATTENUATION

To investigate this question further, consider ultrasonic attenuation and ultrasonic velocity measurements. For ultrasonic attenuation in a metallic polycrystalline engineering material[2-4], the measured attenuation consists of the sum of scattering by flaws, scattering by grains, absorption by various mechanisms, beam spreading, and instrumental effects.

Suppose that the grain size of the metal is the quantity of interest. One would design an experiment or set of experiments in which grain scattering was a dominant factor. The experiment could be designed such that the instrumental effects were negligible or correctable[5]. (These include absorption of energy in the transducer of course.) A correction could be made for beam spreading[6] as long as the velocity anisotropy as well as the velocity in the specimen were known along with the operating frequency and the transducer active diameter. One would have to know that the transducer was a good piston source, too[7]. One can see that even the corrections are complex. Beyond these considerations, one would have to correct for absorption mechanisms in the metal[8]. There is always thermoelastic heat flow[9] and dislocation damping[10]. There may be magnetic effects[11]. These may be negligible at the frequencies where the grain scattering is high enough to measure easily. However, one must measure over a range of frequencies to determine grain size[2-4].

This gets to the heart of the matter. The absorption mechanisms generally yield attenuation dependent upon the square of the frequency. The grain scattering is a function of frequency dependent upon the ratio of the ultrasonic wavelength to the grain diameter. For small grains, the dependence is upon the fourth power of the frequency[12-15]. For large grains, the attenuation is independent of frequency[16]. In the intervening range there is a continuous change in functional dependence.

What variables influence ultrasonic grain scattering?

o   Frequency[12-15]

o   Grain size[12-16]

o   Grain size distribution[17,18]

o   Crystallographic anisotropy of the grain[12-15]

o   Microstructure which may indicate grains which are not single crystals[2-4,19-22]

o   Preferred orientation[23]

o   Elongated or flattened grains[24,25]

    Let us take some specifics.

o   In the Rayleigh region[26] where the wavelength is much greater than the grain diameter, the frequency dependence is f to the fourth power (f^4).

o   When the stochastic region[27] is reached where the wavelength is comparable to the grain diameter, the frequency dependence is f-squared (f^2).

o   In between there is a continuous change. The intermediate region is often where the experiment falls[2-4,28,29].

o   In the Rayleigh region the scattering power of a grain is proportional to its volume, while in the Stochastic region the scattering power is proportional to the grain diameter[16,27].

o   The grain size parameters are not simply the ASTM numbers or the average diameter raised to some power, but are computed from the grain size distribution by more complex formulas[17,18]. In the Rayleigh region the parameter proportional to grain volume is π over 6 times the average of the sixth power of the grain diameter divided by the average of the third power of the grain diameter, that is, $(\pi/6)\langle D^6 \rangle_{av}/\langle D^3 \rangle_{av}$. The average grain diameter for the stochastic region is simply the average grain diameter, that is, $\langle D \rangle_{av}$. The averages are taken over the grain size distribution.

o For cubic crystallites, the crystallographic anisotropy[15] of the grain is simply C_{11} minus C_{12} minus $2C_{44}$. Aluminum is at least ten times lower in anisotropy than most other metals. And the anisotropy enters the scattering formulas to the second power.

o Microstructure resulting in non-single-crystal grains, that is, in transformation products within a prior grain volume, can have two effects:[18–22]

   (1) It can change the average anisotropy of the prior grain volume.

   (2) It can supply a multiplicity of new, smaller scattering centers.

As an example, the martensite transformation in steel and other alloys can lower the scattering by a factor of 100 relative to pure polycrystalline iron of the same prior anstenitic grain size[19,30]. Precipitation reactions can also lower the grain scattering[2]. The martensite in carbon steel can contain absorption mechanisms with attenuation proportional to the square of the frequency[20]. This attenuation can be removed by tempering.

o Preferred orientation can lead to anisotropic attenuation. Drawn zinc[23] has very little grain scattering along the bar axis because the ultrasonic wave travels along only a-axes of the grains in that direction. Scattering, of course, depends on differences in elasticity from grain to grain which are provided by the randomicity of the crystalline axes from grain to grain[16]. The $[111]$ direction is different in compressibility from the $[100]$ direction, and so on.

o Cylinders cause scattering attenuation proportional to the square of their diameter in the Rayleigh region, so it follows that elongated grains or flattened grains will not behave as equiaxed grains[25,26].

An added complication is the possibility of scattering by flaws. If these are comparable to grains in size, they are indistinguishable from them. Larger flaws may provide separate echoes, of course. The small flaws may be segregated, making the attenuation inhomogeneous. Of course, grain growth may be segregated, too, by some vagary of chemistry, cold work, or heat treatment. So, how does one find the grain size in this mix of variables?

One might want to adopt X-ray powder pattern line broadening. That, however, is not the subject of this paper.

Probably a family of measurements will ultimately be necessary unless the number of variables can be pared down drastically. Again, one runs the risk of assuming that he can go back to the simple "forward problem." Probably a better assumption is that ultrasonic attenuation measurements will be one of a family of measurements needed to characterize any material.

With ultrasonic attenuation, one has the magnitude and the frequency dependence to work with. The instrumental contributions can be eliminated experimentally, and the beam spreading can be corrected. Two or three velocity measurements may be needed to prepare for the beam spreading calculation if the material is more than a little anistropic[31]. Then there is the question of the absorption and the flaws. If flaws can be discounted and absorption can be proved to be relatively very small, then the remainder of the attenuation can be ascribed to grain scattering. Then the frequency dependence of the attenuation indicates whether the grains are in the Rayleigh region, the stochastic region, or in between[28,29]. The magnitude of the attenuation should agree with the theory in the region found, as long as the grains are single crystals[28,29]. If not, then the magnitude of the attenuation will not agree with theory. Conversely, if the magnitude of the attenuation and the theoretical expression for it are used with the wrong effective grain anisotropy, the wrong grain size will be calculated, maybe by a factor as high as five (5). As you see, the entire question is complex. The inverse problem can be impacted but not completely solved by ultrasonic attenuation measurements.

III.   ULTRASONIC VELOCITY

Like attenuation, velocity is affected by a multiplicity of factors. Grain size is not one of them. Grain microstructure is, to a degree[18]. Precipitates, flaws, alternate phases[32-37] and so on affect the velocity. Processes commonly classed as "heat treatment" can modify velocities[38] through the above mechanisms. One very interesting example is cast iron. It is an iron alloy like steel but with a huge excess of carbon precipitated as graphite. The graphite occupies 11 to 14 percent of the volume. The morphology of this graphite is the principle source[39] of velocity variations in cast iron, i.e., by up to a ratio of three-to-two (3:2). The morphology varies from large thin flakes to tight little spheres. The spherical or "nodular" form of graphite lowers the modulus of the iron the least because the spheres permit the iron the highest degree of continuity[32].

Where is the "inverse problem?" The "inverse problem" resides in the fact that the shape of the graphite is one of the two principal determinants of the strength of iron[40]. The strength is the quantity one would like to assure by an appropriate measurement. The other principal determinant of strength is a metallurgical property of the iron phase, the pearlite-to-ferrite ratio. A high ratio, i.e., eighty to ninety percent (80-90%) pearlite, implies a stronger metal[33]. The pearlite-to-ferrite ratio perturbs the ultrasonic velocity not in the least, however, when viewed on a scale practical for quality control measurements, plus-or-minus one-half percent (±1/2%). A third factor which influences strength, carbide content, has the effect of increasing the velocity while decreasing the strength[34]. This effect is opposite to the graphite morphology effect, and can complicate the "inverse problem." Proper cooling of the casting from the melt eliminates the formation of carbides, and is striven for when nodular iron (ductile iron) castings are made. Thus, one is tending toward the "forward problem." Similarly, the pearlite-to-ferrite ratio is held approximately at a desired level principally by the alloy chemistry and secondarily by the cooling rate immediately upon casting. Hence, the casting practice tends to return the "inverse problem" toward the monotonic "forward problem." The reason there is any problem at all is that the two processes for making nodular iron both have failure modes for which the effect is a modification of the shape of the graphite away from spherical. This modification would tend to lower the strength, so the measurement of graphite morphology is important. It is important to be able to do it on 100% of production by a rapid nondestructive means, so ultrasonic velocity is one necessary method[30,40]. The pearlite-to-ferrite ratio and the carbide content influence hardness, electrical conductivity, and permeability, so indentation measurements[41] or eddy current measurements[32,42] would be used if there were a question concerning the percentage of pearlite or of carbide.

The ultrasonic velocity varies with average graphite morphology fairly linearly over its entire range. Sorting out iron with ninety to one hundred percent (90-100%) nodularity is relatively easy[39], as the highest nodularity produces the highest velocity. A more difficult question as an "inverse problem" is assuring that the metal has very low nodularity but also contains no flake graphite. This intermediate morphology, known as "compacted graphite iron" displays intermediate strength and favorable thermal conductivity and acoustic damping properties. The "inverse problem" is not straight-forward, as two test pieces might yield the same velocity although one was 100% compacted graphite while the other was approximately 1/3 flake and 2/3 nodular, mixed. Probably such abnormal mixes would not occur, but cast iron is complicated. The pearlite-to-ferrite ratio and the carbide percentage are still questions, too, if strength is at issue.

In general, as seen above, one would need a mix of measure- ments, including ultrasonic velocity, to solve an "inverse prob- lem" in which velocity was a principal discriminant.

IV. SUMMARY

In any inverse problem, several experimental methods will be needed for a complete solution. Ultrasonic attenuation and vel- ocity are useful items in the set of available methods.

REFERENCES

1. ARPA/AFML Program on Quantitative NDE: Various reports issued by WADC/WPAFB, Ohio, 1969 -- date.
2. Papadakis, E. P., "Ultrasonic Attenuation Caused by Scattering in Polycrystalline Media," in Physical Acoustics: Principles and Methods, Vol. IV Part B, W. P. Mason, editor, Academic Press, New York, 1968, pp. 269-328.
3. Papadakis, E. P., "Scattering in Polycrystalline Media," in Methods of Experimental Physics - Ultrasonics, P. D. Edmonds, editor, Academic Press, New York, 1981, pp. 237-298.
4. Papadakis, E. P. "Physical Acoustics and the Microstructure of Iron Alloys," Internat. Metals Rev. (to be published).
5. Papadakis, E. P., "Ultrasonic Attenuation by Spectrum Analysis of Pulses in Buffer Rods: Method and Diffraction Correc- tions," J. Acoust. Soc. Amer. 53, 1336-1343 (1973).
6. Papadakis, E. P., "Ultrasonic Diffraction Loss and Phase Change in Anisotropic Materials," J. Acoust. Soc. Amer. 40, 863-876 (1966).
7. Mansour, T. M., "Evaluation of Ultrasonic Transducers by Cross-Sectional Mapping of the Near Field Using A Point Reflector," Mater. Eval. 37 (7), 50-54 (June 1979).
8. Zener, C., Elasticity and Anelasticity of Metals, U. of Chicago Press, Chicago, Ill., 1948.
9. Lucke, K., "Ultrasonic Attenuation Caused by Thermoelastic Heat Flow," J. Appl. Phys. 27, 1433-1438 (1956).
10. Granato, A. and Lucke, K., "Theory of Mechanical Damping Due to Dislocations," J. Appl. Phys. 27, 583-593 (1956).
11. Mason, W. P., Physical Acoustics and the Properties of Solids, Van Nostrand, Princeton, 1959, pp. 209-224.
12. Lifshitz, E. M. and Parkhomovskii, G. D., Zh. Eksperim. i Teoret., Fiz. 20, 175-182 (1950).

13. Bhatia, A. B., "Scattering of High-Frequency Sound Waves in Polycrystalline Materials," J. Acoust. Soc. Amer. 31, 16-23 (1959).

14. Bhatia, A. B. and Moore, R. A., "Scattering of High-Frequency Sound Waves in Polycrystalline Materials II," J. Acoust. Soc. Amer. 31, 1140-1142 (1959).

15. Merkulov, L. G., "Investigation of Ultrasonic Scattering in Metals," Sov. Phys. - Tech. Phys. 1 (1), 59-69 (Oct. 1956), from J. Tech. Phys. (USSR) 26, 64-73 (1956).

16. Mason, W. P. and McSkimin, H. J., "Energy Losses of Sound Waves in Metals Due to Scattering and Diffusion," J. Appl. Phys. 19, 940-946 (1948).

17. Papadakis, E. P., "Grain Size Distribution in Metals and Its Influence on Ultrasonic Attenuation Measurements," J. Acoust. Soc. Amer. 33, 1616-1621 (1961).

18. Papadakis, E. P., "From Micrograph to Grain Size Distribution With Ultrasonic Applications," J. Appl. Phys. 35, 1586-1594 (1964).

19. Kamigaki, K., "Ultrasonic Attenuation in Steel and Cast Iron," Sci. Rep.   RITU Tohoku Univ., Sendai, Japan, A-9, 48-77 (1957).

20. Papadakis, E. P., "Ultrasonic Attenuation and Velocity in Three Transformation Products in Steel," J. Appl. Phys. 35, 1474-1482 (1964).

21. Papadakis, E. P., "Ultrasonic Attenuation and Velocity in S.A.E. 52100 Steel Quenched from Various Temperatures," Met. Trans. 1, 1053-1057 (1970).

22. Papadakis, E. P. and Reed, E. L., "Ultrasonic Detection of Changes in the Elastic Properties of a 70-30 Iron-Nickel Alloy Upon Heat Treatment," J. Appl. Phys. 32, 682-687 (1961).

23. Papadakis, E. P., "Influence of Preferred Orientation on Ultrasonic Grain Scattering," J. Appl. Phys. 36, 1738-1740 (1965).

24. Morse, P. M., Vibration and Sound, 2nd Edition, McGraw-Hill, New York, 1948, pp. 347-352.

25. Morse, P. M. and Ingard, K. U., Theoretical Acoustics, McGraw-Hill, New York, 1968, pp. 401-407.

26. Lord Rayleigh, Theory of Sound, Macmillan Co., New York, 1929, Vol. II, p. 152.

27. Huntington, H. B., "On Ultrasonic Scattering by Polycrystals," J. Acoust. Soc. Amer. 22, 362-364 (1950).

28. Papadakis, E. P., "Revised Grain Scattering Formulas and Tables," J. Acoust. Soc. Amer. 37, 703-710 (1965).

29. Papadakis, E. P., "Ultrasonic Attenuation Caused by Scattering in Polycrystalline Metals," J. Acoust. Soc. Amer. 37, 711-717 (1965).

30. Papadakis, E. P., "Ultrasonic Attenuation in S.A.E. 3140 and 4150 Steel," J. Acoust. Soc. Amer. 32, 1628-1639 (1960).

31. Papadakis, E. P., "Ultrasonic Study of Simulated Crystal Symmetries in Polycrystalline Aggregates," IEEE Trans. SU-11, 9-29 (1964).

32. Plenard, E. "The Elastic Behavior of Cast Iron," National Metal Congress, Cleveland, 1964.

33. Fuller, A. G., Emerson, P. J. and Sergeant, G. F., "A Report on the Effect Upon Mechanical Properties of Variations in Graphite Form in Irons Having Varying Amounts of Ferrite and Pearlite in the Matrix Structure....," AFS Trans. 88, 21-50 1980).

34. Sergeant, G. F. and Fuller, A. G., "The Effect Upon Mechanical Properties of Variation in Graphite Form in Irons Having Varying Amounts of Carbide in the Matrix Structure....., AFS Trans. 88, 545-574 (1980).

35. Patterson, B. R. and Bates, C. E., "Nondestructive Property Prediction in Grey Cast Iron Using Ultrasonic Techniques," AFS Trans. 89, 369-378 (1981).

36. Henderson, H. E., "Ultrasonic Velocity Technique for Quality Assurance of Ductile Iron Castings," The Iron Worker 37 (3), Summer 1973.

37. Kovacs, B. V. and Cole, G. "On the Interaction of Acoustic Waves With S. G. Iron Castings," Trans. AFS 83, 497-510 (1977).

38. Henderson, H. E., "The Effect of Heat Treatment on Ultrasonic Velocity of Ductile Iron Castings," The Iron Worker 40 (3), Summer 1976.

39. Papadakis, E. P., "Morphological Severity Factor for Graphite Shape in Cast Iron and Its Relation to Ultrasonic Velocity and Tensile Properties," AFS Trans. (to be published).

40. Papadakis, E. P., "Ultrasonic Velocity and Attenuation: Measurement Methods With Scientific and Industrial Applications," in Physical Acoustics: Principles and Methods, Vol. XII, W. P. Mason and R. N. Thurston, editors, Academic Press, New York, 1976, pp. 343-348.

41. Lysaght, V. E., Indentation Hardness Testing, Reinhold, N.Y., 1949, pp. 134-135 and 150-153.

42. Giza, P. and Papadakis, E. P., "Eddy Current Tests for Hardness Certification of Gray Iron Castings," Mater. Eval. 37(8), 45-50,55 (July 1979).

INFERENCE OF FATIGUE CRACK CLOSURE STRESSES FROM

ULTRASONIC TRANSMISSION MEASUREMENTS

R. G. Thompson, C. J. Fiedler, and O. Buck

Ames Laboratory
Iowa State University
Ames, Iowa

INTRODUCTION

The nondestructive determination of stress is usually attempted by measuring stress (or strain) related changes in material properties. Included are direct x-ray measurements of lattice constants and indirect inferences of stress or strain from changes in ultrasonic velocities or magnetic properties. However, in problems in which one wishes to determine localized stresses across an interface between two materials, neither approach is satisfactory. X-rays typically measure surface stresses and do not sample the stress condition near an internal interface. Ultrasonic velocity measurements and magnetic measurements can sense interior conditions, but also average over the properties of the intervening material. Hence, they cannot determine the localized stresses at the interface without sophisticated data reduction techniques such as those employed in holographic reconstructions.

An alternate approach is based on the spectral content of ultrasonic waves scattered from the interface. It is postulated that, because of roughness, the two sides of the interface are not in complete contact. The interface is then viewed as a planar array of alternating contacting and noncontacting regions. The stress level controls the intimacy of contact, which in turn can be related to the resulting changes in the spectral content of ultrasonic waves reflected from, or transmitted through the interface. This paper reviews the physical basis of the phenomenon and presents preliminary results of its application to the problem of determining closure stresses in fatigue cracks. A

161

second application to the problem of determining joining stresses in Nitinol tube couplers is discussed in a companion paper.

STRESS DEPENDENCE OF ULTRASONIC SCATTERING FROM AN INTERFACE

A quasi-static model for the interaction of an ultrasonic wave with the partially contacting interface has been developed. It is assumed that, when the ultrasonic wavelength is large with respect to the dimensions and separation of the contacts, their influence on an ultrasonic wave can be modeled by a pair of effective boundary conditions of the form

$$\sigma_1 = \sigma_2 \tag{1}$$

$$\sigma_1 = \kappa(U_2 - U_1) \tag{2}$$

Here the subscripts 1 and 2 refer to the value of the stress σ and displacement U on either side of the interface, and the proportionality constant, κ, may be thought of as the stiffness of a distributed spring.[1-4] Here only the components of displacement and stress normal to the interface will be considered and hence equations (1) and (2) are written in scalar form. By defining κ to be a third-rank tensor, a more general approach could be presented.

By applying standard analysis, these boundary conditions can be used to compute the reflection R and transmission T of energy at the interface.[5] For normal incidence, the results are

$$R = \frac{j\pi f \rho v / \kappa}{1 + j(\pi f \rho v / \kappa)} \tag{3}$$

and

$$T = \frac{1}{1 + j(\pi f \rho v / \kappa)}, \tag{4}$$

where f is the frequency, ρ is the density, and v is the ultrasonic velocity. The magnitude of the reflection coefficient increases with frequency while the transmission coefficient decreases. The dimensionless parameter governing R and T is $\pi f \rho v / \kappa$. Hence, the interfacial stiffness, κ, can be deduced from an experimental determination of the scale of plots of R or T versus f.

The stress dependence of the ultrasonic scattering enters through the variation of κ with the degree of contact. For periodic arrays of contacts, e.g., a two-dimensional array of alternating contacts and slits, analytical expressions for κ as a function of contact parameters have been obtained.[5] These show that κ varies inversely with the contact separation S and directly with the fractional contact area W/S, where W is the contact width. Increases in a compressive stress would be expected to decrease S,

increase W and thereby increase κ. Alternatively, the case of
randomly spaced, non-interacting contacts can be considered. Using
a slightly different approach, Haines[6] has derived expressions for
reflection and transmission at the interface which are equal to
equations (3) and (4) if one assumes a value of

$$\kappa = \frac{kE}{2} = \frac{(-\sigma^0/\sigma_m)}{\bar{r}} , \qquad (5)$$

where k is a constant of order two, E is Young's modulus, $(-\sigma^0)$ is
the compressive stress at the interface, σ_m is the flow stress of
the material, and $\bar{r}$ is the mean radius of the contact area. The
stress dependence of κ is explicitly shown in equation (5). It
should be noted that $\bar{r}$ also has a stress dependence, but his is
relatively weak. However, $\bar{r}$ does depend strongly on the initial
roughness of the surfaces. Haines' model has been found to be in
semi-quantitative agreement with measurements of Woolridge.[7,8]

A closely related conclusion has been obtained by Buck et al.[9]
They speculate that each area of contact will carry a local stress
equal to the Meyer hardness of the material, which should be
considered equivalent to the flow stress in the notation of Haines.
An ultrasonic determination of the area fraction of contact could
then be used to predict the average stress, equal to fractional
area times Meyer hardness.

APPLICATION TO STUDIES OF CLOSURE OF FATIGUE CRACKS

It is now well accepted that, in the region near the tip of a
fatigue crack, the faces can be in partial contact due to
compressive "closure" stresses which can be developed by a variety
of mechanisms depending on the conditions under which the crack
grew.[10-13] This closure can reduce the ultrasonic detectability
of the crack and influence its rate of growth. The stress
dependence of the ultrasonic transmission has been used to study
this closure condition.

The experiment is conceptually sketched in Fig. 1. A fatigue
crack was grown in a 7075 aluminum compact tension specimen, using
load shedding techniques to maintain a constant $\Delta K_I/K_{Ic} \simeq 25\%$.
The geometry of this sample had been modified so that, in the
plane of the measurement, only the shape shown in Fig. 2 remained.
This allowed illumination of the crack with well-focussed
ultrasonic beams. The broad band probes used had 10 MHz nominal
center frequency, 1.8 cm (3/4 in) diameter and 10 cm (4 in) focal
length in water. At 10 MHz, the diffraction limited focal spot
size was estimated to be 1.5 mm (0.060 in) which was always
adjusted to be in the plane of the crack. In the experiments
reported in this paper, the probes were always aligned coaxially,

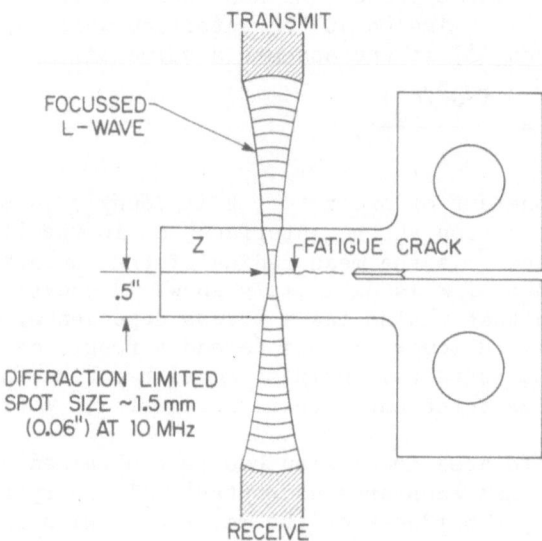

Fig. 1.   Configuration for measuring ultrasonic transmission
          through a fatigue crack in a modified compact tension
          specimen.

so that the direct transmission through the crack was measured.
In work to be reported elsewhere,[14] the orientation of the
receiving transducer was changed so that either longitudinal or
transverse waves diffracted by the tip of the crack were detected.

In preliminary experiments,[5] it was found that the ultrasonic
focal spot width was comparable to the width of the closure region
and hence it was difficult to differentiate one from the other.
In order to overcome this problem and provide a basis to
quantitatively interpret the experiments, a model relating $\kappa(x)$ to
the measured signals was developed. In this model, the focal spot
was assumed to have a Gaussian profile, equations (1) and (2) were
used to model the interface, and the Kirchhoff approximation was
made in the scattering solution. The result was the expression

$$\Gamma^N(f,x_1) \;=\; \int_{-\infty}^{\infty} \left(\frac{1}{1 + j(\pi f \rho v/\kappa(x))} \right) e^{-2(x-x_1)^2/w(f)} . \tag{6}$$

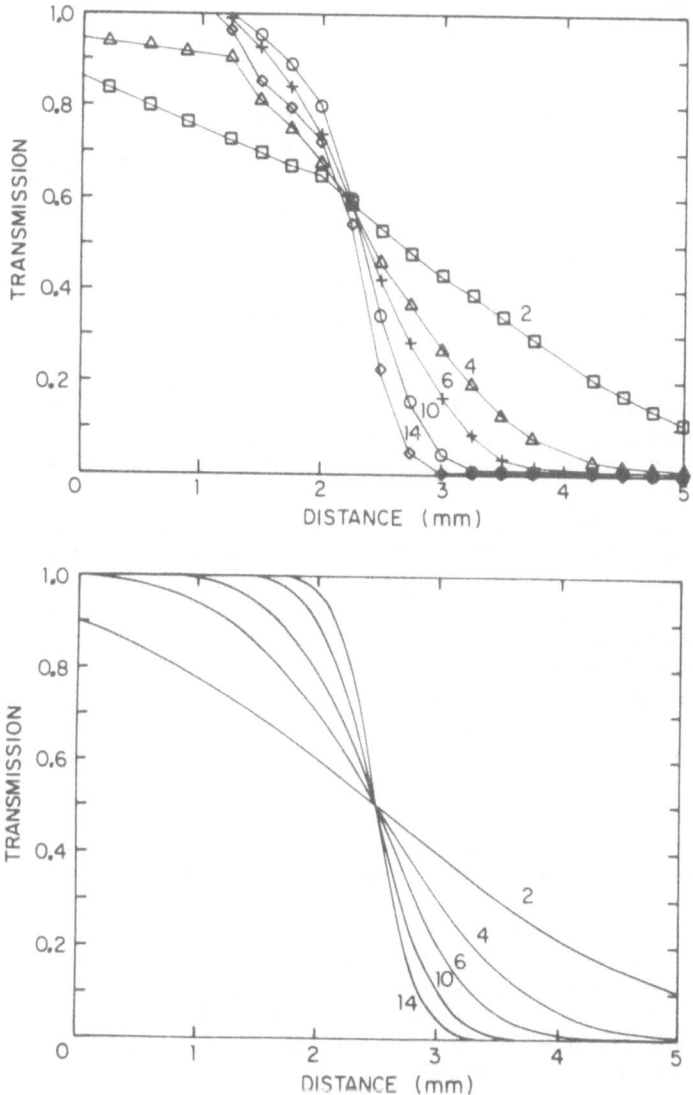

Fig. 2. Ultrasonic transmission past an open crack (saw slot) at 2, 4, 6 10 and 14 MHz. Top: experiment; bottom: theory.

Here $\Gamma^N(f,x_1)$ is the transmission at beam position x_1, normalized by that when the beam totally misses the crack ($x_1 = -\infty$). $\kappa(x)$ is the interface compliance which varies from a value of ∞ at the crack tip ($x = 0$) to zero when the crack is fully open, and $w(f)$ is the frequency dependent width of the ultrasonic beam. Note that equation (6) is simply the overlap of the plane wave transmission factor, equation (4), and the two Gaussian beam profiles. To verify this expression and provide an experimental basis for determining $w(f)$, measurements were performed on a saw slot simulating a fully open crack. The results are shown in Fig. 2. The top diagram shows the experimental data and the bottom diagram shows the prediction of equation (5) when $w(f)$ was selected to have the value (7.9 mm-MHz)/f(MHz) and $\kappa(x)$ was ∞ for $x < 0$ and 0 for $x > 0$. The fit is considered to be quite good.

Two features of Fig. 2 should be noted for future comparison to the fatigue crack data. First, the transition from full to zero transmission is more steep at higher frequencies due to the smaller beam diameter. Second, 50% transmission occurs at the same beam position for all frequencies since, when the beam center is at the crack tip, half of it is blocked for any beam radius.

This procedure was then repeated for the previously described fatigue crack in aluminum. Figure 3 (top) presents the experimental data. Note two major differences from the saw slot response. First, except for the 2 MHz response, none of the curves cross. The effective crack length, defined in terms of the position for 50% transmission, increases with frequency. This is consistent with the prediction of equation (4) that, for fixed κ, transmission decreases with frequency. Hence, regions of the crack which appear transparent at 2 MHz will be nearly opaque at 14 MHz. Second, at high frequencies, the transition of the fatigue crack from complete to zero transmission is not nearly as steep as that of the saw slot. This provides further evidence for the existence of a region of gradual change from the uncracked to the fully open condition.

To test the quantitative accuracy of these comments, trial values of $\kappa(x)$ were substituted into equation (6). Comparison to the experimental data showed semi-quantitative agreement for a range of values of κ, as shown in equation (4). Figure 3 (bottom) presents the theoretical prediction when κ had the exponential form indicated by the solid line in Fig. 4. All the major features of the experiment are reproduced. When the dashed value for κ shown in Fig. 4 was used, the steepness of the theoretically predicted curves were in better agreement with experiment, but their separation at the 50% point was greater than that observed.

Given $\kappa(x)$, one would like to estimate the closure stresses, $-\sigma^0(x)$, holding the crack faces in contact. This step has not been completed, but the philosophical direction can be indicated.

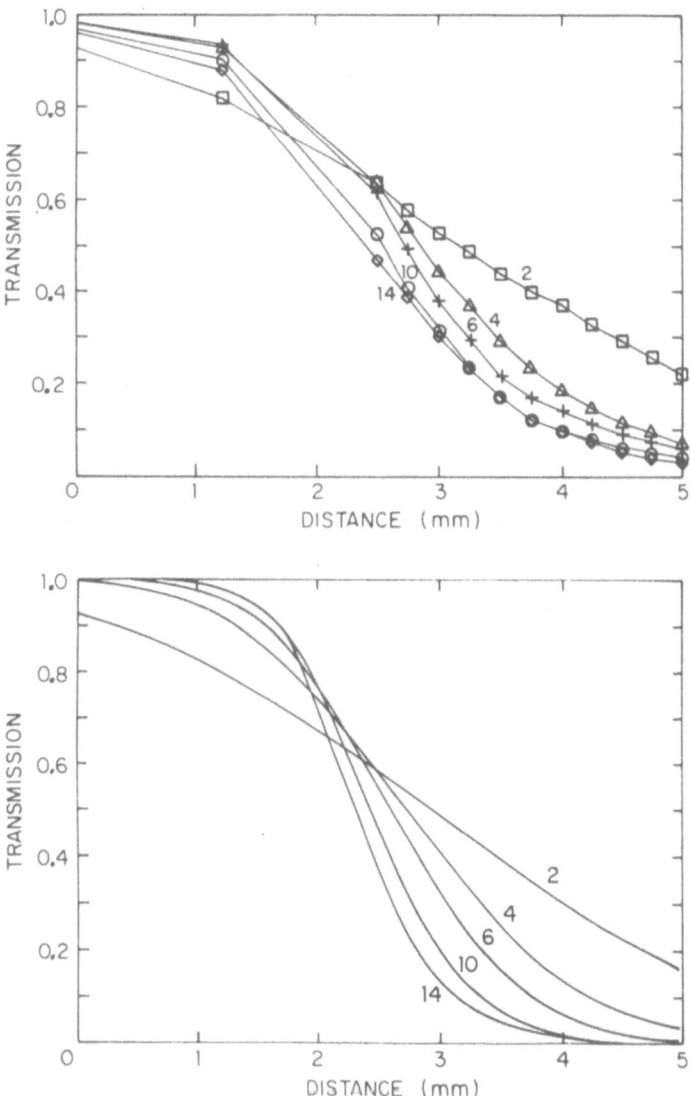

Fig. 3.   Ultrasonic transmission past a fatigue crack at 2, 4, 6, 10 and 14 MHz.   Top: experiment; bottom: theory.

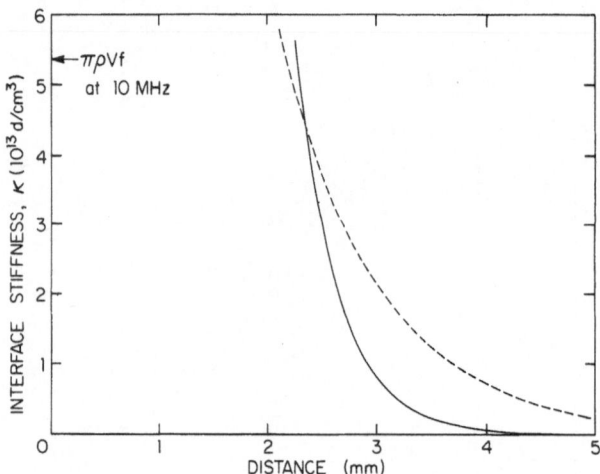

Fig. 4.   Values of $\kappa(x)$ used in modeling fatigue crack closure.

Equation (5) must be combined with an expression relating $\bar{r}$ to σ^o. Haines[6] notes that

$$\text{Fractional contact area} \ = \ \frac{(-\sigma^o)}{\sigma_m} \ = \ N\pi r^{-2} \ , \tag{8}$$

where N is the number of contact points per unit area. Solving for $\bar{r}$ and substituting in equation (5), one finds that

$$(-\sigma^o) \ = \ \sigma_m \ \frac{\pi}{N} \left[\frac{2\kappa}{kE}\right]^2 \ , \tag{9}$$

which is the desired expression for closure stress. Evaluation of equation (9) requires an independent knowledge of N which is the subject of present research.

DISCUSSION

       The details of the above analysis must be improved in several ways. Problems to be addressed include determining the unique functional form of κ which best fits the data and correction for the fact that equation (9) breaks down for large κ since $(-\sigma^0)$ should be bounded by σ_m. Despite these present shortcomings, the authors believe that the approach taken will bear considerable fruit. This possibility of deducing quantitative crack closure stress values in the future is strengthened by the successful application of the model of Haines to the measurement of joining stresses in Nitinol tube couplers.[15] The continuum view adopted here is believed to provide a useful complement to the model of Golan,[16] who discusses the estimation of closure stress from the strengths of signals diffracted from individual, isolated contacts.

ACKNOWLEDGEMENT

The Ames Laboratory is operated for the U.S. Department of Energy by Iowa State University under Contract No. W-7405-ENG-82. This work was supported by the Director of Energy Research, Office of Basic Energy Sciences.

REFERENCES

1. H. G. Tattersall, "The ultrasonic pulse-echo techniques as applied to adhesion testing," J. Phys. D., Appl. Phys. 6, 819 (1973).
2. R. B. Thompson, "A quasi-static model for the interaction of elastic waves with a planar array of voids," unpublished technical report of Rockwell International Science Center (1973).
3. G. A. Alers and L. J. Graham, "Reflection of ultrasonic waves by thin interfaces," 1975 Ultrasonics Symp. Proc., (IEEE, New York, 1975), pp. 579-582.
4. G. A. Alers and R. B. Thompson, "Application of trapped modes in layered media to the testing of adhesive bonds," 1976 Ultrasonics Symp. Proc. (IEEE, New York, 1976), pp. 138-142.
5. R. B. Thompson, B. J. Skillings, L. W. Zachary, L. W. Schmerr, and O. Buck, "Effects of crack closure on ultrasonic transmission," Review of Progress in Quantitative Nondestructive Evaluation 2, D. O. Thompson and D. E. Chimenti, eds. (Plenum Press, New York, in press).
6. N. F. Haines, "The theory of sound transmission and reflection at contacting surfaces," Report RD/B/N4711 (Central Electricity Generating Board, Research Division, Berkeley Nuclear Laboratories, Berkeley, England, 1980).
7. A. B. Wooldridge, "The effects of compressive stress on the ultrasonic response of steel-steel interfaces and of fatigue cracks," Report NW/SSD/RR/42/79 (Central Electricity Generating Board, Northwestern Region, Manchester, England, 1979).

8. A. B. Wooldridge, "The effects of compressive stress and
    contaminating liquids on the ultrasonic detection of
    fatigue cracks," Revue du Cethedec, 17^{e}annee, 4^{e}trimestre
    1980-NS80-2, 233-244 (1980).

9. O. Buck, B. J. Skillings, and L. K. Reed, "Simulation of
    closure: Effects on crack detection probability and stress
    distributions," Review of Progress in Quantitative
    Nondestructive Evaluation 2, D. O. Thompson and D. E.
    Chimenti, eds. (Plenum Press, New York, in press).

10. W. Elber, "Fatigue crack closure under cyclic tension," Engr.
    Frac. Mech. 2, 37-45 (1970).

11. O. Buck, C. L. Ho, and H. L. Marcus, "Plasticity effects in
    crack propagation," Engr. Frac. 5, 23-34 (1973).

12. S. Suresh, G. F. Zaminski, and R. O. Ritchie, "Oxide-induced
    crack closure: An explantion for near-threshold corrosion
    fatigue crack growth behavior," Metal. Trans. A 12A,
    1435-1443 (1981).

13. S. Suresh and R. O. Ritchie, "A geometrical model for fatigue
    crack closure induced by fracture surface roughness,"
    Metal. Trans A 13A, 1627-1631 (1982).

14. R. B. Thompson and C. Fiedler, "Effects of crack closure on
    the transmission and mode conversion of ultrasonic waves,"
    to be submitted to J. Nondestructrive Evaluation.

15. D. O. Thompson, D. K. Rehbein, B. J. Skillings, and J. F.
    Smith, "Inference of compressive stresses at joined
    interfaces using ultrasonic reflectivity," these
    proceedings.

16. S. Golan, "Measuring of closure forces with ultrasonic
    diffracted waves," in New Procedures in Nondestructive
    Testing, P. Höller, ed. (Springer-Verlag, Berlin, in press).

INFERENCE OF COMPRESSIVE STRESSES AT JOINED

INTERFACES USING ULTRASONIC REFLECTIVITY

D. O. Thompson, D. K. Rehbein, B. J. Skillings,
and J. F. Smith

Ames Laboratory
Iowa State University
Ames, Iowa

INTRODUCTION

The nondestructive determination of the compressive component
of stress in various kinds of interference fasteners is a problem
of considerable interest. As noted in the previous paper,[1] none
of the usual techniques employed for stress measurement is
appropriate for this case. This paper reports results obtained
upon Nitinol couplers utilizing ultrasonic reflectivity and
interpretive concepts described in that paper as advanced by
Haines.[2] The reader is referred to a paper in the Proceedings
of the 14th Symposium, Southwest Research Institute, for full
details of the present work.[3]

Nitinol, a nickel-titanium alloy, is a shape memory material
which undergoes a rather special type of martensitic transformation
that results from the weakening of an elastic constant with
changing temperature.[4-10] At a temperature at which some small
value of the constant is reached, the alloy becomes mechanically
and thermodynamically unstable and a phase change occurs. The
temperature at which the phase change occurs can be controlled
by deviation from stoichiometry or by alloying additions,[11] e.g.,
iron. Fasteners used in this work were fabricated by the Raychem
Corporation using a 5% iron addition which places the martensitic
phase change near -47°C.

The Nitinol fastener is a short, thick-walled hollow cylinder,
and is illustrated in Fig. 1. This figure shows a longitudinal
cross-section of the fastener which has been cut in half. Note
the lands on the inner surface of the fastener which form regions
of maximum compressive stress with the joined tubing. The

171

Fig. 1.   Longitudinal cut through a coupler showing the positions
of the four lands.

fastener join is made in the martensitic phase and the assembly
warmed back to the austenitic phase. The Nitinol fastener then
"remembers" its previous shape and, if matters are properly
controlled, develops a compressive stress in the components being
joined. These stresses are the only means used to make the join.

SAMPLE PREPARATION AND STRESS CHARACTERIZATION

    A number of assembled samples using stainless steel 316
tubing and Nitinol couplers described above were fabricated. The
wall thickness of the stainless tubing was 0.050° and the nominal
outer diameter was 0.5". In order to vary the values of
compressive stress introduced into the fabrication, a set of
tubing samples was prepared by reducing the tubing OD's by
specified amounts. Since the Nitinol fasteners seek an equilibrium
inner diameter at room temperature, a reduction in the tube OD
serves to reduce the magnitude of the compressive stress generated
at the interface.

    Compressive interfacial stress values were estimated utilizing
an elastic 2-D stress analysis for thick cylinders and measurements
of the deformation produced on the ID of the tube. It is realized
that the analysis used is not completely appropriate inasmuch as
the stresses generated exceed the elastic limits of the materials
used, especially the stainless tubing. However, the results
obtained using the current analysis and dimensional measurements

are sufficient to establish the nondestructive technique. These analyses are currently being extended utilizing 3-D finite element elastic-plastic analyses.

ULTRASONIC MEASUREMENTS

In Fig. 2 are shown typical ultrasonic responses obtained for a Nitinol fastener-tube assembly at three different positions along a generator of the Nitinol cylinder. As noted on the figure, the left trace was obtained between lands (cf. Fig. 1), the center trace was obtained at a position just adjacent to a land, and the right trace directly over a land. The first signal in each train is the echo from the interface between the fastener and tubing, the second is the reflection from the inner surface (ID) of the tubing, and the remaining signals result from multiple bounces from within both the fastener and tubing. Note the absence of the second echo in the position adjacent to the land. This suggests that the fastener and tubing are separated at this position due to the sharpness of the land and the stiffness of the stainless tubing. The magnitude of the interface echo received at this position was used to normalize all other interface echoes obtained in the measurements. This procedure serves to cancel diffraction and attenuation losses expected in the measurement process. All measurements were made using a nominal 15 MHz focussed transducer in water immersion and with both pulsed and tone burst excitation.

EXPERIMENTAL RESULTS AND INTERPRETATION

As discussed in the previous paper[1] and by Haines,[2] it is postulated that ultrasonic reflections which occur from two surfaces in contact are determined by the amount of surface in contact, and that because of surface asperities, this contact is never complete. The results are sensitive to stress because the compressive stresses at the interface produce deformation of the asperities and thereby a change in the metal-metal contact area. Haines' formulation of this problem has been used extensively in this work. In this formulation, Haines discusses both ultrasonic reflectivity and transmissivity for cases that correspond to first contact and repeated contact. Haines' reflectivity results for the first contact case and for similar materials separated by an interface are given in equation (3) of the previous paper using the identification made in equation (5).

In Fig. 3 are shown ultrasonic reflectivity results obtained on a set of six Nitinol fastener-tube assemblies (points and dashed joining curve). The normalized reflection coefficient is plotted on the vertical axis and the estimated interfacial stress

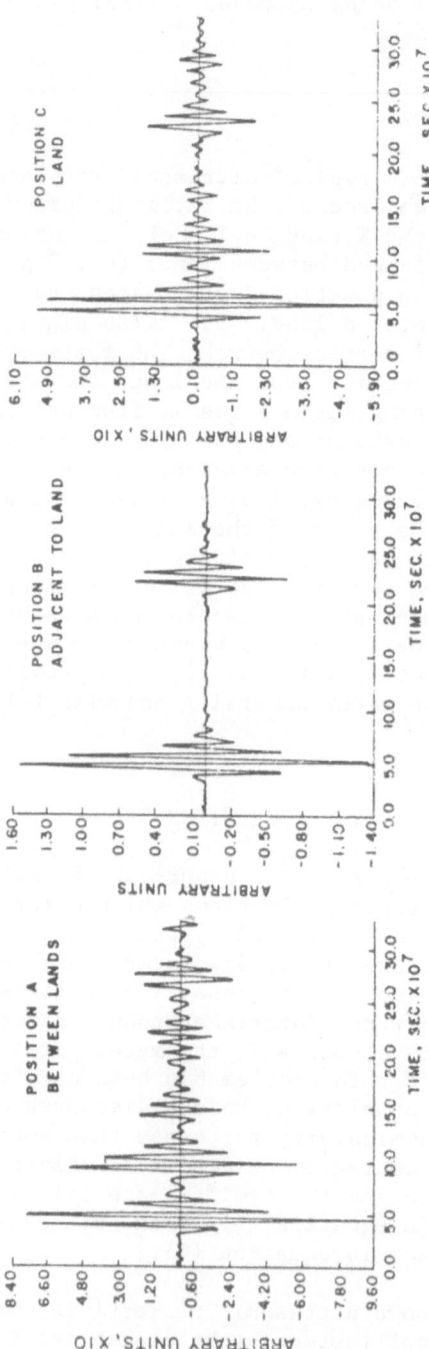

Fig. 2. Ultrasonic echo patterns from area between lands (Position A), adjacent to land (Position B) and on a land (Position C). Position B illustrates sonic disbond between the coupler and tube.

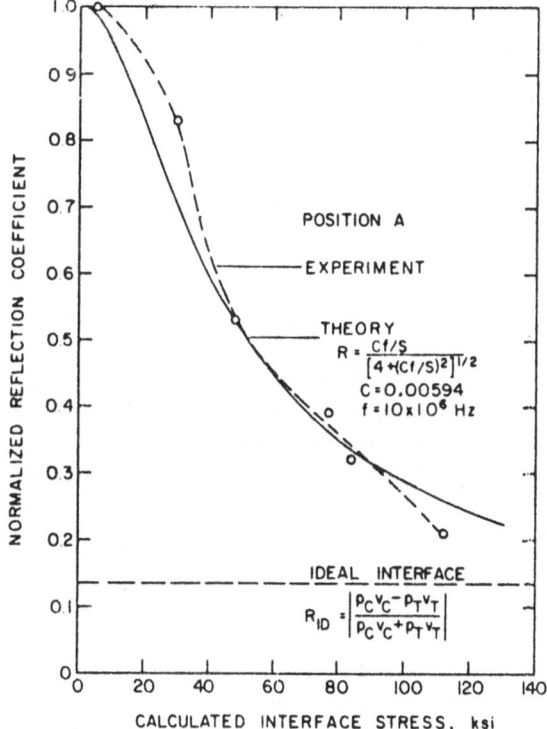

Fig. 3. Normalized reflection coefficient vs interface stress.
Comparison of experimental data with theoretical fit.

is given on the abscissa. The OD's of the tubing used in this set
varied from full size (as received) to 0.019" undersize. As will
be seen, this range was sufficient to produce a broad range of
variation in both the reflectivity and the interfacial stress.
The dashed line across the figure represents the reflectivity to
be expected from an interfacial contact between Nitinol and
stainless steel assuming perfect interfacial contact.

The results of these experiments have been interpreted
successfully utilizing Haines' model. To compare the experimental
results with this model, Haines' theoretical prediction was fitted
to the data at the 53 percent reflectivity point by selecting a
value for the parameter C (noted in Fig. 3) where C is a material
constant (cf. preceding paper and reference to Haines, equations
(3) and (5)). This single constant was then sufficient to develop
the rest of the theoretical curve which is shown as a solid line.
It is seen that the experimental results and theoretical
predictions are in good agreement.

DISCUSSION

The authors believe that the experimental results confirm
Haines' model very well. Additional experiments (not reported
here) in which samples were prepared with a significantly
different surface roughness are also in at least qualitative
agreement with model predictions. It is realized that improvement
is needed in the way that the interfacial compressive stresses
are determined. Work is in progress to replace the analytical
plane elastic stress analysis used in this work with a 3-D finite
element elastic plastic analysis. Results obtained from this
effort are expected to provide even better agreement with
Haines' model inasmuch as the inclusion of plastic effects will
tend to compress the high end of the stress scale shown in Fig. 3
to smaller values. A further refinement of Haines' model is also
necessary in order to compare further with experimental results.
The current model treats the case of contact between similar
metals, whereas this work has used assemblies of dissimilar metals.
It is expected that this refinement will have the most effect at
high stress levels.

ACKNOWLEDGEMENT

This work was sponsored by the Center for Advanced
Nondestructive Evaluation, operated by the Ames Laboratory, USDOE,
for the Naval Sea Systems Command and the Defense Advanced
Research Projects agency under Contract No. W-7405-ENG-82 with Iowa
State University and the Center for Advanced Nondestructive
Evaluation, operated by the Ames Laboratory, USDOE, for the Air
Force Wright Aeronautical Laboratories/Materials Laboratory and the
Defense Advanced Research Projects Agency under Contract No.
W-7405-ENG-82 with Iowa State University.

REFERENCES

1. R. B. Thompson, C. F. Fiedler and O. Buck, "Inference of
    fatigue crack closure stresses from ultrasonic
    transmissivity," these proceedings.
2. N. F. Haines, "The theory of sound transmission and
    reflection at contacting surfaces," Report RD/B/N4744,
    Central Electricity Generating Board, Berkeley Nuclear
    Laboratories.
3. D. K. Rehbein, B. J. Skillings, J. F. Smith, and D. O.
    Thompson, Proceedings of the 14th Symposium on NDE,
    Southwest Research Institute.
4. R. J. Wasilewski, "Elastic-modulus anomaly in TiNi," Trans.
    Met. Soc. Am. Inst. Mining Met. Engrs. $\underline{233}$, 1691 (1965).

5. N. G. Pace and G. A. Saunders, "Ultrasonic study of the martensitic phase change in TiNi," Phil. Mag. 22, 73 (1970).

6. N. G. Pace and G. A. Saunders, "Lattice properties in the vicinity of the martensitic transformation in TiNi," Solid State Commun. 9, 331 (1971).

7. N. G. Pace and G. A. Saunders, "Ultrasonic study of lattice stability in In-Ti alloys," Proc. Roy. Soc. A 325, 521 (1972).

8. D. B. Novotny and J. F. Smith, "Single crystalline elastic constants of F.C.C. Thallium-Indium alloys," Acta Met. 13, 881 (1965).

9. D. J. Gunton and G. A. Saunders, "The elastic behaviour of In-Ti alloys in the vicinity of the martensitic transformation," Solid State Commun. 14, 865 (1974).

10. M. R. Madhava and G. A. Saunders, "An ultrasonic study of the elastic phase transition in In-Cd alloys," Phil. Mag. 36, 777 (1977).

11. C. M. Jackson, H. J. Wagner, and R. J. Wasilewski, "55-Nitinol: the alloy with a memory: its physical metallurgy, properties, and applications," Report NASA-SP 5110, National Aeronautics and Space Administration, Washington, DC (1972).

FREQUENCY DEPENDENT PROPERTIES OF MATERIALS CONTAINING

A DISTRIBUTION OF PORES AND/OR INCLUSIONS

V. K. Varadan, V. V. Varadan and L. Adler

Wave Propagation Group
Departments of Engineering Mechanics and
  Welding Engineering
Ohio State University
Columbus, Ohio 43210

INTRODUCTION

The properties of a material are significantly changed by the presence of pores and/or inclusions. The dynamic behavior of such materials depends on the concentration, type, size and distribution of the pores and inclusions. Such materials behave effectively like lossy materials not only due to geometric dispersion or multiple scattering, but also real losses, if any, associated with the scatterers. A study of wave propagation in such materials is a convenient way to experimentally and theoretically predict the phase velocity and effective dynamic properties such as elastic moduli and dielectric constants.

We consider an elastic medium containing a random distribution of cavities (pores) with uniform or Gaussian size distributions. The waves incident on such inhomogeneous media undergo multiple scattering due to the presence of pores thus reducing the scattering amplitude or cross section by attenuation of waves. The attenuation of waves depends critically on the material properties of the host medium (matrix), the distribution of the pores and the frequency of the incident wave. The problem is very difficult and to our knowledge, rigorous theories with numerical results are not available in the literature. In this paper, a T-matrix theory is presented for studying the multiple scattering[1] of both longitudinal and transverse elastic waves by the pores. A statistical analysis with quasi-crystalline approximation (QCA) and Percus-Yevick (PY) or self-consistent (SC) pair correlation function is then employed to obtain expressions for the average amplitudes of the coherent fields which

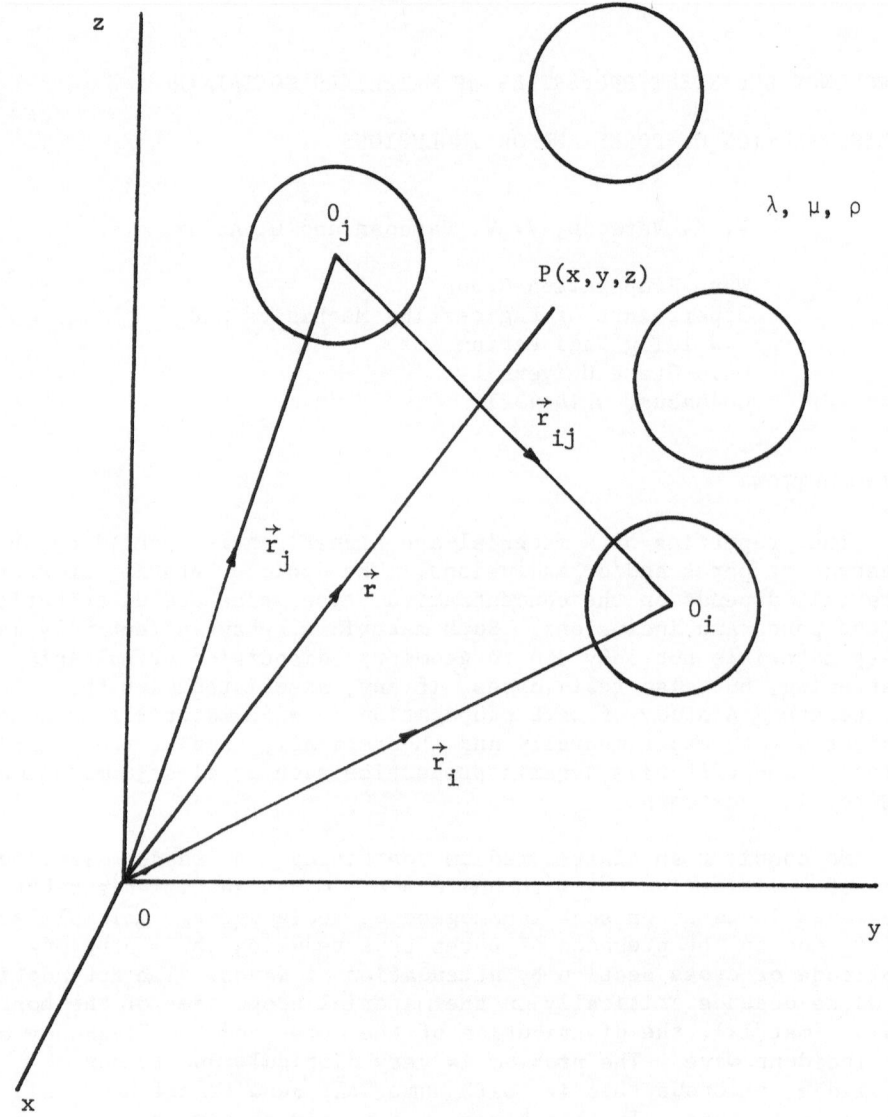

Fig. 1. Scattering Geometry.

may be solved to yield the phase velocity and coherent attenuation
of the waves.

WAVES IN DISCRETE RANDOM MEDIA - A SELF CONSISTENT MULTIPLE SCATTERING FORMALISM

The distribution of pores in the host medium (matrix) is depicted
in Fig. 1. The properties of the elastic medium are given by Lames'
constants λ and μ and density ρ. In Fig. 1, 0_i and 0_j refer to the
center of the i-th and j-th pores, respectively and they are referred
to the origin 0 by the spherical polar coordinates $(r_i, \phi_i, \emptyset_i)$. P
is any point in the medium outside the pores (the matrix medium).

A time harmonic plane wave of unit amplitude and frequency
is incident on the medium such that the direction of propagation of
the incident waves is along the z-axis, which may be written in terms
of displacement field vector $\vec{u}^0$:

$$\vec{u}^0(\vec{r}) = \hat{z}e^{i(k_p z-\omega t)\hat{z}} + \hat{x}\ ^{i(k_s z-\omega t)\hat{z}} \tag{1}$$

where k_p and k_s are the compressional and shear wave numbers given by
$k_p = \omega/c_p$; $c_p = \sqrt{(\lambda + 2\mu)/\rho}$ and $k_s = \omega/c_s$; $c_s = \sqrt{\mu/\rho}$. The incident
wave undergoes multiple scattering. Let $u_i^s(r)$ be the field scattered
by the i-th scatterer. The problem at hand reduces to computing the
total total wave field at any point in the matrix medium and hence
the bulk properties, satisfying the appropriate boundary condition
on the surface of the scatterers and radiation conditions at infinity.

The total field at any point in the matrix medium can be interpreted as the sum of the incident field and the fields scattered by
all the scatterers, which can be written as

$$\vec{u}(\vec{r}) = \vec{u}^0(\vec{r}) + \sum_{i=1}^{N} \vec{u}_i^s(\rho_i) \quad ; \quad \vec{\rho}_i = \vec{r}-\vec{r}_i. \tag{2}$$

However, the field that excites the i-th scatterer is the incident
field $\vec{u}^0$ plus the fields scattered from all other scatterers except
the i-th. The term exciting field $\vec{u}^e$ is used to distinguish between
the field actually incident on a scatterer and the external incident
field $\vec{u}^0$ produced by a source at infinity. Thus, at a point $\vec{r}$ in the
vicinity of the i-th scatterer, we write

$$\vec{u}_i^e(r) = \vec{u}^0(\vec{r}) + \sum_{j \neq i}^{N} \vec{u}_j^s(\vec{\rho}_j) \quad ; \quad a \leq |\rho_j| < 2a \tag{3}$$

where 'a' is a typical dimension of the scatterer.

The exciting and scattered fields for each scatterer can be expanded in terms of vector spherical functions with respect to an origin at the center of that scatterer:

$$\vec{u}_i^e(\vec{r}) = \sum_{\tau=1}^{2} \sum_{\ell=1}^{\infty} \sum_{n=0}^{\ell} \sum_{\sigma=1}^{2} b_{\tau\ell n\sigma}^i \text{ Re } \vec{\psi}_{\tau\ell n}(\vec{\rho}_i) = \sum_{\tau n} b_{\tau n}^i \text{ Re } \vec{\psi}_{\tau n}^i \quad (4)$$

$$\vec{u}_i^s(\vec{r}) = \sum_{\tau n} B_{\tau n}^i \text{ Ou } \psi_{\tau n}^i \quad (5)$$

where $\vec{\psi}_{\tau\ell n\sigma}$ ($\tau = 1,2,3$) are vector spherical basis functions [2]. Field quantities that are regular at the origin are expanded in terms of the regular (Re) basis set, and outgoing (Ou) functions satisfy the radiation conditions at infinity for the scattered field. The superscript i on the basis functions refers to expansions with respect to O_i, and b_n^i and B_n^i are the unknown exciting and scattered field coefficients. We also expand the incident field in terms of vector spherical functions:

$$\vec{u}^0 = \sum_{\tau n} a_{\tau n} \text{ Re } \vec{\psi}_{\tau n}^i \ e^{i \vec{k}_\tau \cdot \vec{r}_i} \quad (6)$$

where $a_{\tau n}$ are the known incident field coefficients.

The unknown coefficients $b_{\tau n}^i$ can be related to $B_{\tau n}^i$ by means of the T-matrix, see Ref. [3].

$$B_{\tau n}^i = \sum_{\tau'n'} T_{\tau n, \tau'n'}^i \ b_{\tau'n'}^i \quad . \quad (7)$$

Substituting Eqs. (4), (5) and (6) in (3), we obtain

$$\sum_{\tau n} b_{\tau n}^i \text{ Re } \vec{\psi}_{\tau n}^i = e^{ik_\tau \hat{z} \cdot \vec{r}_i} \sum_{\tau n} a_{\tau n} \text{ Re } \vec{\psi}_{\tau n}^i + \sum_{j \neq i}^{N} \sum_{\tau n} B_{\tau n}^j \text{ Ou } \vec{\psi}_{\tau n}^j \quad (8)$$

In order to express the fields in Eq. (8) with respect to a common origin at O_i, we employ the translation and addition theorems for the vector spherical functions [2] which may be written in a compact form as follows:

$$\text{Ou } \vec{\psi}_{\tau n}^j = \text{Ou } \vec{\psi}_{\tau n}(\vec{r}-\vec{r}_j) = \sum_{\tau'n'} \sigma_{\tau n \ \tau'n'}(\vec{r}_i-\vec{r}_j) \text{ Re } \vec{\psi}_{\tau'n'}^i \quad : \quad (9)$$

Employing Eqs. (7) and (9) in (8) and using the orthogonality of the vector spherical basis functions, we obtain the following set of coupled algebraic equations for the exciting field coefficients $b_{\tau n}^{i}$

$$b_{\tau n}^{i} = a_{\tau n} \, e^{i \vec{k} \cdot \vec{r}_i} + \sum_{j \neq i}^{N} \sum_{\tau' n'} \sum_{\tau'' n''} T_{\tau' n', \tau'' n''}^{j} \, b_{\tau'' n''}^{j} \, \sigma_{\tau'' n'', \tau n}(\vec{r}_i - \vec{r}_j) \quad (10)$$

From Eq. (10), it can be seen that the exciting field coefficients of the i-th scatterer explicitly depend on the position and orientation of all scatterers. In this paper, we consider a random distribution of identical scatterers and the case when $N \to \infty$ and the volume occupied by the scatterers $V \to \infty$ such that $N/V = n_0$ is a finite number density. For such distribution, a configurational average of Eq. (10) can be made over the positions of all scatterers [1,2] with QCA [4] to arrive at an equation for the configurational average $\langle b_{\tau n}^{i} \rangle_i$ of the exciting field coefficients with one scatterer fixed:

$$\langle b_{\tau n}^{i} \rangle_i = a_{\tau'' n''} \, e^{i \vec{k}_\tau \cdot \vec{r}_i} \quad (11)$$

$$+ (N-1) \sum_{\tau' n'} \sum_{\tau'' n''} T_{\tau' n', \tau'' n''} \int_V p(\vec{r}_j | \vec{r}_i) \langle b_{\tau'' n''}^{j} \rangle_j \, \sigma_{\tau' n' \, \tau n} \, d\vec{r}_j$$

where $p(\vec{r}_j | \vec{r}_i)$ is the two particle joint probability density.

The joint probability density is defined as

$$p(\vec{r}_j | \vec{r}_i) = \begin{cases} \dfrac{1}{V} g(|\vec{r}_j - \vec{r}_i|) & ; \quad |\vec{r}_j - \vec{r}_i| > 2a \\[2ex] 0 & ; \quad |\vec{r}_j - \vec{r}_i| < 2a \end{cases} \quad (12)$$

Equation (12) implies that the particles are hard (no-interpenetrations) and the excluded volume is a sphere of radius '2a' although the particles themselves may be non-spherical. The function $g(|\vec{r}_j - \vec{r}_i|)$ is called the pair correlation function and depends only on $|\vec{r}_j - \vec{r}_i|$ since spherical statistics are used. If the particles are not identical, then 'a' must be replaced by the mean radius and the T-matrix must be replaced by one averaged over size. The pair correlation function for an ensemble of particles depends on the nature and range of the interparticle forces. Percus and Yevick [5] have obtained an approximate integral equation for the pair correlation function of a classical fluid in equilibrium. Wertheim [6] has obtained a series solution of the integral equation for an ensemble of 'hard' spheres. The statistics of the fluid are then the same as

those of the ensemble of discrete hard particles that we are considering.

To solve the integral equations given by (11), we consider the
inhomogeneous medium with discrete scatterers as a homogeneous
continuum and assume that the average coherent wave is a plane wave
propagating with an effective wave number K in the same direction as
the incident plane wave. We can thus write

$$\langle b_{\tau n}^i \rangle = X_{\tau n}\ e^{i\ K\hat{z}\cdot\vec{r}_i} \tag{13}$$

where $X_{\tau n}$ is the amplitude of the coherent wave.

Substituting Eq. (13) in (11) employing the joint probability
function as defined before and the divergence theorem to convert the
volume integral in (11) to surface integrals and using the extinction
theorem which cancels the incident wave, we obtain a set of simultaneous coupled homogeneous equations for the coefficients $X_{\tau n}$ given by

$$X_{\tau n} = c \sum_{\tau''n''} \sum_{\tau'n'} \sum_{q=|n'-n|}^{|n'+n|} X_{\tau''n''} T_{\tau'n',\tau''n''}\ C_{\tau'n',\tau n}^q\ \frac{I_q}{(k_\tau^2 - K^2)a^2} \tag{14}$$

where $c = 4\pi a^3 n_0/3$ is the effective spherical concentration of the
scatterers per unit volume, C^q is an expression containing Wigner
coefficients, and

$$I_q(k,k_\tau,c) = \frac{6c}{(k_\tau a)^2 - (Ka)^2}\ [2k_\tau a\ j_q(2Ka)\ h_q'(2K_\tau a)$$

$$-2Ka\ h_q(2k_\tau a)\ j_q(2Ka)] \tag{15}$$

$$+ 24c \int_{x=1}^{\infty} x^2[g(x)-1]\ h_q(k_\tau x)\ j_q(Kx)\ dx$$

Equation (14) is a system of simultaneous linear homogeneous
equations for the unknown amplitudes $X_{\tau n}$. For a nontrivial solution,
we require that the determinant of the truncated coefficient matrix
vanishes, which yields an equation for the effective wave number K
in terms of k_τ and the T-matrix of the scatterer. This is the dispersion relation for the scatterer filled medium. Equation (14) is
a general expression valid for any arbitrary shaped scatterer, since
the T-matrix is the only factor that contains information about the

exact shape and boundary conditions at the scatterer. The effective
wave number K obtained in the analysis is a complex quantity, the
real part of which relates to the phase velocity, while the imaginary part relates to the attenuation of coherent waves in the medium.

RESULTS AND CONCLUSIONS

In the Rayleigh or low frequency limit, the size of the scatterers is considered to be small when compared to the incident wavelength. It is then sufficient to take only the lowest order coefficient in the expansion of the fields. In this limit, the elements
of the T-matrix can be obtained in closed form for various simple
shapes. It can be shown that at low frequencies, only $X_{\tau 0}$, $X_{\tau 1}$ and
$X_{\tau, -1}$ of Eq. (14) make a contribution. The value of K as determined
by the above dispersion relations is a real quantity for lossless
(elastic) material and a complex quantity for lossy (viscoelastic)
material, and relates to phase velocity $V_p = \omega/K$. In this limit, we
normally study the dependence of phase velocity on concentration,
angle of incidence and aspect ratio of the pores. The general
tendency of the phase velocity is to increase (for inclusion) and
decrease (for cracks and cavities) as concentration increases. Thus,
the phase velocity vs. concentration in this limit is not very useful
both from theoretical and experimental point of view. The plots of
absorption and coherent attenuation due to multiple scattering vs.
frequency for various concentrations carry more information [1,7].

To study the response at higher frequencies, we must consider
higher powers of $k_\tau a$, and this implies that a larger number of terms
$(X_{\tau n})$ must be kept in the expansion of the average field. This is
best done numerically. For a given value of $k_\tau a$, the T-matrix of
the pore is computed. Next, the coefficient matrix of $X_{\tau n}$ [Eq.(14)]
is formed. The complex determinant of the coefficient matrix is
computed using standard Gauss elimination techniques. For a given
$k_\tau a$, the root of the equation set M = 0 is searched in the complex
K plane $(K_1 + iK_2)$ using Muller's method. Good initial guesses were
provided by the Rayleigh limit solutions at low values of $k_\tau a$ and
these could be used systematically to obtain convergence of roots
at increasingly higher values of $k_\tau a$. The real part K_1 determines
the phase velocity, while the imaginary part K_2 determines the
coherent wave attenuation.

Here, we present some sample numerical calculations of spherical pores in graphite and aluminum and graphite inclusions in an
iron casting. The longitudinal and shear wave velocities of graphite
are taken as c_p=3200 m/sec and c_s=1800 m/sec, for aluminum, c_p =
5900 m/sec and c_s=3200 m/sec; and for iron c_p=5800 m/sec and c_s =
3100 m/sec.

Numerical computations have been performed for a random dis-

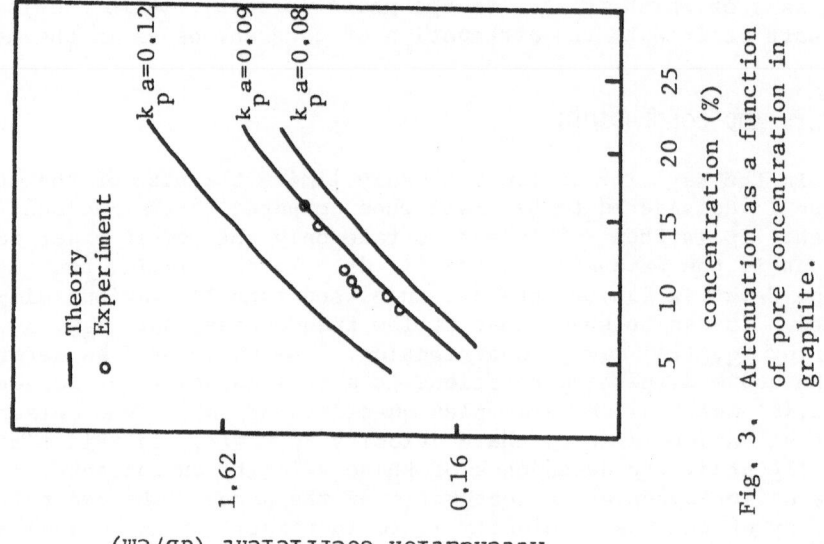

Fig. 3. Attenuation as a function of pore concentration in graphite.

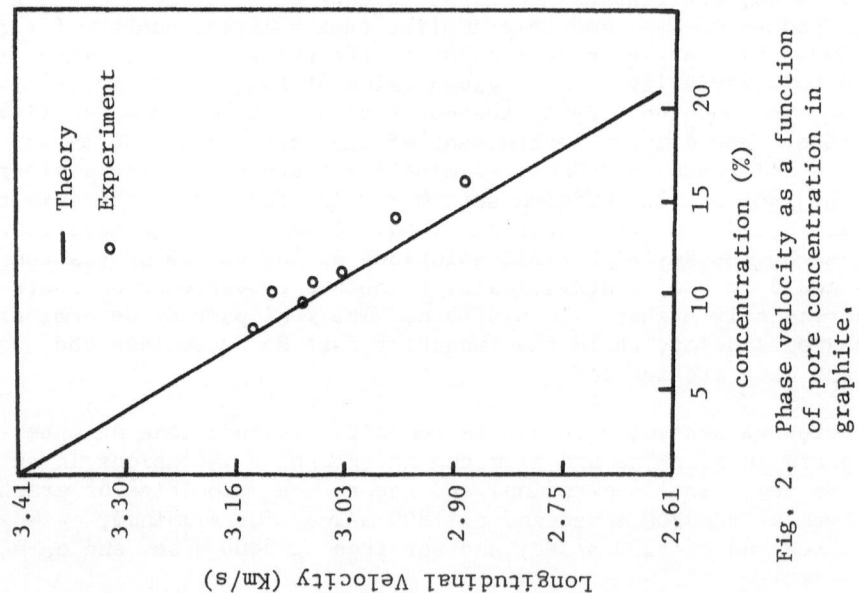

Fig. 2. Phase velocity as a function of pore concentration in graphite.

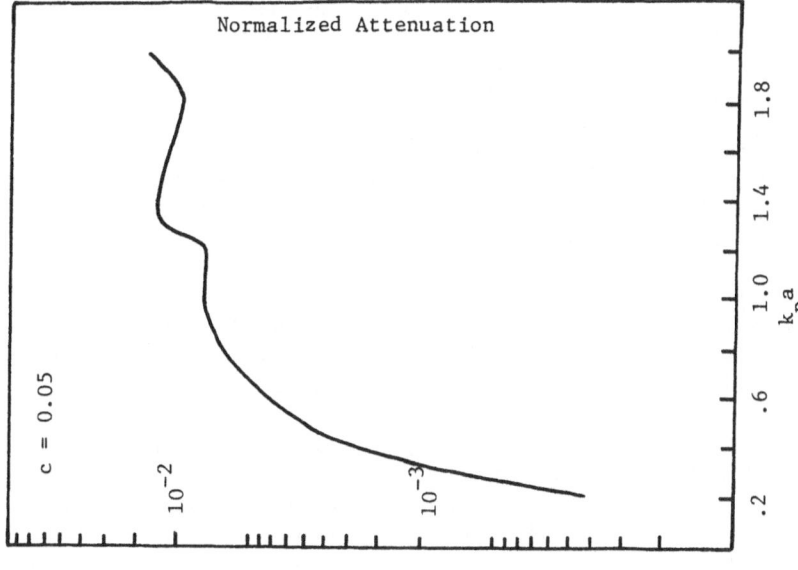

Fig. 5. Attenuation as a function of non-dimensional wavenumber for graphite particles in iron casting.

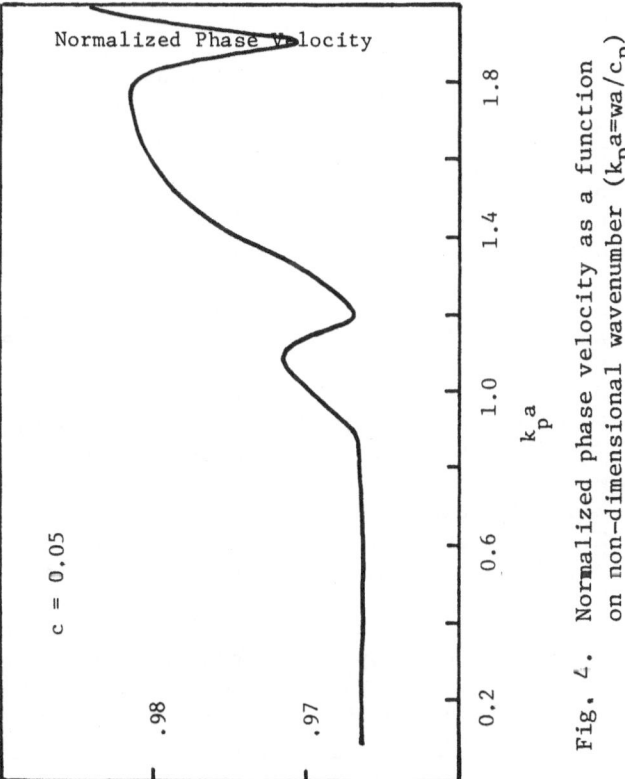

Fig. 4. Normalized phase velocity as a function on non-dimensional wavenumber ($k_p a = wa/c_p$) for graphite particles in iron casting.

bution of spherical pores in graphite and aluminum and steel. Although the results are presented only for longitudinal wave incidence, calculations for shear wave incidence are quite straightforward. In Fig. 2 the phase velocity is plotted as a function of concentration for a Gaussian size distribution of spherical pores for various values of $k_p a$. The phase velocity decreases with concentration and frequency. The plot shows good agreement with the measurements performed by S. Wang and L. Adler. In Fig. 3, calculated values of the attenuation are compared with experimental results. In Figs. 4 and 5, the phase velocity and attenuation of longitudinal waves in an iron casting containing 5% by volume of spherical graphite particles is plotted as a function of frequency. At higher values of $k_p a$, there are oscillations in both plots which are due to resonances of the individual graphite particles.

It is relatively easy at this stage to compute the effective modulii of such materials as a function of frequency and scatterer concentration. Wave propagation experiments and calculations as described above are an effective means of non-destructively characterizing a material containing an unknown distribution of pores and inclusions.

REFERENCES

1. V.K. Varadan, Y. Ma, and V.V. Varadan, A multiple scattering theory for elastic wave propagation in discrete random media, J. Acoust. Soc. Am., submitted.
2. V.K. Varadan, Multiple scattering of acoustic, electromagnetic and elastic waves, in: "Acoustic, Electromagnetic and Elastic Wave Scattering-Focus on the T-matrix Approach," V.K. Varadan and V.V. Varadan, eds., Pergamon Press, New York (1980).
3. V.V. Varadan, Elastic wave scattering, in: "Acoustic, Electromagnetic and Elastic Wave Scattering-Focus on the T-matrix Approach," V.K. Varadan and V.V. Varadan, eds., Pergamon Press, New York (1980).
4. V.N. Bringi, T.A. Seliga, V.K. Varadan, and V.V. Varadan, Bulk Propagation Characteristics of Discrete Random Media, in: "Multiple Scattering and Waves in Random Media," P.L. Chow, W.E. Kohler and G.C. Papanicolaou, eds., North-Holland Publishing Company, Amsterdam, (1981).
5. J.K. Percus and G.J. Yevick, Analysis of classical statistical mechanics by means of collective coordinates, Phys. Rev. 110:1 (1958).
6. M.S. Wertheim, Exact solution of the Percus-Yevick Integral Equation for Hard Spheres, Phys. Rev. Lett. 10:321 (1963).
7. V.K. Varadan, V.V. Varadan and Y. Ma, Frequency dependent dynamic properties of rubber materials with a random distribution of voids, J. Acoust. Soc. Am., submitted.

MEASUREMENT OF THIN CASE DEPTH IN HARDENED STEEL

BY ULTRASONIC PULSE-ECHO ANGULATION TECHNIQUES

Morris S. Good* and Joseph L. Rose

Sigma Research Incorporated
Richland, Washington 99352

Drexel University
Philadelphia, Pennsylvania 19104

ABSTRACT

This study evaluated ultrasonic pulse-echo angulation as a nondestructive technique for measuring thin case depths (less than 2 mm) in hardened steel. In concept, this method has been used to measure case depths greater than 2 mm and has also been suggested as a technique for determining nominal grain size. It has not been applied to case depths less than 2 mm because impediments, such as the obscuring effect of the front interface echo, the extremely low signal-to-noise ratio of grain backscatter, and the small area localization usually required by industrial inspection procedures prevent accurate measurement.

Study results indicate that improved signal processing techniques (i.e., RF signal averaging, and smoothing) can overcome these obstacles. The final algorithm predicted effective case depth at 0.91, 1.40, and 1.98 mm at ultrasonically calculated values of 0.94, 1.40, and 2.11 mm, with respective standard deviations of 0.25, 0.23, and 0.81 mm. The extremely localized fluctuations in the case depth specimen contributed significantly to the measured standard deviation.

KEY WORDS: shear wave, backscatter, case depth, ultrasonic, pulse-echo, RF signal averaging, spatial averaging, Hilbert transform, moving average window, martensite, pearlite.

*Currently with Sigma Research Inc., Richland, WA 99352.

INTRODUCTION

The ultrasonic pulse-echo technique (illustrated in Fig. 1) localizes ares measurements by sonic resolution (approximately 1 mm) and may be effective for case depths of 0.5 mm and greater.

This range reflects an extension of angulation methods previously reported in literature (Emerson, 1976; Perennes et al., 1979). Angulation techniques direct the reflected wave front away from the transducer. This action has a dual function. First, the front wall surface echo returning to the transducer is a portion of the spherical wave and sidelobe pattern; therefore, energy levels are smaller than the reflected plane wave front by several orders of magnitude. This reduces the masking effect of the front wall echo on superficial scatters. Second, the incident angle is such that a shear wave is formed in the steel sample of interest.

CASE-HARDENED SPECIMEN PREPARATION AND MATERIAL CHARACTERISTICS

A block of 1060 steel 2.5 by 2.5 by 10.2 cm was induction-hardened on one side to create a planar case depth of 2 mm. A Knoop microhardness indentor produced transverse profiles along six marked indications (see Figs. 2 and 3).

Figure 4 shows micrographs of the case, transition zone, and core areas (Taylor, 1978). The left micrograph is characteristic of nearly 100% martensite. The center micrograph displays the transition between a predominately martensitic structure to a predominately pearlitic structure. The difference in ultrasonic scattering properties is related to these changes in the internal grain structure.

A hardness value of 30, Rockwell scale C, was selected as the effective case depth specification value. Nominal case depth was measured at three points. These measurements were determined as the average among the case depths measured along the lines M1-M4, M2-M5, and M3-M6.

A tapered wedge was ground from the hardened surface, as shown in Fig. 5, and the surface was polished. This resulted in mean effective case depths for lines M1-M4, M2-M5, and M3-M6 of 0.91, 1.40, and 1.98 mm, respectively.

DEVELOPMENT OF DATA ACQUISITION SYSTEM

The computer-controlled data acquisition, as shown in Fig. 6, was used to evaluate various techniques that would be contained in the data acquisition protocol. These studies assessed transducer angulation and RF signal-averaging.

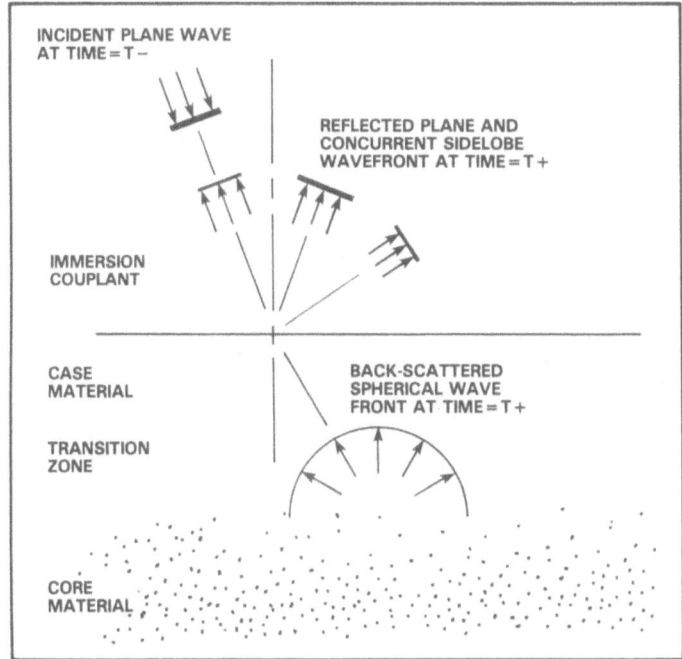

Figure 1.   Incident, reflected, refracted, and back-scattered wave
phenomena resulting from wave-/case-hardened specimen
interaction.

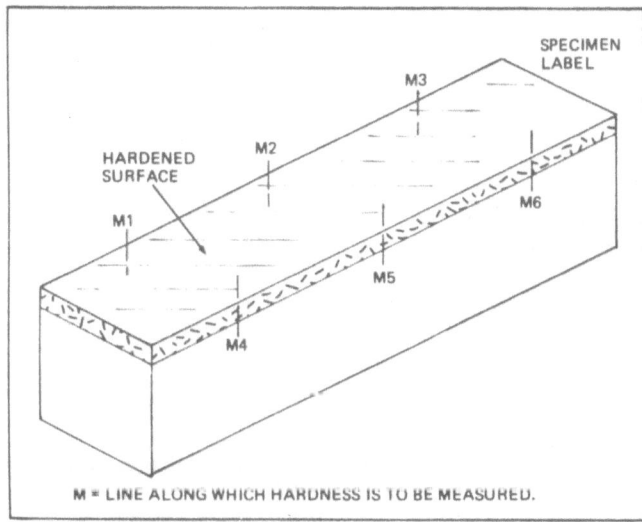

Figure 2.   Case-hardened specimen (indications M1 through M6
display position of microhardness traverses).

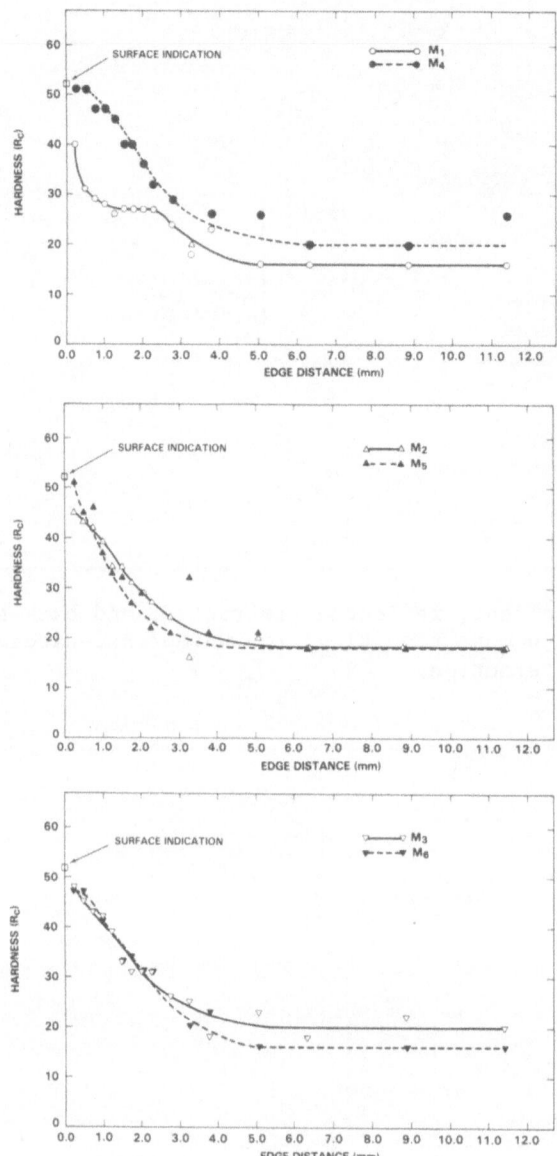

Figure 3.   Transverse microhardness profile along paired sets,
             M1–M4, M2–M5, and M3–M6.

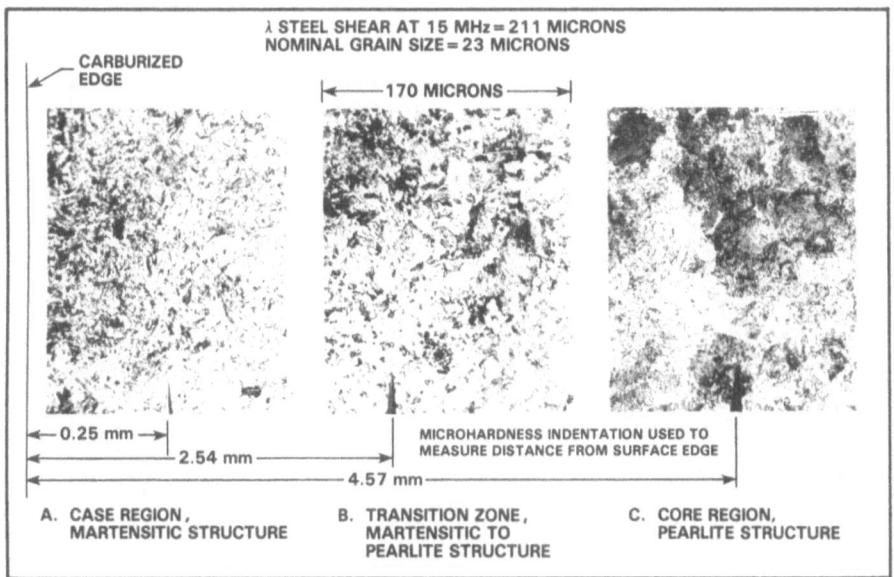

Figure 4.   Micrographs of steel specimen.

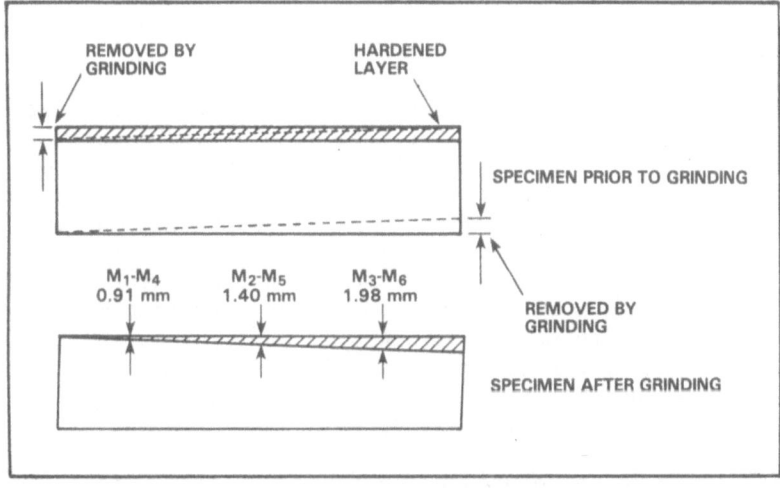

Figure 5.   Preparation of steel specimen for variable case-
hardened depth.

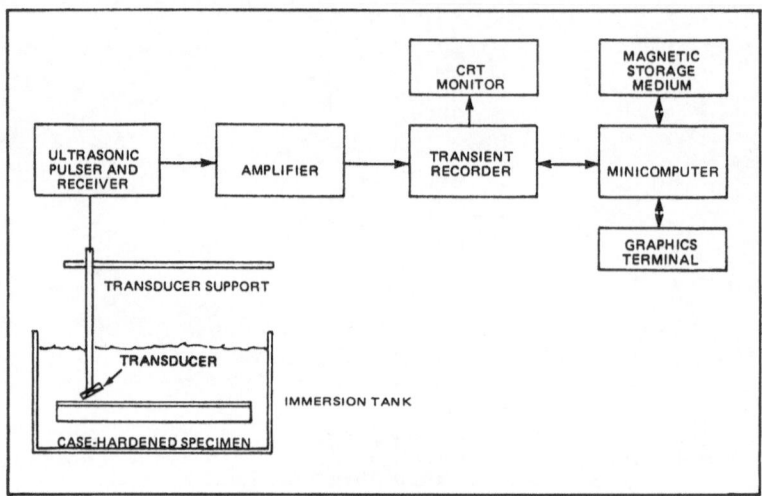

Figure 6.   Block diagram of computer-controlled data acquisition
            and analysis system.

TRANSDUCER ANGULATION AND SIGNAL AVERAGING

     Sonic beam angulation allows the user to select refracted
longitudinal or shear waves. Perennes et al. (1979) used
20-degree inclination angle in water; and Goebbels (1980) discusses
ultrasonic grain size measurement and elaborates on the relative
amplitude difference (approximately 34 dB) between shear and
longitudinal backscattered energy in steel. In his study,
Goebbels selected frequency values for each vibrational mode so
that the wave lengths of the two respective vibration modes were
equal. The signal-to-noise ratio was largest for an inclination
of 50 degrees shear excitation in steel. This compares with
optimal pulse-echo condition for shear wave excitation
(Krautkramer and Krautkramer, 1977).

     Signal averaging reduced noise to acceptable levels. As
illustrated in Fig. 7, a single RF transient response is
characteristic of signal-to-noise ratio of approximately one-tenth.
Affects of random RF spurts were minimized by signal-averaging
techniques. All data henceforth were averaged 255 times.

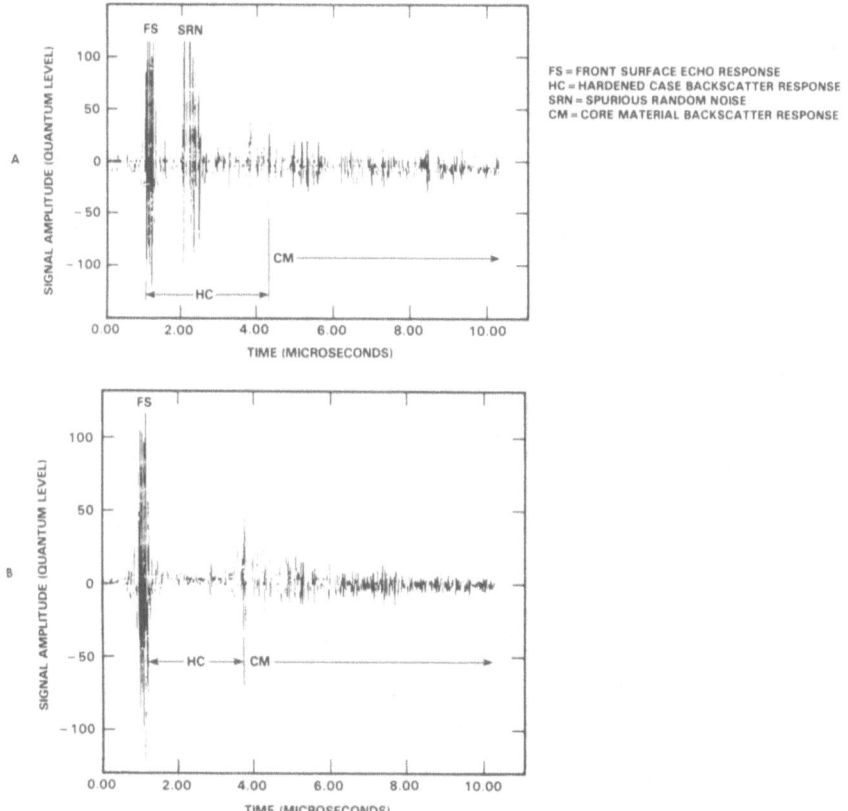

Figure 7.   Case-hardened RF signal response (sampling rate 100 MHz).
            (a) Single-signal response; (b) signal averaged 255 times.

DATA COLLECTION AND POSTPROCESSING

    Three sequential processes were used to measure case depth:
envelope determination (by the Hilbert transform), spatial
averaging of several stored envelopes, and envelope smoothing using
a moving average window (MAW). The RF signal envelope contained
localized amplitude perturbations that are extremely sensitive
to probe movement relative to the steel specimen. Spatial
averaging of the processed envelopes gave an envelope characteristic of an area rather than a point. Signal envelopes were averaged
and smoothed as illustrated in Fig. 8. Eight unique envelopes
were used in all subsequent spatial averaging procedures.

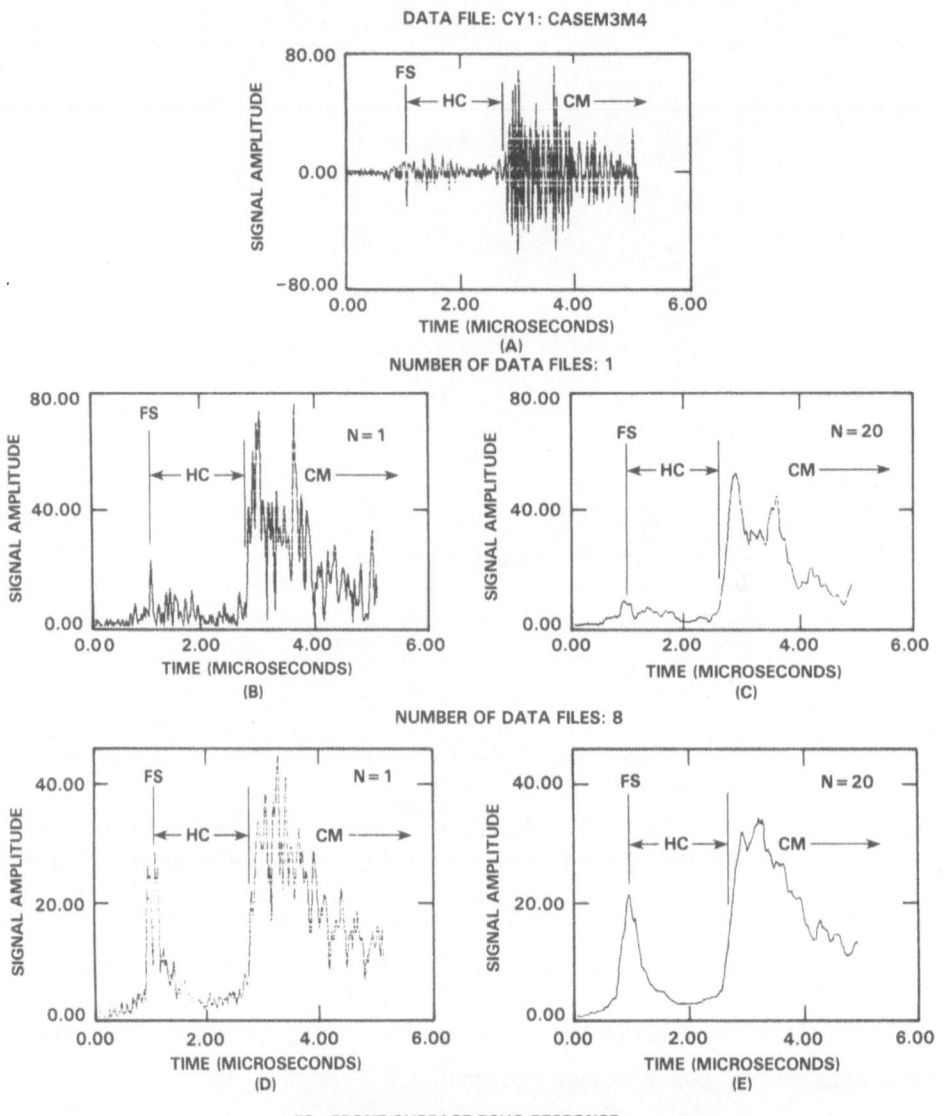

Figure 8.   Effects of spatial averaging combined with the moving
average window (MAW).   (a) RF signal, averaged 255 times;
(b) signal envelope via Hilbert transformation;
(c) smoothed envelope using a 0.2 μs MAW; (d) spatial
average of eight unique envelopes; (e) smoothed spatial
averaged envelope using a 0.2 μs MAW.

MODEL ANALYSIS

Past efforts have attempted to measure case depth by an extension of pulse-echo techniques where depth was determined from the interim between the arrival times of the front surface echo and the detected onset of increased backscatter. Inherent weaknesses in this approach are the accurate spatial determination of the front surface and the assumption that the leading edge of increased backscatter concurs with the boundary layer of the specified effective case hardness value.

To relate effective case depth with the leading edge of increased backscatter, let us assume that the refracted shear wave is entering a case-hardened specimen. After propagating through the hardened surface layer a distance S, it encounters diffuse scatter characteristics of the core area (Fig. 9). In addition, as the wave propagated through the steel specimen, it traversed a plane characteristic of a given hardness value, which segregates the effective case layer from all other material. If we measure the distance along the propagational path between the discriminating hardness plane and the point where significant backscattering is initiated and label this offset U, we can determine the effective case depth by the following equation:

$$d = (S - U) \cos \theta . \tag{1}$$

By using a calibration procedure that acquires an ultrasonic signature at a known case-hardened depth d', a simple relation can be formulated:

$$d = \left(\frac{V_{ss} (T_{bs} - T'_{bs})}{2} \right) \cos \theta + d' , \tag{2}$$

where primed values represent values obtained during a calibration procedure. The expression relates effective case depth to a relative time change and simplifies the measuring process. The accuracy of this model was tested by a series of ultrasonic measurements on the case-hardened specimen.

EXPECTED TRENDS

Four graphs in Fig. 10 evaluate case depth as a function of arrival time. The three graphs plotted to the left represent, from top to bottom, ultrasonic signatures of progressively thinner effective case depths. Shifting of the leading edge relative to a reference point confirms the basic premise that relative changes in case depth are related to shift of backscatter arrival time.

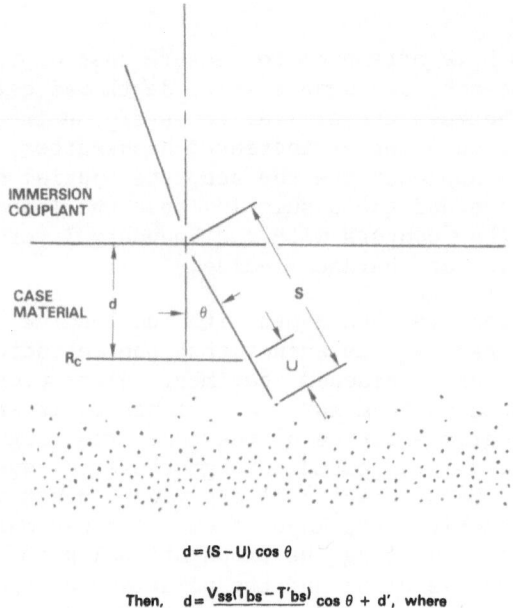

$$d = (S - U) \cos \theta$$

Then, $d = \dfrac{V_{ss}(T_{bs} - T'_{bs})}{2} \cos \theta + d',$ where

θ = REFRACTION ANGLE IN STEEL
V_{ss} = SHEAR WAVE VELOCITY IN STEEL
T_{bs} = ARRIVAL TIME OF
        BACKSCATTER ENERGY
d = EFFECTIVE CASE DEPTH
S = STEEL PROPAGATIONAL PATH LENGTH
      TRAVERSED WITH ONSET OF
      INCREASED BACKSCATTER
U = STEEL PROPAGATIONAL PATH LENGTH
      BETWEEN R_c HARDNESS PLANE AND
      ONSET OF INCREASED BACKSCATTER

Figure 9. Model representation of case–hardened specimen for
           ultrasonic pulse–echo angulation determination of
           effective case depth.

The fourth graph displays an ultrasonic signature from the
nonhardened surface opposite the case-hardened layer.

LEADING EDGE DETERMINATION

     In determining a consistent manner of leading edge detection,
a MAW gate width and the threshold must be selected. This was
accomplished by using a data base of twenty-four sets of signals,
eight sets for each of the three known effective case depths.

     To evaluate the effects of the MAW as a function of gate size,
selected values were used in conjunction with a preselected
threshold value of 3 dB. (Figure 11 illustrates the effect of a
0.3 µs MAW gate.) Within-class scatter decreased as the gate size

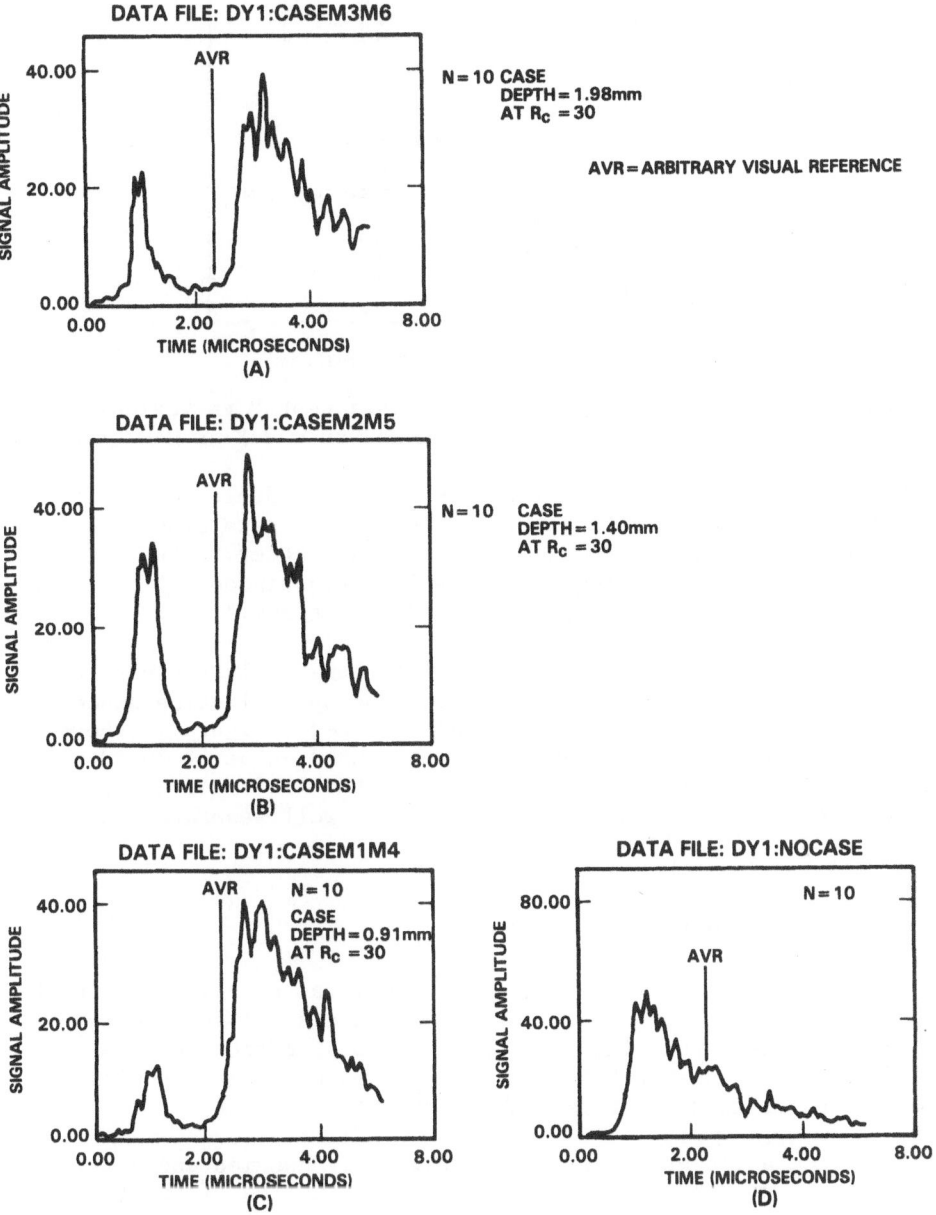

Figure 10.   Comparison of backscattered energy at different
effective case depths (all envelopes determined by
spatial average of 8 signal envelopes and smoothed with
0.1 μsec MAW.   (a) Effective case depth equals 1.98 mm;
(b) effective case depth equals 1.40 mm; (c) effective
case depth equals 0.91 mm; (d) unhardened surface.

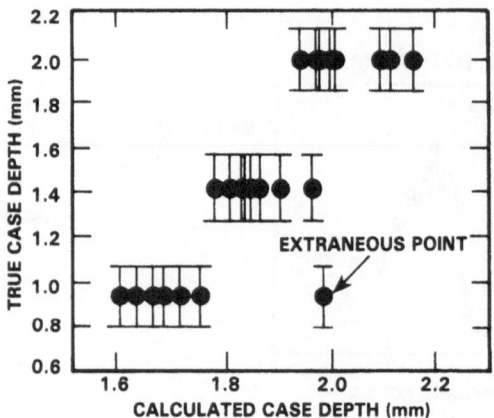

Figure 11. Graphic display of "true case depth" vs "calculated
            case depth," 3 dB threshold for a 0.3 μs MAW gate size.

increased; however, increasing gate size also decreases spatial
resolution. Our selection criteria becomes a gate size
sufficiently large to reduce within-class scatter to an acceptable
level without progressing to an overly large window. A gate size
of 0.3 μs was subjectively determined as optimum.

    A selected set of threshold values were evaluated with a
0.3 μs MAW gate duration to determine the optimal value based on
the criteria of minimum within-class scatter. Figure 12 shows
graphic results for the respective threshold value of 20 dB. In
using this technique, only one point of the 1.40-mm set overlaps
the 1.98-mm regime. All other points are well separated into
their respective class sets.

MODEL REEVALUATION

    A reevaluation of the previous model was performed since the
calculated values of effective case depth did not seem to form a
strictly linear relation with true values of effective case depth.
Governing reasons for this trend were thought to be associated
with changes in the lateral beam resolution.

    Since a linear relation was previously assumed and only three
known case depths are given, a quadratic equation of the form
given below was selected as a curve-fitting technique.

$$d = W_1 (P^2) + W_2 (P) + W_3 , \tag{3}$$

where W_1, W_2, and W_3 are equal to their respective weighting
functions and $P = T_{bs} - T'_{bs}$.

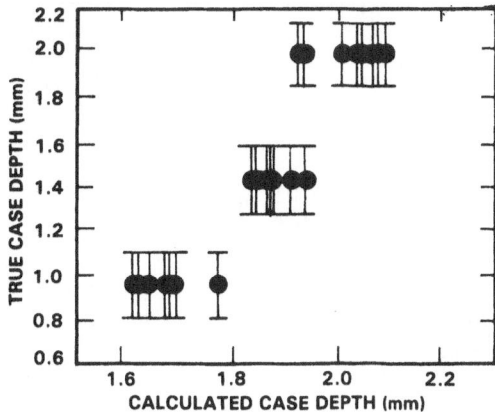

Figure 12. Graphic display of "true case depth" vs "calculated case depth," using 0.3 μs MAW gate size and a 20 dB threshold.

Weighting coefficients were determined by using the mean arrival time values determined by the 0.3 μs MAW and a 20 dB threshold (see Fig. 13). This method allowed for empirical fluctuations, such as lateral beam resolution and beam divergence. In analyzing the data, statistical information, such as the standard deviation at each of the three known case depths, measurement can evaluate algorithm stability (see Table 1).

CONCLUSIONS

An algorithm was formed to determine effective case depth nondestructively by the ultrasonic pulse–echo angulation technique. This method integrated various signal-processing techniques in a relatively consistent and accurate procedure. Moreover, results demonstrate that a strictly monatomically increasing relation with respect to the true effective case depth exists in the regime of 1.0 to 2.0 mm and backscatter arrival time. This can be used as an extension of the linear relation

Table 1. Statistical Data on Values Determined from Quadratic Algorithm

True Effective Case Depth	Calculated Mean Value	Calculated Standard Deviation
0.91 mm	0.94 mm	0.25 mm
0.40 mm	1.40 mm	0.24 mm
1.98 mm	2.11 mm	0.81 mm

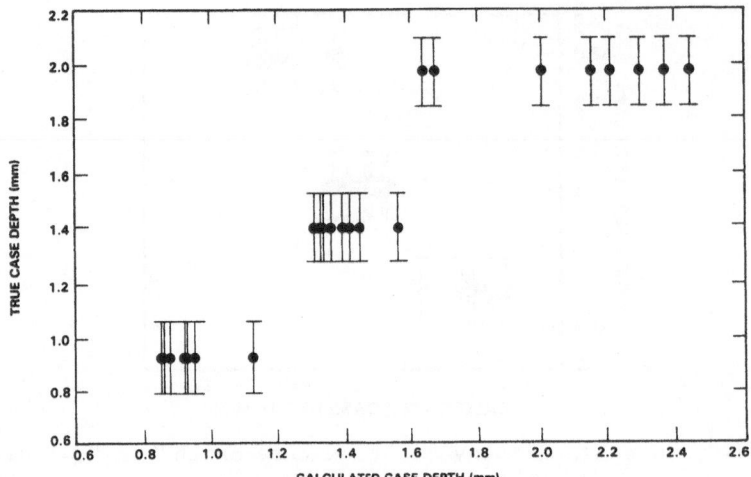

Figure 13. Graphic display of "true case depth" vs "calculated
            case depth," MAW gate size 0.3 μs, 20 dB threshold,
            quadratic equation with weights.

established for large case depths of 2 mm and greater. There are
also reasons to believe that even thinner effective case depths,
such as 0.5 mm, are feasible.

     Accurate measurements can be made on thinner values due to
the offset that exists between the respective planes of critical
hardness and initiation of backscatter. This offset is seen as
the large time difference between the front wall surface echo and
the small relative time shifts corresponding to changes in
effective case depth. This indicates that backscatter is
initiated posterior to the discriminating hardness plane for the
scanned specimen.

     Prior restrictions of ultrasonically determined effective
case depth were due to poor spatial resolution, extremely low
signal-to-noise ratio, and the obscuring effect of the front
surface echo. With improved signal-processing techniques, many of
these restrictions were relaxed. The final algorithm predicted
effective case depth at 0.91, 1.40, and 1.98 mm at ultrasonically
calculated values of 0.94, 1.40, and 2.11 mm with respective
standard deviations of 0.25, 0.23, and 0.81 mm. Error sources are
small changes in the relative position of transducer and specimen,
and those relating to the ultrasonic technique used. The
localized case depth fluctuation is a significant source
contributing to the standard deviation (Fig. 3).

     More work is needed to reduce the scatter of measurements of
thin case depths and to firmly establish the previously defined

relations. This work can be defined in two parallel efforts: first, the acquisition, integration, and customization of electronic instrumentation; and second, increased signal processing by software to establish optimum methods. This would enhance the signal-to-noise ratio, improving analysis, and possibly lead to a purely analog instrument that would greatly expedite testing procedures. Extensive studies regarding different steels, hardening procedures, and degrees of hardness are recommended for future technique refinement in consistency, calibration, and defining limiting conditions.

ACKNOWLEDGEMENTS

This research project was carried out between January and June 1982 with sponsorship from the Office of Naval Research in Arlington, Virginia. The authors gratefully acknowledge Mr. M.K. Premkumar, a Drexel University graduate student, for technical services concerning extensive metallurgical tests on the case-hardened specimens.

REFERENCES

1. Emerson, P. J., 1976, Alternative Methods for Measuring Case Depth in Ferrous Components, British J. of NDT 18, 2.
2. Perennes, C., Chretien, N., Jussiaume, A., and Gaillot, D., Nondestructive Testing of Hardening Depth in Steel Cylinders by Two Ultrasonic Methods, "Proceedings of Ninth World Conference on Nondestructive Testing," Session 1B-2, November 1979.
3. Taylor, J. L., 1978, An Introduction to Case Hardening Processes, British J. of NDT 18, 2.
4. Goebbels, K., 1980, Structure Analysis by Scattered Ultrasonic Radiation, "Research Techniques in Nondestructive Testing, Vol. IV," Academic Press, New York.
5. Krautkramer, J., and Krautkramer, H., 1977, "Ultrasonic Testing of Materials," trans. Springer-Verlag, New York.

ULTRASONIC MICROSTRUCTURAL CHARACTERIZATION

OF CRYSTALLINE METALS

ACOUSTIC EMISSION FOR IN-PROCESS MONITORING AND MICROSTRUCTURE CONTROL?

H. N. G. Wadley and R. Mehrabian

Center for Materials Science
National Bureau of Standards
Washington, DC 20234

ABSTRACT

Acoustic emissions are the elastic waves emitted by sudden localized changes of stress by, for example, the formation of cracks, plasticity, and phase transformations. It is beginning to be considered a potential in-process monitoring technique for quality and productivity improvement as a sensor for closed loop feedback control systems. Applications of the technique are held back because of its complicated nature and because the signals are controlled rather subtly by microstructure. In this review we describe the theoretical framework that has begun to emerge and which now provides a physical understanding of acoustic emission. We then reconsider the results of laboratory studies and recent applications to assess, in the light of this understanding, the contribution acoustic emission methods might make toward in-process monitoring and microstructure control during metals processing.

INTRODUCTION

The metals and ceramics industries have recognized the need for increased monitoring of their manufacturing processes to both measure microstructure evolution and detect the presence of defects. Such monitoring, in conjunction with microprocessor-based feedback control loops, would result in optimized microstructures and products with fewer defects with resulting improvements in productivity, quality, and competitivity of products that exploit this new technology.

There are several potential techniques that show promise for
in-process control. Techniques such as eddy currents and ultrasonics are already under investigation. One drawback with these
approaches however is the need for introduction of some radiation
(electromagnetic or mechanical) into the product and then measurement of its propagation characteristic. This is usually performed
on a point-by-point basis and an image formed. Such scanning
techniques tend to be slow, do not necessarily monitor the entire
product volume, and frequently interfere with the production process.

One technique which shows considerable promise of overcoming
these particular criticisms is that of acoustic emission. Acoustic
emission is the term used for the elastic waves generated by abrupt
localized changes of stress in a solid [1]. The waves propagate
away from the source and cause transient (ns-ms time scale) surface
displacements that may be measured with transducers (some of which
could be noncontacting) [2]. The acoustic emission signal is, as
it were, the dynamic displacement response of a solid body to
stress. A strain-time curve, obtained from a mechanical properties
test, would be the analogous static response to the same deformation.

Acoustic emission has been used fairly extensively as a laboratory tool to investigate the micromechanisms of deformation and
fracture by which a material responds to stress during mechanical
testing [1,5-8]. It has also found increasing application as a
nondestructive evaluation (NDE) technique for detecting and locating flaws in structures subjected to stress [3,4]. More recently,
it is beginning to be considered a candidate technique for in-
process monitoring because acoustic emission signals are emitted
by the very mechanisms a material normally responds to processing.
It is also suspected that emission of a different form will also
be emitted by defects that form during processing.

There are two issues that need to be addressed for each potential application of acoustic emission to in-process monitoring.
First, it is always the case that not every mechanism by which a
product responds to processing will generate detectable signals.
The "detectability" of the processing mechanisms will depend upon
the amplitude of the elastic waves emitted by the source and the
background noise over the frequency range of the detecting instrumentation. During processing, much of the background noise is
confined to low frequencies (10 to 50 kHz) and thus processing
mechanisms that are abrupt (high frequency) are more readily detectable. This, together with the elastic wave amplitude itself, is
likely to be very material dependent. For example, the formation
of a brittle microcrack in a ceramic during cooling is much more
likely to be detected than the plastic deformation, during rolling,
of already cold worked aluminum. At the present time it is unclear
which materials/processes are suitable for acoustic emission monitoring of microstructure evolution and defect detection.

Second, for defect detection applications, methods are required to unambiguously separate defect signals from others which are non-critical and to deduce the severity of the defect. All products contain defects and "severity" is a factor determined by the use to which a product is ultimately put. Some method is required that allows the parameters of a defect that determine its impact on fitness for purpose to be deduced from the actual acoustic emission signal itself.

It is the purpose of this paper to review the existing understanding of relationships between source mechanisms and acoustic emission signals and to use this to assess the strengths and weaknesses of the acoustic emission technique for in-process monitoring and control. It will emerge that while the technique is not perfect for all material mechanisms, processes, it has much to offer for certain specific procedures, particularly those in which rapid phase transitions occur or where the detection of brittle cracking is necessary.

SOURCE-SIGNAL RELATIONSHIPS

Acoustic emission signals are generated by changes of stress in a body. During materials processing, changes of internal stress are accommodated by:
- The motion of dislocations.
- The formation of cracks.
- Phase transformations.
- Thermal expansions.
- Changes of externally applied loads.
- Magnetic domain motion

Elastic waves radiate away from the site of such processes and ultimately reach the surface of the component causing transient surface displacements that can be detected with piezoelectric crystal or even laser interferometer transducers. The signals from these transducers can then be amplified and displayed or recorded for later analysis [10], figure 1. We should like to predict these signals and to assess their potential for monitoring materials processing.

Acoustic emission transducers measure the displacement (or in some cases velocity) as a function of time over some small area of the surface of a sample. Since the voltage waveform of the new generation of transducers is only slightly distorted from the actual surface displacement waveform (i.e., their transfer functions are almost constant, say κ volts/meter, between 10 kHz and 1 MHz) we shall evaluate this displacement waveform itself. Convolution (or just multiplication by κ for a new generation of

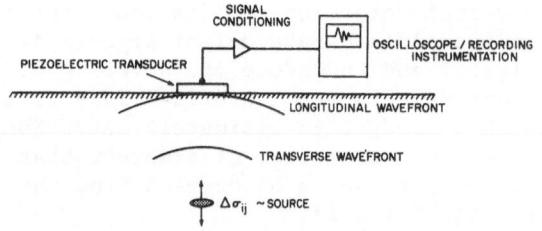

Fig. 1   The longitudinal and transverse elastic waves radiated
         by a stress change source propagate to the surface of a
         body causing transient surface displacements. These are
         converted, by a piezoelectric crystal, to a voltage-time
         waveform that can be recorded or observed on an oscillo-
         scope.

wideband transducers) of the transfer function of a particular
transducer with the displacement waveform will give the detected
acoustic emission signal.

      Elastodynamic techniques using what has become known as the
Green's function approach have recently been quite successfully
applied to the calculation of surface displacement waveforms due
to simple defect sources provided they are localized both in
space and time [5,9,11]. The basic method of calculation consists
of:

   • Deduction of a representation of the source as a local
     strain rate, stress change rate, or force dipole density.
     These three representations are equivalent and are
     simply related using elasticity theory. An example of
     the force dipole representation for various defects in
     an isotropic elastic body (Lamé constants λ and μ) is
     shown in figure 2. The application of such dipoles
     (force separation products) would cause the propagation
     of an identical elastic disturbance to that of the
     defect itself [11].

   • Evaluation of the surface displacement waveform for
     each (tensor) component of the source (i.e., for each
     dipole), figure 3. This amounts to solution of the
     elastic wave equation for a unit force dipole; the
     so-called dynamic elastic Green's function. The actual
     displacement due to a given orientation dipole of
     strength other than unity is found by simple linear
     scaling.

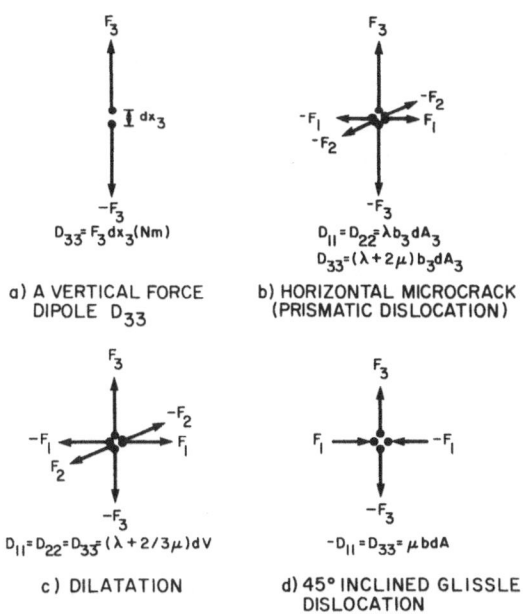

Fig. 2.   (a) A change in a single component stress ($\Delta\sigma_{33}$) can be
          modelled as the creation of an infinitesimal force dipole
          D_{33}. The change in stress due to acoustic emission
          sources are modelled as equivalent force dipole combina-
          tions. The strengths of the dipole components are
          related to the product of the strain created by the
          appearance of the source and the elastic moduli of the
          material. Examples are given for; (b) horizontal penny
          shaped microcrack that grows in a vertical stress field,
          (c) an isotropic dilatation, and (d) a glissile disloca-
          tion loop inclined at 45° to axis 3 [11].

• Calculation of the displacement waveform for the defect
  is then achieved by simple addition of the appropriate
  dipole component waveforms and convolution with the
  appropriate source time dependence, figure 4.

     The most difficult of the three steps is undoubtedly calcula-
tion of the dynamic elastic Green's function. While wave propagation in a body containing no interfaces or any free surfaces (an
infinite body) is simple to calculate and depends only upon the
bodies elastic properties and density, wave propagation in bodies
with free surfaces is greatly complicated by reflections/mode
conversions of elastic waves at free surfaces or interfaces.
Thus, the dynamic elastic Green's function of a body with free

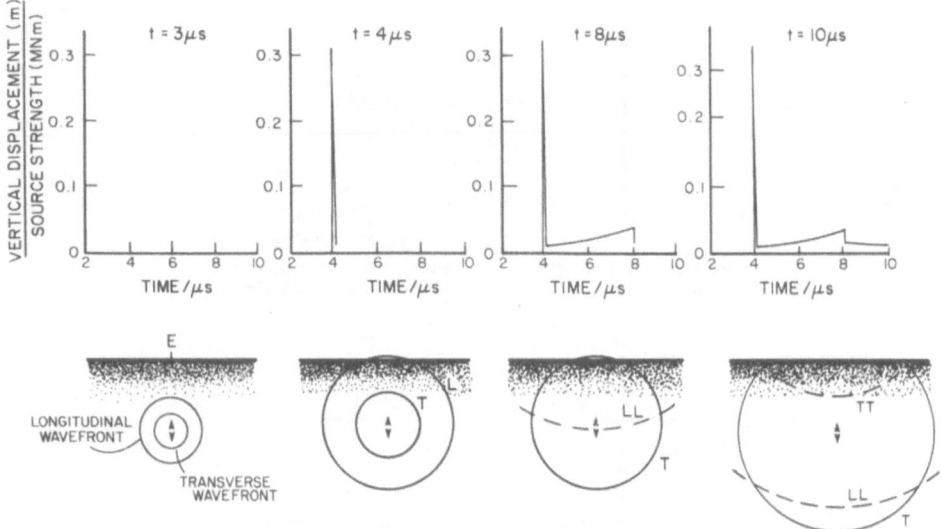

Fig. 3   The application of an internal vertical dipole (at t=o)
         causes radiation of two elastic waves, L and T, that
         propagate spherically from the source at differing veloc-
         ities. The arrival of the wavefronts at the surface
         causes displacements. The time dependence of these dis-
         placements and the wavefront positions, at various times
         after application of a dipole of a step-function time
         dependence with 30 ns risetime are shown in (a) and (b)
         respectively. It was assumed $C_1 = 2C_2$. Mode converted
         wavefronts are not shown for simplicity.

surfaces depends both on material properties (elastic constants
and density), the precise geometry of the body and the location
of the source within the body. Even when all of these parameters
have been defined, Green's functions can currently only be calculated for infinite bodies, infinite half spaces (one flat surface),
and infinite plates (two that coplanar surfaces). Fortunately,
many components can be approximated to these shapes for short
periods of time following action of the source allowing estimation
of the acoustic emission signals.

        To illustrate the Green's function approach, we show, in
figure 3, the displacement waveform at the epicenter of an elastic
half-space due to the vertical dipole source shown in figure 2(a).
Only the vertical (out-of-surface) displacement is shown, horizontal (or in-plane) displacements also occur but are more difficult

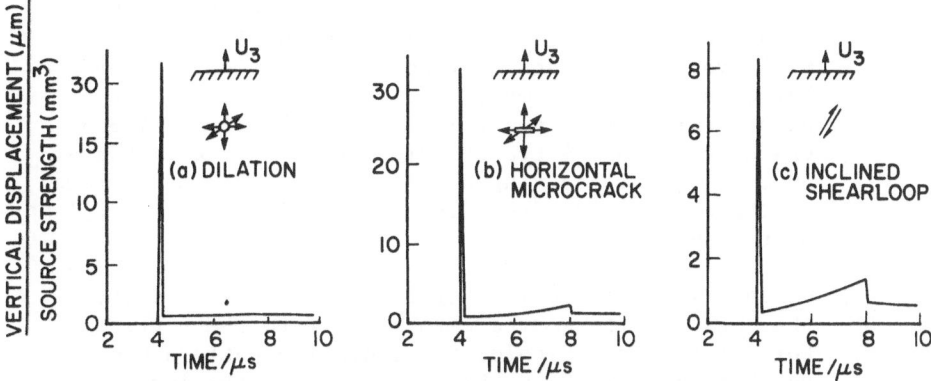

Fig. 4.   Time dependence of epicenter displacement due to the
application of combinations of force dipoles representing
an isotropic dilatation, a horizontal mode I loaded
penny-crack and a 45° inclined circular dislocation loop.
The source risetime was 30 ns, the source-epicenter distance 25 mm, $C_1 = 6250$ ms^{-1} and $C_2 = 3125$ ms^{-1} [11].

to measure. The waveforms show, for example, that the signal
from a vertical dipole consists of a pulse (that arrives 4 μs
after operation of the source) corresponding to the propagation
of a spherical longitudinal wavefront (velocity $c_1 = (\lambda + 2\mu/\rho)^{\frac{1}{2}} =$
6.25 mmμs^{-1}) from source to epicenter (a distance of 25 mm) and a
negative step (at 8 μs) corresponding to a transverse wavefront
(velocity $c_2 = (\mu/\rho)^{\frac{1}{2}} = 3.125$ mmμs^{-1}) propagating along
the same path. The amplitude and direction of waveform displacements varies with dipole orientation [11] but, at the epicenter
for the vertical dipole the longitudinal displacement is in the
direction of the force (out-of-the-surface) while the transverse
wave tends to restore the surface toward its original position.
It can be seen there is a finite static displacement for $t \geqq 8$ μs;
a consequence of the continued presence of the dipole within the
body. This is the static displacement that would be registered
by a strain gauge in a mechanical properties test.

Displacement waveforms due to the simplest forms of fracture,
plastic deformation and thermal expansion (dilatation) are shown
in figure 4. At epicenter, the amplitudes of the elastic displacements of the longitudinal arrival are controlled by four factors:

1. Source Strength. The strength of the dipole components
   (units of Nm) scale, linearly, the displacement waveform.

Thus, a 1Nm strength vertical dipole causes a peak
displacement in figure 3 of ∿ 0.3 μm while a 2Nm dipole
would cause an identically shaped displacement waveform
but with twice the amplitude. For a microcrack, the
dipole strength is controlled by crack volume (area x
crack opening). A typical microcrack with a radius of
30 μm and crack opening displacement of 0.3 μm (volume
∿ 1000 μm^3) would, from figure 4(b) give a peak displacement of ∿ 30 x 10^{-12} m.

2. Orientation. The epicenter displacement varies with
   the angle of the source with respect to the surface.
   For the vertical dipole, the radiation pattern has
   $\cos^2\theta$ form where θ is the angle between the force and
   the vertical axis. Thus a horizontal dipole (D_{22}, D_{11}),
   for which θ = 90° would have zero strength vertical
   displacements in the epicenter waveform. The radiation
   pattern for the dislocation source is of the form sin2θ
   while that for the microcrack would be $\cos^2\theta + \dfrac{\lambda}{\lambda+2\mu} \sin^2\theta$.
   The radiation from the dilatation is isotropic [11].

3. Time Scale. The area under the L arrival is proportional
   to the strength of the source. If the strength is
   maintained constant but the time taken to fully apply
   it (risetime) varied, then the width (in time) of the
   longitudinal arrival also varies in the same way and
   the amplitude varies inversely so that the area (units
   of meters x seconds) under the arrival remains constant.
   The effect of extending the risetime upon the acoustic
   emission waveform, while every other variable remains
   fixed, is shown in figure 5. From these figures we see
   that only rapid events are likely to generate detectable
   dynamic displacements.

4. Distance. The energy contained in a wavefront as it
   propagates away from the source is conserved, provided
   internal friction and scattering are absent. Since the
   energy per unit area of wavefront is inversely proportional to r^2 (where r is the source-wavefront distance)
   and the energy in the wavefront is proportional to the
   square of the particle displacement (u^2), the amplitude
   of the L arrival varies as 1/r, as can be seen in
   figure 5. Furthermore, since the signal between the L
   and T arrivals is also stretched out in time as the
   propagation distance increases, this part of the waveform
   between wave arrivals apparently decays as $1/r^2$. Thus,
   for propagation distances of more than a few centimeters
   the epicenter signal will appear as a single pulse.

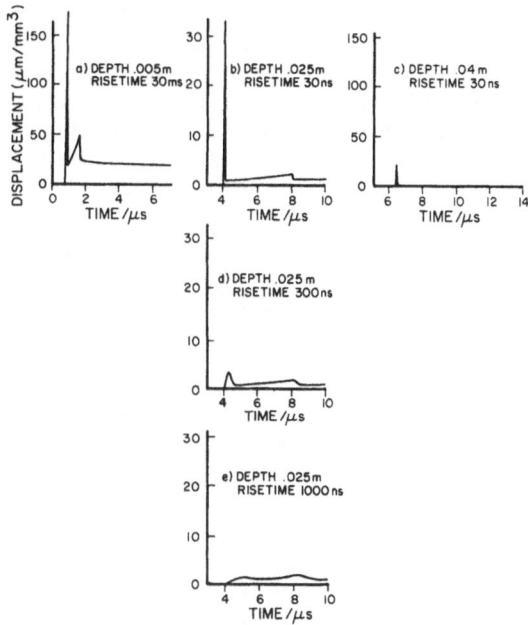

Fig. 5.   The effect of carrying the depth of the source and the source risetime for the microcrack source [11].

It has been shown [11] that for defect sources the amplitude of these single longitudinal pulses is related to the properties of the defect and the elastic behavior of the body;

$$U_3 = \frac{bavc_2^2}{rc_1^3}$$ 45° dislocation loop

$$U_3 = \frac{8(1-\nu)\sigma a v}{3Ec_1 r}$$ Horizontal microcrack

where c_1 = longitudinal wavespeed (ms^{-1})
   c_2 = transverse wavespeed (ms^{-1})
   b = Burger's vector of a dislocation (m)
   v = radial velocity of circular defect (ms^{-1})
   a = circular defect radius (m)
   r = source-epicenter distance (m)
   σ = applied stress (Nm^{-2})
   E = Young's modulus (Nm^{-2})

These relations enable prediction of the acoustic emission displacement for defects of known size (a), velocity (v), depth (r),

and orientation (θ). It can be seen that this emission is dependent upon both elastic properties (λ, μ) of the material and
microstructure which controls a and v. The microstructure effects,
it turns out, can be very subtle, and to illustrate this we shall
review a series of experiments where the microstructure has been
systematically varied and a standardized parameter of the acoustic
emission response measured during uniaxial tensile deformation at
constant strain rate.

MICROSTRUCTURE EFFECTS ON ACOUSTIC EMISSION FROM PLASTIC DEFORMATION

Single Crystal Aluminum

Perhaps the simplest case of plastic deformation would be
that for single crystal high purity aluminum oriented for initially
single slip. A typical experimental result showing the acoustic
emission power (intensity) variation as a function of strain is
shown in figure 6 [5]. In the experiment, the sample was deformed
at a constant slow strain rate ($\sim 10^{-4}$ s^{-1}) and acoustic emission
in the frequency range 0.1 to 1.0 MHz measured. It is observed
that the emission quickly reaches a maximum shortly after the
initiation of plastic flow and that less power is detected as the
strain increases beyond this point, even though the total amount
of slip per unit time (strain rate) was more or less constant.

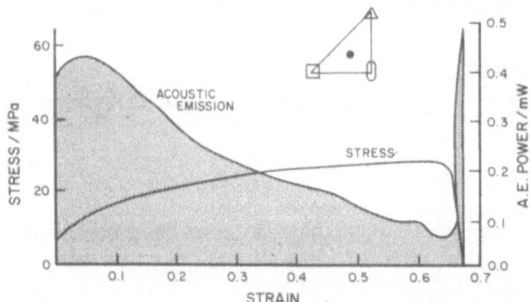

Fig. 6.   The acoustic emission intensity (power) detected during
           the tensile deformation of [149] single crystal aluminum
           decreases with increasing plastic strain during parabolic
           work hardening. The measurements were made over a fre-
           quency band from 0.1 to 1.0 MHz with an amplification of
           98 dB. A smooth relationship is observed because many,
           individually small signals are emitted within the
           response time of recording instrumentation resulting in
           observation of a mean emission intensity [5].

For an explanation of this behavior we must consider the microscopic nature of slip. The imposed strain rate is satisfied by the formation and propagation of dislocations on slip planes at $\sim 45°$ to the tensile axis. We can approximate each event in such a process by the formation of individual dislocation loops. A single loop expanding at a velocity v to a final radius a generates, at the epicenter of a half-space, a maximum displacement:

$$U_3 = \frac{c_2^2 bav}{rc_1^3}$$

Using values of $c_1 = 6.4$ mmμs^{-1}, $c_2 = 3.2$ mmμs^{-1}, $b = 2.9 \times 10^{-10}$ m for aluminum and assuming $r = 40$ mm we find:

$$U_3 = 2.8 \times 10^{-13} \, av \quad (m)$$

Now, the smallest displacement detectable under laboratory conditions (low noise) using modern acoustic emission transducers is around 10^{-14} m. Setting this as the threshold for detectability gives the following condition for a microscopic plastic deformation to generate a detectable signal;

$$av \geq 0.035 \, m^2 s^{-1}$$

where a is the radius to which the loop expanded before arrest at a pinning point, and v the radial velocity. If two loops, very close to each other (c.f. shortest observed wavelength of emission) had moved together, their signals would be in phase (additive). Thus for slip involving the cooperative motion of n closely spaced dislocations, the criterion for detectability is:

$$nav \geq 0.035 \, m^2 s^{-1} .$$

Suppose a were a large fraction of the crystal radius (a reasonable supposition at the initiation of plastic deformation of a perfect single crystal), say ~ 1 mm, then for a detectable signal from a single loop, $v \geq 35$ ms^{-1}, not an unreasonable value to expect. In practice, both a and v for slip events are statistically distributed, figure 7, and only those events whose nav products are at the extreme of the distribution generate detectable events. At the initiation of deformation, a substantial fraction of slip events may propagate sufficiently so that even moderately slow dislocation groups would give detectable signals. Workhardening, however, reduces the mean free path between sessile dislocations, restricting a and resulting in fewer events that radiate detectable elastic waves; thus the emission power decreases.

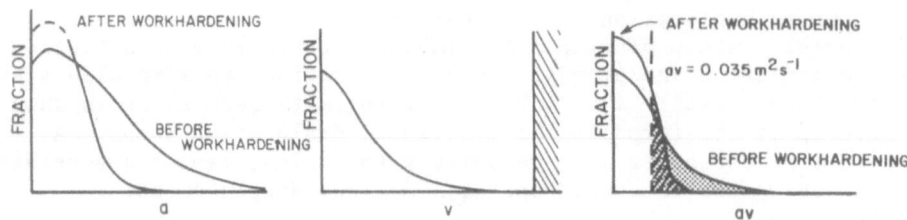

Fig. 7.   Schematic diagrams of hypothetical slip distance and
           velocity distributions and the distribution of the av
           product. Only events for which av > 0.035 m^2s^{-1} gener-
           ate detectable elastic waves. In single crystals of
           pure metals this occurs more frequently at small strain
           (low dislocation density).

Grain Size Effects

    The effect of grain size upon acoustic emission of high
purity aluminum is shown in figure 8 [5]. It is well known that
grain boundaries are effective barriers to moving dislocations.
Thus, in a polycrystalline metal, the distance a dislocation loop
can expand over will be, at most, that of the radius of the
largest grain in the sample. The distribution of slip distances
over some range of plastic strains will now be a complicated
function of the grain size distribution and internal grain dislo-
cation-dislocation interactions. It seems reasonable to suppose
that, at the initiation of slip, a substantial fraction of moving
dislocations will propagate over a distance of about the mean
grain diameter.

    As the grain size is reduced, the critical velocity of a
single dislocation required to give a detectable signal increases
sharply. For example, for a mean grain diameter of 1 mm,
$v \sim 35$ ms^{-1} while for a diameter of ~ 35 μm, $v \sim 1000$ ms^{-1}. In
fcc pure metals we do not expect dislocation velocities much
above about 100 ms^{-1} and so emission from small grain material
must come from groups of cooperatively moving dislocations. The
probability that enough dislocations move together to give a
detectable signal decreases with grain size and is considered the
cause for a loss of acoustic emission with decreasing grain size.

    A note of caution should be made however. Some workers have
apparently observed an increase, at very small grain size, in the
acoustic emission. Up to the present time, this controversial
observation has not been fully resolved [12].

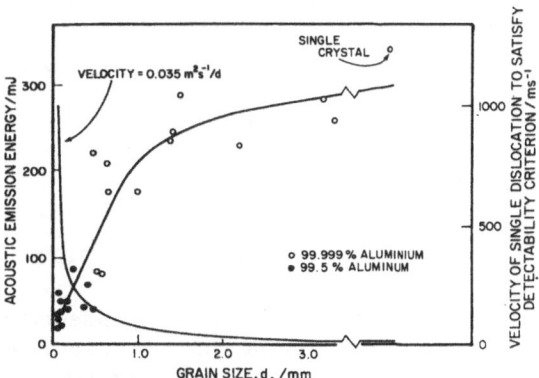

Fig. 8.   The integrated acoustic emission power (the energy) from
           polycrystalline aluminum decreases with decreasing grain
           size because the average dislocation propagation distance
           decreases resulting in an increase in necessary disloca-
           tion velocity (for generation of detectable signals) to
           values above those attainable during plastic deformation
           [5].

In alloys of aluminum, substitutional impurity segregation
to dislocations can occur, even when the alloy element concentration is below the solid solubility. This drastically alters the
dynamics of dislocations and leads to effects, during mechanical
testing, such as yield points, Luders bands, and dynamic strain
aging. The occurrence of these phenomena is almost invariably
accompanied by intense acoustic emission signals [5]. Both the
incidence of the effects themselves and their associated acoustic
emission signals vary greatly with grain size; see for example
figure 9 where the acoustic emission energy is plotted as a function of grain size for an Al-1.3% Mg alloy solid solution and compared with the result shown in figure 8 for high purity aluminum.

The peak in the acoustic emission of the solid solution at a
grain size of ∿ 80 μm arises from a competition between two effects:

- As the grain size increases a increases.
- As the grain size increases v decreases.

The velocity decrease occurs because the flow stress decreases with
grain size and, for grain sizes greater than about 500 μm the yield
stress is insufficient to separate mobile dislocations from the
cloud of impurity atoms (Cotterell atmosphere) segregated to their
core. This results in drift controlled motion of dislocations at

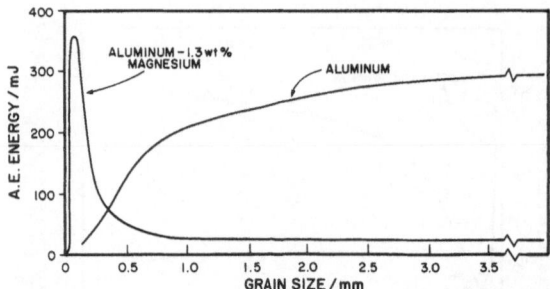

Fig. 9.   A comparison of the grain size dependence of acoustic
           emission from pure aluminum and an aluminum-magnesium
           solid solution alloy shows that for small grain size,
           solute atoms enhance the emission from dislocations,
           while for large grain size the emission from dislocations
           is suppressed [5].

velocities too low to give acoustic emission even though propagation distances are greater than adequate for radiation of detectable signals in pure aluminum.

Precipitation Effects

      There are two extreme classes of dislocation-precipitate
interaction:

      • When the precipitates are strong, widely spaced and inco-
        herent with the matrix, dislocations bow between partic-
        les leaving behind a dislocation loop around the particle.

      • When the precipitates are weak, closely spaced, and
        coherent with the matrix, dislocations may penetrate
        the precipitates and shear them.

      The former process encourages uniformly distributed slip
while the latter leads to the formation of intense slip bands.
These bands form because the successive passage of dislocations
through shearable precipitates reduces their strength and causes
a local strain instability.

      Studies in which the precipitate strength, distribution, and
coherency have been systematically varied in aluminum alloys [13-15]
indicate the uniform slip process of dispersion hardened alloys
(strong, relatively widely spaced incoherent precipitates) give
only moderate levels of detectable acoustic emission because a is

limited to the interparticle spacing (a few μm). On the other
hand, when precipitate shear occurs, intense acoustic emission
signals are observed, figure 10. Similar results are also obtained
in quenched and tempered ferritic steels [16], figure 11.

These studies serve to illustrate the complimentary role
acoustic emission plays in the monitoring of plastic flow. In
contrast to stress-strain measurements and hardness testing where
the integrated static response of the material to some load is
measured, acoustic emission measures that part of the slip distribution for which the criterion $av > 0.035$ m^2s^{-1} is satisfied (the
high speed events), and this is very sensitive to metallurgical
variables, often much more so than the stress-strain curve.

From a processing viewpoint, the passive nature of the
acoustic emission monitoring instrument together with the sensitivity of the emission to microstructure would indicate a potential
nondestructive method for monitoring and online feedback control
of microstructure. The background noise level, which determines
the constant in the detectability criterion, controls the range of
microstructures observable in this way.

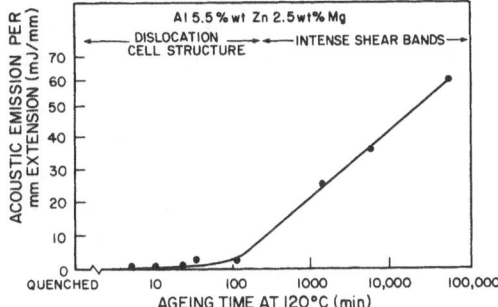

Fig. 10. Aging supersaturated aluminum alloys results in "shear-
         able" precipitates which "focus" dislocation motion onto
         closely spaced slip planes (intense shear bands). The
         acoustic emission from this mechanism of dislocation
         motion is much more readily detected than that leading
         to a cellular network in solid solution or dispersion
         strengthened microstructures.

ACOUSTIC EMISSION FROM FRACTURE

Following a similar procedure to that used for plastic
deformation, the formation of a horizontal "penny shaped" crack

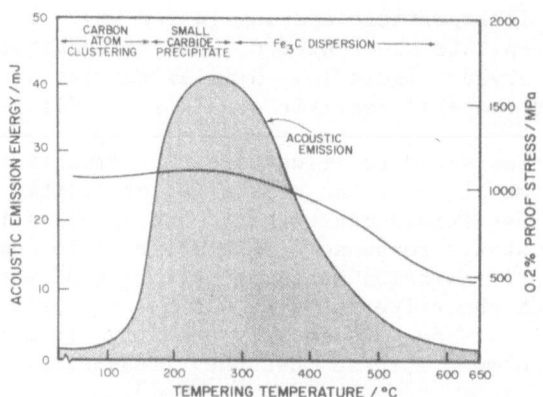

Fig. 11. The acoustic emission emitted during plastic deformation
of quenched and tempered (for 100 minutes) Fe 3.25 wt.%
Ni 0.21 wt.% C steel is greatest from samples containing
small carbide precipitates. The development of a dispersion of strong Fe_3C [16].

of radius a at a velocity v in a half-space with preexisting vertical stress σ will give detectable acoustic emission only if:

$$\sigma a^2 v \geqq 5 \times 10^{14} \, rT \text{ (watts)}$$

where r is the source-epicenter distance and T the displacement
threshold above which a signal exceeds background noise [17].
Using $r = 2 \times 10^{-2}$ m and T (determined by noise) $\sim 10^{-14}$ m gives
the condition (in MKS) units that:

$$\sigma a^2 v \geqq 0.10 \text{ watts}$$

Consider a ferritic steel undergoing cleavage microfracture
for which a is about half the grain size (say $a \sim 10$ μm),
$v \sim 10^3$ ms^{-1}, and $\sigma \sim 500$ MPa then

$$\sigma a^2 v = 500 \times 10^6 \cdot 10^{-10} \cdot 10^3 \text{ (watts)}$$

$$= 50 \text{ watts}$$

Thus cleavage microfracture is likely to generate signals 500
times larger than the detection threshold and should be easily
detectable.

Other micromechanisms of fracture, that might occur during
processing, will have different distributions of a and v. Some,

depending upon noise background, may be detectable, others may
not. A convenient way to consider this is to map, on a plot of v
against crack area, the regions covered by various fracture mechanisms and to superimpose upon this a detectability criterion.
This has been done in figure 12 using a threshold for detectability
of 10^{-13} m (more typical of noise in a working environment).

Because there are distributions of a and v, for each fracture
process there is a field of detectability, the boundaries of
which are controlled by extremal statistics. Only those events
that are to the right of the shaded line give detectable events.
In steel, brittle intergranular fracture and cleavage fracture are
easily detected while microvoid coalescence will be undetectable.

Interestingly, it is likely that a fraction, at least, of
the inclusion distribution generates detectable signals and
during processing listening to their fragmentation might provide
a quality control method to control dense inclusion segregates.

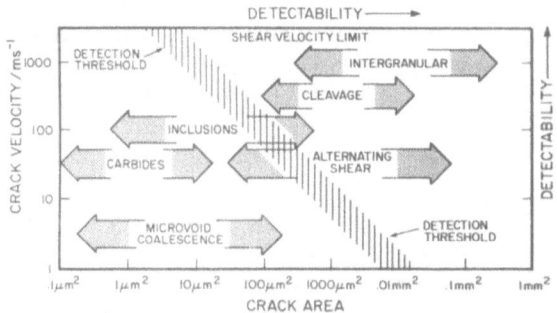

Fig. 12. Detectability of fracture events in steels. Detection
threshold assumes transducer sensitivity of 10^{-13} m,
source-transducer distance of 0.1 m, stress of 500 MNm^{-2}.

ACOUSTIC EMISSION FROM PHASE TRANSITIONS

Many of the thermomechanical treatments to which materials
are subjected during processing are designed to induce controlled
phase transformations that result in optimal microstructures. It
has become recognized that sensors are required to monitor (in
situ) phase transformations as part of a closed loop feedback
control system that would improve both material quality and
industrial productivity. Acoustic emission methods have begun to
be applied to the monitoring of phase transformations with varying
degrees of success.

Solid State Phase Transformations

Solid state phase transformations almost invariably cause
development of an internal stress due to density and crystal
structure differences between the different phases of an alloy
system. Consequently, one expects that the development of the
stress field itself or, perhaps, the mechanisms by which the
stress field is relaxed (loss of coherency) are potential sources
of acoustic emission.

Speich and Schwoeble [18] have systematically investigated
the effect of cooling rate on the acoustic emission accompaning
the austenite-ferrite transformation in plain carbon steels. As
the cooling rate was increased the reaction products changed from
pearlite through bainite to martensite. Their observations are
summarized in table 1.

From this it is clear that the diffusion controlled nucleation
and growth of ferrite and carbides, while causing the development
of appreciable stresses, fails to generate elastic waves in the
acoustic emission range of frequencies. However, the diffusionless
transformation of austenite to martensite, which occurs at velocities $\sim 10^3$ ms^{-1} is easily detectable and provides a simple method
of deducing the M_s and M_f temperatures for a steel and other
materials [19,20].

Table 1. Acoustic Emission During Continuous Cooling of Plain
          Carbon Steel

Transformation Product	Phase Transformations	Acoustic Emission Activity
Pearlite	Diffusion controlled simultaneous growth of lamellar ferrite and cementite	None detectable
Bainite	Diffusion controlled growth of small carbides (< 1 μm) and lath ferrite.	None detectable
Martensite	Diffusionless transformation in which laths or plates typically 20 μm in diameter and several microns thick transform from fcc to bct structure at $\sim 0.3 \, C_2$.	Very energetic signals detected

There are essentially three mechanisms by which the formation of martensitic from austenite may radiate elastic waves [9]:

1. The change of elastic constants.
2. Dynamic momentum associated with density changes.
3. Volumetric and shape changes.

For small applied or preexisting stresses, and for laths that propagate at less than about 10 percent the shear wave speed, the first two mechanisms are likely to make only a small contribution to the signal. It is therefore reasonable to assume the detectability of acoustic emission signals is controlled by shape and volumetric changes that occur as a fixed number of atoms in the fcc austenite rearrange their crystal structure to form a bct martensite lath or plate.

To estimate the acoustic emission amplitude consider a region of austenite about 25 mm below the surface of an isotropically elastic half space transforming to martensite. There are two principal components to the shape change (ignoring associated plastic effects) a volume expansion (dilatation) and a shear; both of which generate elastic waves.

The emission from the volume expansion can be estimated as follows. For a 1 percent carbon concentration, the austenite lattice parameter is 3.59 Å while those of martensite are 2.848 and 2.977 Å giving a volume dilation of 4.3 percent. Assuming that a 30 μm radius penny shaped region 2 μm thick transformed, the volume change would be 240 μm^3. If the transformation occurs at say 10^3 ms^{-1} (30 percent of the shear speed), the duration of the source is 30 ns and an acoustic emission signal with the temporal form of figure 3(c) is radiated. Its peak amplitude will be 30 μm mm^{-3} x 240 x 10^{-9} mm^3 = 8 x 10^{-12} m, a very large signal, (about 800 times greater than the detection threshold of laboratory instrumentation).

Typically shears of about 20 percent also occur. These contribute no displacement if they are aligned parallel or perpendicular to the vertical axis. The maximum displacement occurs at 45° when the temporal waveform of figure 3(d) must be added to the dilatation signal. To estimate the amplitude we note the shear is roughly equivalent to about 10^4 dislocations gliding across the disc. Therefore, the peak displacement, from figure 3(d), is (8 μm per mm^3) x bπa^2 x 10^4 = 8 x 10^{-11} m, an order of magnitude greater than that of the dilatation.

Therefore, we see that very energetic signals are predicted for martensitic transformations and the monitoring of these ought

to be relatively straightforward. The signal strength will
depend upon microstructure. The model above would suggest a
great sensitivity to grain size, martensite morphology (which
controls the volume of each microtransformation) and carbon
concentration (which controls the shape change).

Acoustic emission techniques are very useful for following
the kinetics and measurement of the dynamics of martensitic
transformations [19,20]. For example, in figure 13, the acoustic
emission emitted during the continuous cooling of a high carbon
steel is shown [18]. The martensitic start temperature (M_s) for
this steel was $\sim$ 200 °C; only a few acoustic emissions were generated above this temperature, possibly associated with a local
stress assisted transformation. As the temperature continued to
decrease below M_s, increasing rates of emission were observed,
each emission presumably associated with the rapid growth of a
martensitic plate across an austenite grain. The maximum transformation rate appeared to be about 60 °C below M_s and emission
ceased entirely by about 100 °C below M_s.

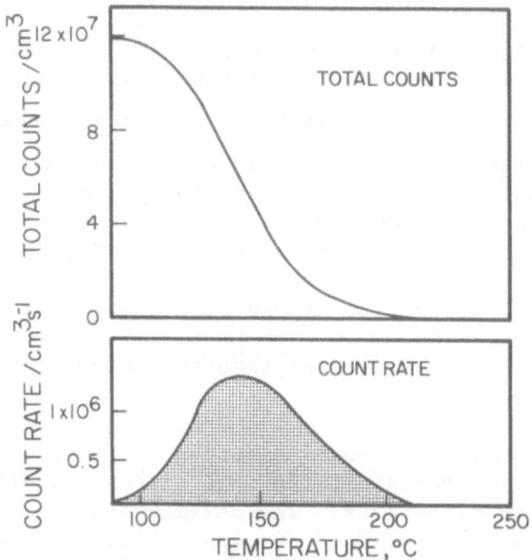

Fig. 13.   Upon cooling a high carbon steel ($M_s \sim$ 200 °C) acoustic
emission signals (counts) are detected near the M_s temperature. The acoustic emission rates goes through a
maximum and then decreases about 100 °C below M_s as the
transformation goes to completion [18].

The carbon concentration of 4300 series of low alloy steels has a very strong effect upon both the temperature dependence of the emission during continuous cooling and the number of detectable signals per unit volume, figure 14 [18]. The temperature dependence of the emission is consistent with the decrease in M_s temperature with increasing carbon concentration. The effect of carbon concentration upon the number of detectable signals however is more likely to be a manifestation of changes in the dynamics and morphology of the martensite transformation.

While almost all solid state phase transformations result in the development of internal stresses, only those from the rapid martensitic transformations generate detectable elastic waves directly. However, mechanisms of plastic deformation and even fracture are induced by these internal stresses and relaxation processes (or secondary stress changes).

For example, rapidly cooled high carbon steels have a very high internal stress which is normally relieved by a tempering treatment that allows dislocation/impurity atom migration. Shea and Harvey [21] have observed, using acoustic emission methods, that isothermal tempering at too low a temperature will result in subsequent cracking of the steel; it is believed because of impurity atom locking of dislocations.

Similarly, while the primary stress developed in and around a second phase precipitate has not proven detectable by acoustic emission methods, Cannelli and Cantelli [22] have observed signals

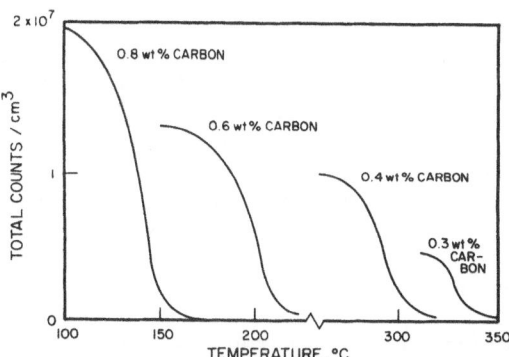

Fig. 14.   The acoustic emission activity during the austenite-
           martensite transformation in 4300 series low alloy steel
           increases with increasing carbon concentration. The
           temperature shift of the curves is due to the lowering
           of the M_s with increasing carbon concentration [18].

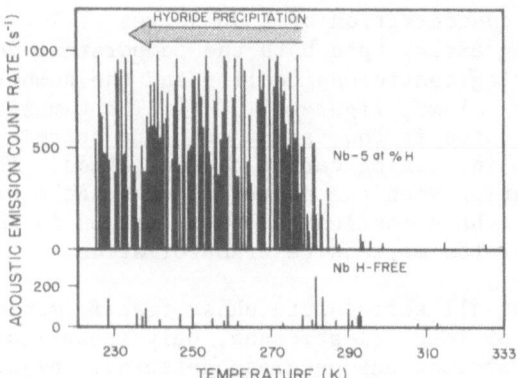

Fig. 15. Showing the generation of acoustic emission during pre-
          cipitation of NbH upon cooling.

from hydride precipitates of Nb, Ta, and V as they subsequently
fracture and cause lattice plastic deformation, figure 15, both
of which relax the primary stress. These results suggest there
may well be some merit in using acoustic emission techniques to
detect the point, during aging, when the coherency stresses of
small precipitates are relaxed by the formation of interfacial
dislocation structures around semicoherent precipitates (e.g.
when $\theta'' \rightarrow \theta'$ in the Al-Cu system). The prismatic punching out of
dislocation loops during precipitation in other alloy systems
(e.g. Mo-C) might also be worthy of investigation.

Liquid-Solid Transformations

     There have been few studies of acoustic emission during
solidification. Work on Pb-Sn and Sn-Bi [23] indicated the
generation of acoustic emission signals when solidification
conditions were such that interdendritic porosity (solidification
shrinkage of the final interdendrite liquid) occurs. The emission
associated with this process was very intense, though the precise
physical mechanism is unclear. This work also indicated the
plastic deformation of primary dendrites could also generate low
intensity acoustic emissions.

     Fuerer and Wunderlin [24] measured the acoustic emission
during solidification of an Al-4.5% Cu-0.2% Ti alloy and observed
acoustic emission generated, they believed, by the formation of
porosity. They varied the volume fraction of porosity by adjusting
the hydrogen content of the melt and found the solidification
acoustic emission was proportional to the volume fraction of
porosity, table 2.

Table 2.   Summary of Results for Solidification Emission
of Al-4.5% Cu-0.2% Ti

H_2content/cm^3 at STP/100 g	Total A.E. Count	Pore Fraction (%)
0.05	1.05×10^4	0.19
0.17	2.75×10^4	0.46
0.23	6.35×10^4	0.63

They considered the emission to be generated by the unstable
formation of hydrogen "bubbles" in the melt close to the liquid-
solid interface.   The lower hydrogen solubility of solid aluminum
results in a hydrogen supersaturation in the melt close to the
liquid-solid interface.   The relief of this supersaturation acts
as the driving force for H_2 bubble formation in an analogous role
to that of strain energy reduction in the formation of cracks.
This dilatation source then radiates longitudinal elastic waves
that are transmitted through the liquid-solid interface and
ultimately detected by a transducer as acoustic emission.

A study of the acoustic emission generated during rapid
solidification has been initiated by DARPA at the National Bureau
of Standards.   The study is directed toward development of acoustic
emission source characterization methodologies and their application to in-process control of rapid solidification processing.
Exploratory studies have initially concentrated on deducing the
origin of the acoustic emission generated during electron beam
surface melting and resolidification of 2219 and 1100 aluminum
alloys.

It has been found that acoustic emission signals are emitted
during solid state heating (due to thermoelastic effects similar
to those utilized for laser generation of elastic waves [25]),
melting and resolidification [26].   The acoustic emission emitted
during resolidification, after attainment of a steady state
temperature field, increases with the electron flux due to the
increased volume of resolidifying metal.   For a given melt depth,
the acoustic emission from 2219 aluminum alloy is up to 100 times
more energetic than that from 1100 aluminum, figure 16.

Metallographic studies have indicated the occurrence of
solidification cracking and course slip bands in the copper
alloyed 2219 alloy.   Both phenomena were absent in the commercial

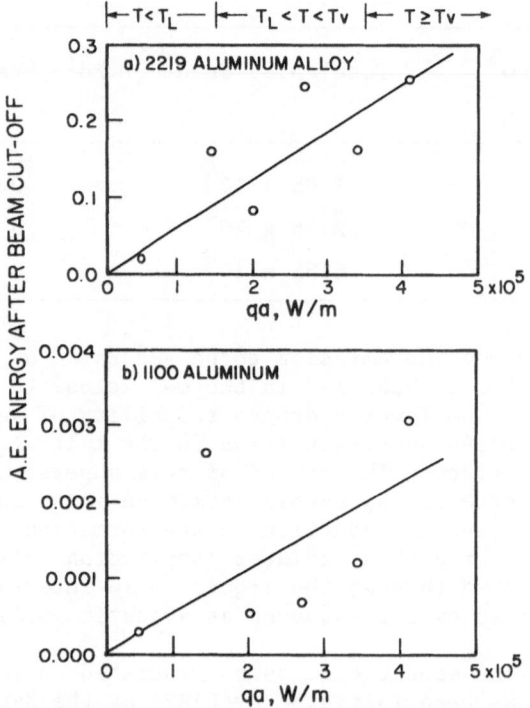

Fig. 16. Acoustic emission energy emitted during rapid solidifica-
          tion of 2219 and 1100 aluminum alloys as a function of
          qa; where q is the electron flux (Wm^{-2}) and a is the
          radius of circular region heated.

purity 1100 aluminum. The results indicate that the large solidification and thermal contraction stresses set up during rapid
solidification are responsible for plastic deformation in both
materials. The acoustic emission from dislocations is very weak
in the fine-grained 1100 aluminum whereas in the alloy it is much
stronger. This, together with the additional emission of hot
tearing results in much greater levels of detectable emission.

    A goal of the NBS program is ultimate development of in-
process monitoring methodologies for advanced metal processing
techniques. Acoustic emission is, however, already being routinely
used for certain, more conventional processing both in the U.S.A.,
but particularly abroad, and we consider these below.

IN-PROCESS MONITORING APPLICATIONS

Up to this point the results of fundamental studies of acoustic emission from deformation, fracture, and phase transformations have been reviewed with a view to understanding the physical basis of phenomena observed to date. Even though the physical understanding is less than complete, there are, nevertheless, important practical examples of the application of acoustic emission to a variety of metal processing procedures.

Spot Welding

Resistance spot welding consists of compressing two metal parts together and passing sufficiently high a current to cause local melting to form a weld. It has been a successful joining method for many years but suffers from the problem that because the spot weld is usually buried between two sheets, inspection is both difficult and costly. Acoustic emission has proven a highly successful method for overcoming this problem [27].

Resistance spot welding consists of the following sequence: electrodes are set-down on a part, a force applied, current flows causing melting and nugget formation, the current is removed and cooling occurs and finally, the electrodes are lifted-off. It has been found that each component of the process generates detectable acoustic emissions, as shown schematically in figure 17.

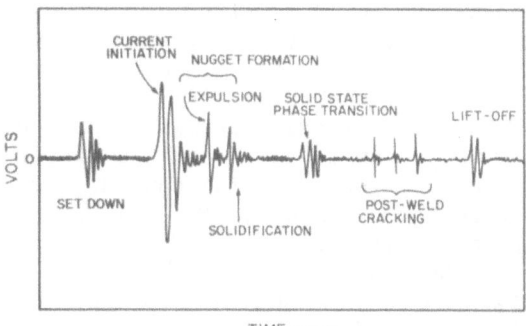

Fig. 17. Schematic example of acoustic emission signals typically detected during resistance spot welding. The signals from the normal sources of welding (such as set down) and the detection of cracking allows closed loop feedback control of the welding and online quality control [27].

The signals from individual components of the process such as set down, initiation of current, nugget formation, etc. are related to electrode condition, surface condition, nugget volume, etc. and provide a basis for closed loop feedback control of processing. Deleterious conditions such as metal expulsion and post weld cracking also generate distinctive acoustic emission signals enabling their detection and control. It is therefore apparent that acoustic emission techniques can be used for optimization of this (and similar) joining processes to increase productivity and quality.

Heavy Section Welding

The most common site of failure in structures fabricated from heavy section steels is the weld. These failures are initiated by defects formed during welding that are often very difficult to detect with conventional nondestructive methods. When defects are detected their repair can often leave a weld with worse problems than that of the original flaw because of damage to weld microstructure and additional residual stresses. The repair of welds is a major cost item in the production of components manufactured from heavy section steel.

The removal of flaws through in-process detection and repair would both improve repair quality and, provided only one or two weld passes are removed and replaced, result in considerable cost savings over conventional heavy section post weld repair. The obstacle to this has been lack of an effective in-process sensor to detect flaws dynamically as they form. In-process monitoring using acoustic emission shows great promise for providing the much needed tool [28,29].

The primary problem is that both flaw formation and the welding process generate acoustic emission. However, simple spatial and temporal filtering (cluster analysis) has been quite successful at discriminating defect signals from noise. As an example, Bentley, Dawson, and Prine [29] recently conducted a trial during which they monitored automatic and submerged arc welding of plates of pressure vessel and 316 stainless steel into which 20 defects were deliberately introduced. Two sets of instrumentation were used to monitor the welding--the primary difference being the positioning of the transducer pairs on opposite surfaces. Conventional NDE methods were used to establish that the intended defects were indeed produced; these methods also detected some unintended, natural defects.

It was found that one or other instrumentation sets detected 15 of the 20 intended defects and that each alone, detected 10. Cracks and slag inclusions were more readily detectable than other defect types, table 3.

Table 3. Welding Defect Detection with Acoustic Emission

Plate Material	Welding Method	Intended Defect Type Detected	Defect Type
Mild steel	Manual metal arc	2 hot cracks Slag inclusion Porosity Lack of fusion	2 cracks Slag inclusion Porosity Lack of fusion Lack of fusion/slag*
Mild steel	Automatic submerged arc	Hot crack Slag inclusion 2 areas porosity Lack of fusion	Crack Slag inclusion 1 area of porosity Lack of fusion/slag Crack/slag*
Stainless steel	Automatic submerged arc	2 hot cracks Slag inclusion Porosity Lack of fusion	2 cracks Lack of fusion/slag Lack of fusion/crack

*Natural (unintended) defect.

Other studies have also indicated that cracking during post-weld treatments can also be sensitively detected in reactor steels [30].

Studies such as these clearly demonstrate the potential of in-process weld monitoring by acoustic emission. The major remaining obstacle is a reliable technique of signal analysis to:

1. More reliably separate defect signals from noise.
2. Characterize flaw severity for input into accept-reject criteria.

Other supporting studies of the role of metallurgical and welding variables ought also be carried out to determine the bounds of "quiet" flaw growth.

SUMMARY

Acoustic emission has already become a recognized technique for at least one in-process monitoring application. The reasons for this, and why it may be considered a candidate technique for other in-process monitoring and microstructure characterization applications in part are:

- Some natural material responses to processing (e.g., martensitic phase transformations and plasticity) emit detectable acoustic emission enabling possible feedback control of processing procedures.

- Abnormal material responses (e.g., brittle microcracking) may emit extra highly energetic emission signaling poor quality.

- The passive nature of the technique (it requires no artificial energy sources), its 100 percent volume coverage (possibly with non-contacting sensors), and the short response time (a few ms) result in short feedback times and minimal interference with production line processing techniques.

However, much research is still required both at the basic and more applied levels. Particularly, the problems for the future that must be addressed are:

- Noise: The material generated emission has to be detected in the presence of noise and some materials/processing procedures may not emit signals with sufficient signal-to-noise ratio.

- Characterization: In practice many acoustic emissions are detected but only a few may be from deleterious defects that must be removed during processing. Signal analysis methods are required which will allow rapid, on-line defect assessment.

- Exploratory studies are required to pioneer developmental research in promising areas of processing.

ACKNOWLEDGMENTS

We wish to express our gratitude to our colleagues Drs. C. B. Scruby, J. E. Sinclair, R. B. Clough, and J. A. Simmons for many helpful discussions. Support from the Defense Advanced Research Projects Agency under DARPA order No. 4275 is gratefully acknowledged.

REFERENCES

1. H. N. G. Wadley, C. B. Scruby, and J. Speake, Int. Met. Rev. 25(2), 41 (1980).
2. W. Sachse and N. Hsu, Physical Acoustics XIV (1980).

3. J. C. Spanner, "Advances in Acoustic Emission," Proc. Int.
    Conf. Acoustic Emission, Anaheim, CA (1979). H. L. Dunegan
    and W. F. Hartman, Eds., Dunhart (1981).
4. J. M. Carlyle, J. Acoustic Soc. Am., Suppl. 1, $\underline{68}$, S104
    (1980).
5. C. Scruby, H. Wadley, and J. Sinclair, Phil Mag A, $\underline{44}$(2),
    249 (1981).
6. N. Hsu, J. A. Simmons, and S. C. Hardy, Mat. Eval. $\underline{35}$, 100
    (1977).
7. H. N. G. Wadley, C. B. Scruby, and G. Shrimpton, Acta
    Metallogr. $\underline{29}$, 399 (1980).
8. H. N. G. Wadley and C. B. Scruby, AERE-R10351, "Elastic Wave
    Radiation from Cleavage Crack Extension" (1982).
9. J. A. Simmons and R. B. Clough, "Dislocation Modeling of
    Physical Systems," Proc. Int. Conf., Gainesville, FL, June
    1980, Pergamon Press (1981).
10. C. B. Scruby, J. C. Collingwood, and H. N. G. Wadley, J.
    Phys. D. $\underline{11}$, 2359 (1978).
11. C. B. Scruby, H. N. G. Wadley, and J. J. Hill, AERE-R10534,
    "Dynamic Elastic Displacements at the Surface of an Elastic
    Half-Space due to Defect Sources" (1982).
12. W. W. Gerberich and K. Jatarallobhula, Proc. Conf. Nondestruc-
    tive Evaluation: Microstructure Characterization and Relia-
    bility Strategies, 5-9 Oct. 1980, Pittsburgh, PA.
13. H. N. G. Wadley and C. B. Scruby, Metal Science, $\underline{12}$, 285
    (1978).
14. K. Rusbridge, C. B. Scruby, and H. N. G. Wadley, AERE-R10363,
    "Origin of Acoustic Emission in Aged Al-Zn-Mg Alloys; Part
    I: The Base Ternary Alloy" (1982).
15. C. B. Scruby, H. N. G. Wadley, and K. Rusbridge, AERE-R10364,
    "Origin of Acoustic Emission in Al-Zn-Mg Alloys; Part II:
    Copper Containing Quarternary Alloys" (1982).
16. H. N. G. Wadley, C. B. Scruby, P. Lane, and J. A. Hudson,
    Metal Science $\underline{15}$, 514, Nov.-Dec. (1981).
17. C. B. Scruby and H. N. C. Wadley, 5th Int. Conf. on Quantita-
    tive NDE in the Nuclear Industry, San Diego, May 1982.
18. G. R. Speich and A. J. Schwoeble, ASTM STP-571, 40 (1975).
19. J. Baram and M. Rosen, Scripta Met. $\underline{13}$, 565 (1979).
20. J. Baram and M. Rosen, Acta Metallogr. $\underline{30}$, 655 (1982).
21. M. M. Shea and D. J. Harvey, Scripta Met. $\underline{16}$, 135 (1982).
22. G. Cannelli and R. Cantelli, "Advances in Acoustic Emission,"
    Proc. Int. Conf. Acoustic Emission, Anaheim, CA (1979).
    H. L. Dunegan and W. F. Hartman, Eds., Dunhart (1981).
23. H. M. Tensi and W. Radtke, Metall. $\underline{32}$(7), 681 (1978).
24. J. Feurer and R. Wunderlin, Int. Conf. Solidification and
    Casting, Metals Society, London $\underline{2}$, 18 (1977).
25. H. N. G. Wadley, J. Simmons, C. K. Stockton, M. Rosen, and
    R. Mehrabian, DARPA/AF Review of Quantitative NDE, Aug. 2-7,
    1981, Boulder, CO.

26. R. B. Clough and H. N. G. Wadley, Proceedings of 2nd Int.
     Conference on Use of Electron and Laser Beams for Metals
     Processing, Los Angeles, 1983.
27. S. J. Yahaviolos, M. F. Carlos, S. J. Slykhouje, and
     S. J. Ternowchek, Mat. Eval. $\underline{39}$, 1057 (1981).
28. D. W. Prine, Fifth International Acoustic Emission Symposium,
     Tokyo (1980).
29. P. G. Bentley, D. G. Dawson, and D. Prine, RNPDE Report
     ND-R-767(R) (1982).
30. P. Jax, Proc. 2nd Int. Conf. on Structural Mechanisms in
     Reactor Technology (Berlin) G614-EUR4820 (1973).

USE OF ULTRASONIC TECHNIQUES TO ASSESS

THE MECHANICAL PROPERTIES OF STEELS

B. E. Droney

Research Department
Bethlehem Steel Corporation
Bethlehem, PA 18016

ABSTRACT

A computer-based ultrasonic test system was assembled to automatically acquire ultrasonic velocity and attenuation data on commercial steel samples. Attenuation data on plain carbon plate steels showed significant statistical correlation to ferrite grain size and to yield strength, but more consistent coupling is needed if the technique is to be developed into an on-line test for steel plates. Surface-wave velocity and attenuation data on sheet steel samples undergoing tensile testing (acoustoelastic data) were found to be weakly correlated to yield strength. Variations in residual and bending stresses precluded consideration of acoustoelasticity as as on-line technique to determine properties of sheet. Our recent work has focused on nondestructively determining the anisotropy in velocity and attenuation in sheet steel samples, and on relating this data to mechanical properties. The potential for developing an on-line technique to determine crystallographic texture looks promising, based on recent tests using electromagnetic acoustic transducers (EMATs) to generate shear-wave and Lamb-wave modes in sheet.

INTRODUCTION

To characterize steel implies measurement of several parameters including steel alloy composition or grade, metallographic determination of grain size, and ASTM standard tests for strength, hardness and toughness. Destructive sampling techniques, such as mechanical tensile tests, continue to be the principal methods by which mechanical properties are characterized.

The determination of material properties by nondestructive
techniques offers significant advantages over the conventional
destructive ones, if it can be accomplished reliably and in the
steel mill environment. Some advantages include:

o   The ability to test 100% of the material rather than
    individual tensile samples.

o   The ability to measure properties on completed steel
    structures where destructive sampling is not feasible.

o   Capability to ship the product without delays for sample
    preparation and testing off-line.

o   The feedback of information for process control to assure
    uniformity of properties.

Of the techniques available to nondestructively characterize
steels, ultrasonics offers considerable promise. For example, it
is known that yield strength is related to grain size, and the
latter can be estimated via the ultrasonic attenuation due to grain
scattering. Likewise, other steel properties are related to
crystallographic texture (preferred orientation), which can be
studied via changes in ultrasonic velocity. However, to assess
mechanical properties over a range of commercial steel grades and
processing conditions requires consideration of variables in
addition to grain size. They include alloy content, dual-phase
structure, carbide size and distribution, and surface condition.
The paper describes experimental work to investigate the influence
of some of these variables on the properties of commercial steel
samples.

ULTRASONIC ATTENUATION IN PLATE STEELS
AS A PREDICTOR OF YIELD STRENGTH

Ultrasonic attenuation measurements on plate steels can be
shown to correlate with ferrite grain size[1]. Use of the attenuation
data of longitudinal waves, along with plate alloy composition, can
can then be used to predict mechanical properties of commercial plate
steels. Under certain conditions (fixed transducer path length,
10 MHz frequency, limited range of grades) one can avoid the need to
extract grain size data, and instead use the attenuation coefficient
directly in a regression equation to estimate mechanical properties of
plate steels. In work at Bethlehem by Klinman and Stephenson[2], the
yield strength measured on tensile samples was correlated to the yield
strength predicted from attenuation data and alloy composition data.
In the statistical analysis, plate alloy composition explained about
50% of the variance, whereas attenuation plus composition accounted
for about 70% of the variance.

Although plate alloy composition is usually well known in the steel mill, and the agreement between predicted and measured yield strengths was relatively good, we have not yet implemented the technique. The reason is that we anticipate significant problems in acquiring highly accurate attenuation data in the steel mill environment, since plates may have high temperatures, surface scale, and a wide range of thickness.

ULTRASONIC TESTS ON SHEET STEELS

The clean and smooth surface condition of sheet products enhances the probability of successful mill applications of ultrasonics. This is one of the reasons for directing the nondestructive measurement of properties research toward sheet steel applications. But as noted above, information besides ferrite grain size in needed. Our initial experimental work to assess the ultrasonic properties of sheet steel was focused in two areas:

   o   Surface-wave acoustoelasticity in sheet samples. One
       motivation for this approach was that potential
       applications on moving sheet steel would, in general,
       involve naturally applied tensile loads.

   o   Velocity anisotropy in as-received sheet samples.
       Ultrasonic velocity is known to be a good measure of
       crystallographic texture -- an important parameter for
       cold-rolled sheet steel.

By acoustoelasticity, we generally mean the study of changes in ultrasonic velocity or attenuation with residual or applied stress[3]. We will describe experiments to determine the surface-wave acoustoelastic constant as the slope of velocity versus applied tensile load. By velocity anisotropy, we refer to the dependence of ultrasonic wave speeds on the direction of propagation. For sheet steel, the velocity measured in the plane of the sheet is of interest since it relates to texture.

Several grades of commercial hot-rolled and cold-rolled sheet samples were selected, and the samples were characterized as to their properties. The samples represent low-carbon and solution strengthened (rephosphorized) and micro-alloyed (Cb+V) grades. Sheet thickness was about 2.5 mm for hot-rolled and 1.0 mm for cold-rolled samples. For studies relating to acoustoelasticity, the hot-rolled sheet samples were sheared into strips at 0-, 45-, and 90-degrees with respect to the rolling direction, and machined to standard tensile samples. Some samples were specially prepared (surface-ground, stress-relief annealed) to study the effects of residual and bending stresses on the ultrasonic data.

Surface-wave Acoustoelasticity in Sheet

     Although it is understood how the acoustoelastic constants
relate to properties of single crystals[4], it is uncertain to what
degree these constants may relate to the microstructure or
mechanical properties of commercial steels. Hence, we used an
empirical approach, and no attempts were made to extract the tensor
components of elastic moduli that would more fully describe
anisotropic sheet steel. For this work we used sheet steel samples
of similar grain size, so Rayleigh scattering was not considered as
a factor relating to yield strength.

     The tensile test procedure, when logging data with the
instruments reported previously[3], was to gradually load the sample
to about half-yield stress, then unload. This procedure permitted
transducer coupling variations to be assessed from repeated
acoustoelastic tests on each sample. Later, each sample was loaded
further and the mechanical properties were determined. Our analysis
procedure was to statistically determine the correlation between
mechanical properties and the small changes in ultrasonic velocity
and attenuation that occur in the elastic region of the engineering
stress versus strain curve.

     To acquire the ultrasonic data during tensile testing,
ultrasonic surface-wave (Rayleigh-wave) transducers were placed in
sliding contact with the tensile sample mounted in the Instron
tensile test machine. Surface waves at 5 MHz or 20 MHz were used in
the initial tests, so that the ultrasonic data would be independent
of sample thickness variations during deformation. Ultrasonic
transmitter and receiver transducers were spaced two inches apart,
so that Rayleigh waves were propagated along one surface over the
two-inch gage length of the sheet tensile samples. Amplitude of the
received ultrasonic signal and the time-of-flight for the wave to
propagate from transmitter to receiver were both acquired
automatically. The variation in amplitude as the sample was
elastically deformed provided a measure of the change in attenuation
versus applied stress. The propagation time provided phase velocity
measurements to be recorded as a function of stress. Phase velocity
was acquired with a Hewlett Packard 5370 time-interval counter,
using a zero-crossing method and 100-times averaging within the
counter. We were able to achieve a typical time resolution of 0.1
nanoseconds for the wave traveling over the two-inch gage length.
Very small changes in velocity (0.0005%) and attenuation (0.1 dB)
could be resolved during the tensile test. The attenuation data
showed poor repeatability due to transducer coupling variations, so
only the ultrasonic velocity data will be described in this paper.

     The slope of the velocity curve shown in Figure 1 determines an
effective surface-wave acoustoelastic "constant," at some specified
stress level, say 10 ksi. The ultrasonic surface-wave velocity

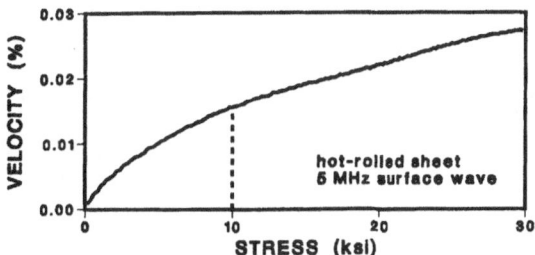

Fig. 1. Surface-wave velocity data acquired during tensile test. The acoustoelastic "constant" was determined from the slope of the velocity versus stress curve at 10 ksi.

varied smoothly with stress, but it is apparent that the slope is not constant as would be expected, based on the assumption of uniaxial tensile stress. This was attributed to variations in bending stresses as the samples were installed in the Instron test machine. A surprising effect noted was the difference in both the initial velocity and the acoustoelastic constant measured on opposite surfaces of most of the samples tested. This difference was likely caused by nonuniform sheet reduction in rolling, and to stresses generated in the strip during the coiling process at manufacture.

TABLE I. ULTRASONIC DATA UNDER APPLIED STRESS

Hot-rolled steel sheet samples were stress relief annealed. The slope of surface-wave velocity versus stress (acoustoelastic constant) at 10 ksi, and the yield strength were both averaged over sample orientation.

GRADE	ACOUSTOELASTIC CONSTANT	YIELD STRENGTH	
1008	$(x\ 10^{-5}/ksi)$	(ksi)	
capped	.7	38	non alloyed
killed	.6	44	"
rephos.	.9	48	"
Cb+V	.5	80	micro alloyed

The acoustoelastic data on hot-rolled sheet appeared to have a
random dependence on sample orientation, so the data from tests at
0, 45 and 90 degrees were averaged, and Table I relates the
acoustoelastic data averaged over each grade to average yield
strength. We found that there were weak correlations between the
average acoustoelastic "constant" at 10 ksi versus yield strength.
In particular, rephosphorized sheet was separable from low-carbon
sheet, but low-alloy sheet was not. Acoustoelastic data measured
for the high-strength, micro-alloyed sheet overlapped those of the
low-carbon sheets, so alloy composition is a much more significant
factor than ultrasonic velocity, in regard to prediction of yield
strength. Indeed, steel alloy composition alone was sufficient to
account for more than 90% of the variation in yield strength. From
this, we conclude the acoustoelastic technique is not sufficiently
repeatable as a predictor of yield strength. But it is possible
that acoustoelasticity may have important on-line applications to
other properties such as "shape", that relate to residual stress.
Before seriously considering on-line applications of ultrasonic
velocity measurements, we feel a more reliable technique is needed
to separate the effects of residual stress and texture.

Velocity Anisotropy versus Sheet Texture

The results described above were for hot-rolled sheet. For
cold-rolled sheet, the velocity change with sample orientation,
called planar velocity anisotropy, was an order of magnitude larger
than any velocity change due to tensile loading. It is known that
velocity anisotropy correlates to texture and to formability in
cold-rolled sheet steel[6]. Recent theoretical work has clarified
how ultrasonic velocity is influenced in polycrystalline steel or
aluminum by several parameters, including texture[7], residual
stress[8], and tensile and bending stresses. Details of the
relationships between velocity and texture have been reviewed in
the literature[9]. Statistical correlations between Young's modulus
and sheet drawing and formability parameters have been reported[10],
and similar destructive techniques are in routine use in the steel
industry to determine texture.

Test Procedure -- Velocity Anisotropy. Hot-rolled and
cold-rolled sheet samples were tested in the as-received, unstressed
state, and at ambient temperature. Piezoelectric transducers and
EMATS were used. Ultrasonic velocity data was obtained as a
transmitter-receiver ultrasonic transducer pair was rotated from 0
to 180 degrees, where the rolling direction was labeled 0 degrees.
Velocity and signal amplitude data were recorded at 15-degree
increments in transducer-pair orientation. Most sheets produced
velocity profiles which exhibited 2-fold or 4-fold symmetry
for rotation about the normal to the sheet surface.

As an independent measure of texture, the bar velocity was

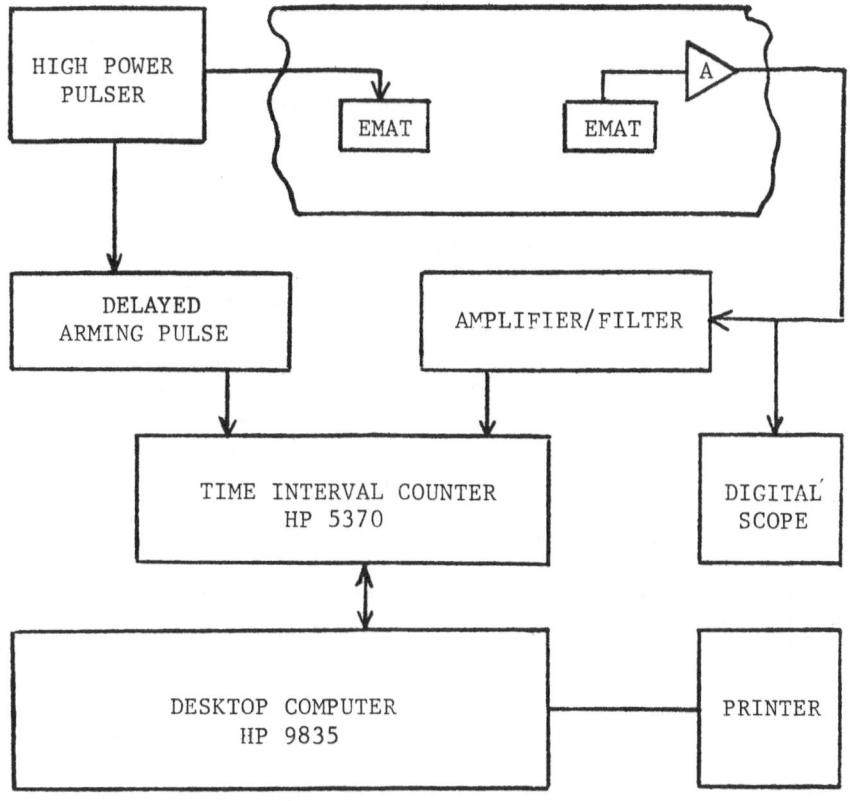

Fig. 2. Block diagram of ultrasonic equipment to measure velocity anisotropy in sheet steel samples, using EMATs.

measured on samples punched from each sheet, using an instrument designed to determine Young's modulus based on a length resonance method. Three 1/4-inch x 4-inch length strips were punched from each sheet, and the bar velocity was calculated from the resonant frequency and length.

A block diagram that illustrates phase velocity measurements on sheet steel is shown in Figure 2. Components of the programmable ultrasonic equipment include: high-power Velonex pulser, non-contact electromagnetic-acoustic transducers (EMATs) with associated impedance matching and preamplifier circuits, digital storage oscilloscope, and the time interval counter described previously for the tensile tests. The counter was programmed to measure frequency of the received signal. The measurement was done on a period-by-period (one-wavelength) basis, using a delayed arming pulse to select a specific cycle of the received EMAT waveform.

Several EMATs were constructed to generate and receive shear-horizontal and Lamb waves in sheet. Shear-horizontal waves were generated with a wavelength of one centimeter, using a two-layer stripline coil. The design utilized permanent magnets intended to support shop tools. The EMAT-pair spacing was about 10 centimeters, and off-axis beam skewing by as much as 15 degrees could be accommodated.

Results -- Velocity Anisotropy in Sheet Steel. The surface-wave velocity versus transducer orientation is compared in Figure 3(a), on the same four hot-rolled sheet grades described previously on acoustoelasticity. The non-alloyed, hot-rolled samples have less than 0.2% change in velocity. They are considered isotropic in regard to mechanical properties. Some hot-rolled grades have nearly 1% change in velocity. For the cold-rolled samples, Figure 3(b) shows the surface-wave velocity change was as much as 3.2%. This relatively high anisotropy is anticipated, assuming typical rolling textures for sheet steel.

Using the EMATs described above, Figure 3(c) shows the velocity anisotropy for two modes of wave propagation in one cold-rolled sheet sample. The phase velocity for each mode is calculated simply from the product of wavelength and frequency. The modes are labeled by S(0) and SH(0) representing the low-order symmetric Lamb and shear-horizontal waves. For the shear-horizontal mode the data is symmetric about 45 degrees, in contrast to the surface-wave data and the Lamb-wave data. Standard deviation of the shear-horizontal and the symmetric Lamb-wave velocity data was about 0.1%. That of the antisymmetric Lamb-wave velocity, not shown in the figure, was about 0.5%. Velocities were repeatable within 0.1% as the EMAT was scanned along the sample surface. The magnitude of the velocity change with orientation can easily be resolved with the ultrasonic equipment, which implies there is good sensitivity to changes in texture. Thus ultrasonic velocity measurements using EMATs could be applied on-line, as a nondestructive technique to determine properties relating to crystallographic texture.

Ultrasonic velocity data has been obtained on other samples of as-received sheet. As an example, Table II summarizes the velocity anisotropy for three nondestructive measurements of velocity determined with EMATs, compared to the resonant bar velocity. Velocities of the lowest-order shear-horizontal, SH(0), and two Lamb-wave modes, S(0) and A(0) were measured versus direction of propagation. EMAT period was one centimeter. For each sheet, the SH(0) velocities were about the same at 0 and 90 degrees, while the other three velocities differed at 0 and 90 degrees, as anticipated for cold-rolled sheet textures. Trends were similar for the S(0) and bar velocity, V(Y), versus direction of propagation. The A(0) mode is highly dispersive at long wavelengths, so its velocity responds to slight variations in sheet thickness. Magnitudes of the

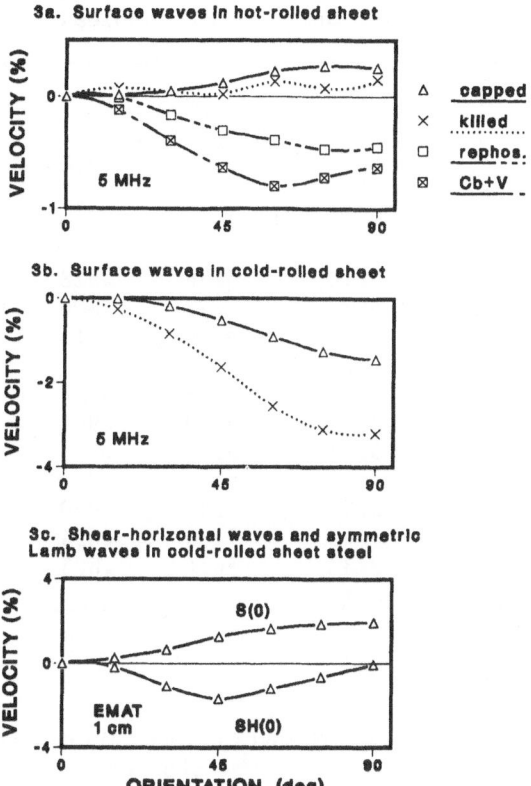

Fig. 3.   Examples of velocity anisotropy in as-rolled sheet steel.

two Lamb-wave velocities were in agreement with approximate values
calculated from the Rayleigh-Lamb frequency equation. As a
nondestructive test for crystallographic texture in sheet, it
appears that SH(0) and S(0) velocity data combined can provide at

TABLE II.   ULTRASONIC VELOCITY ANISOTROPY DATA

Three grades of cold-rolled steel sheet were nondestructively
tested as-received with EMATs, and these velocities were
compared at 0, 45, and 90 degrees to the bar velocity, V(Y),
from destructive tests on the same samples.

| SAMPLE | PHASE VELOCITY (x 10^5 cm/s) | | | |
	SH(0)	S(0)	A(0)	V(Y)
1008 capped				
1 - 0 degrees	3.30	5.36	.828	5.16
1 - 45	3.25	5.43	.806	5.26
1 - 90	3.30	5.46	.816	5.25
rephos.				
3 - 0 "	3.30	5.33	.936	5.17
3 - 45	3.21	5.40	.941	5.37
3 - 90	3.31	5.38	.946	5.20
dual-phase				
10 - 0 "	3.32	5.49	.940	5.07
10 - 45	3.43	5.52	.894	5.31
10 - 90	3.32	5.41	.803	5.18

least as much information as the V(Y) data, where the latter is
determined by destructive tests. The absolute accuracy of the
technique, in terms of assessing the many possible types of sheet
textures, has not yet been established. It is likely that
independent measurement of through-thickness velocity will be
needed, in order to fully characterize texture.

SUMMARY AND CONCLUSIONS

    We experimentally studied several factors that affect
ultrasonic data and their relation to mechanical properties of
steel, and conclude the following:

    o   Ultrasonic attenuation data can be combined with knowledge
         of plate alloy composition, to improve the prediction of
         yield strength in plate steel. However, the technique is
         not in use, in part because of the need for plate surface
         preparation.

o   Coupling variations need to be overcome to permit reliable ultrasonic testing of steel product in motion.

o   The acoustoelastic data in sheet steel has a weak relationship to yield strength, in that various alloys are separable statistically but with a low confidence level. Variations in residual and bending stresses preclude consideration of acoustoelasticity as as on-line technique to determine properties of sheet.

o   Shear velocity anisotropy is a useful indication of crystallographic texture variation between sheet samples. The potential for developing an on-line technique to determine velocity anisotropy looks promising, based on use of EMATs to generate shear-wave and Lamb-wave modes in sheet.

Factors other than mechanical properties can have a large influence on measurements of elastic constants in sheet steels. Sheet anisotropy, residual stress, temperature, and magnetic fields all are very important factors to be considered if measurements are to be made in the steel plant environment.

ACKNOWLEDGMENTS

I would like to thank R. R. Pradhan for characterizing the properties of the sheet steel samples, W. J. Bantz for assistance in developing velocity measurement techniques that are appropriate for non-contact ultrasonic transducers, and G. L. Hunsicker for constructing the EMATs and programming the computer.

REFERENCES

1.   R. Klinman, G. R. Webster, F. J. Marsh and E. T. Stephenson, "Ultrasonic Prediction of Grain Size, Strength, and Toughness in Plain Carbon Steel," Materials Evaluation 38 (1980) 26-32.
2.   R. Klinman and E. T. Stephenson, "Ultrasonic Prediction of Grain Size and Mechanical Properties in Plain Carbon Steel," Materials Evaluation 39 (1981) 1116-1120.
3.   M. Hirao, H. Fukuoka, K. Hori, "Acoustoelastic Effect of Rayleigh Surface Wave in Isotropic Material," J. Appl. Mech. 48 (1981) 119-124.
4.   M. A. Brezeale and J. Philip, "Relation of the Third-Order Elastic Constants to Other Nonlinear Quantities," Proc. IEEE Ultrasonics Symposium (1981) 425-431.
5.   B. E. Droney and R. Klinman, "Ultrasonic Techniques for Determining the Mechanical Properties of Steels," Proc. AIP Conf. Physics in the Steel Industry (1981) 210-228.

6. H. Kitagawa, "Analysis of Elastic Anisotropy in Cold Rolled
    Steel Sheet," Sixth Int. Conf. on Textures of Materials, Tokyo
    (1981) 1166-1178.
7. D. E. MacDonald, "On Determining Stress and Texture Using
    Ultrasonic Velocity Measurements," IEEE Trans. Sonics and
    Ultrasonics, SU-28 (1981) 75-79.
8. G. C. Johnson, "On the Applicability of Acoustoelasticity for
    Residual Stress Determination." J. Appl. Mech. 48 (1981)
    791-795.
9. R. E. Green,"Ultrasonic Investigation of Mechanical
    Properties," Treatise on Materials Science and Technology
    (H. Herman, ed.) Chap. II, Academic Press, New York, 1973.
10. P. R. Mould and T. E. Johnson, Jr., "Rapid Assessment of
    Drawability of Cold-Rolled Low-Carbon Steel Sheets," Sheet
    Metal Ind. (1973) 328-348.

DETERMINATION OF MICROSTRUCTURAL PARAMETERS

BY MAGNETIC AND ULTRASONIC QUANTITATIVE NDE

Werner A. Theiner and Herbert H. Willems

Fraunhofer-Institut
für zerstörungsfreie Prüfverfahren
D-6600 Saarbrücken, FRG

INTRODUCTION

Microstructure parameters like precipitations, grain boundaries, dislocations, grain size and atomic defects (impurities, vacancies) determine the mechanical properties of technical steels in well-known manner[1] and, therefore, the estimation of these parameters is a general aim in the field of quality and production control as well as in the development of new steels. Nondestructive (nd) techniques, which are sensitive to "lattice imperfections" and macro-stresses, should cover the range from atomic up to macroscopic dimensions in the centimeter range. This problem can be solved by a combination of different and independent nd quantities as listed in Table 1 for ultrasonic and electro-magnetic methods.

ULTRASONIC ND TECHNIQUES

Structure characterization of polycrystalline materials (i.e., steels) by means of ultrasound is essentially based on attenuation measurements. Inelastic scattering (absorption), as well as elastic scattering of the ultrasonic energy, contribute to the total attenuation, which is quantitatively described by the attenuation coefficient α. Confining us to room temperature and frequencies in the lower MHz range, ultrasonic absorption is caused mainly by dislocation damping and magnetoelastic effects, whereas elastic scattering takes place due to the acoustic mismatch at grain boundaries or phase boundaries. In steels, the overall frequency dependence of the absorption coefficient α_A is generally well described by a linear relation.[2] The scattering

249

Table 1.   Linear Dimensions of Structure Parameters and
           Examination Methods

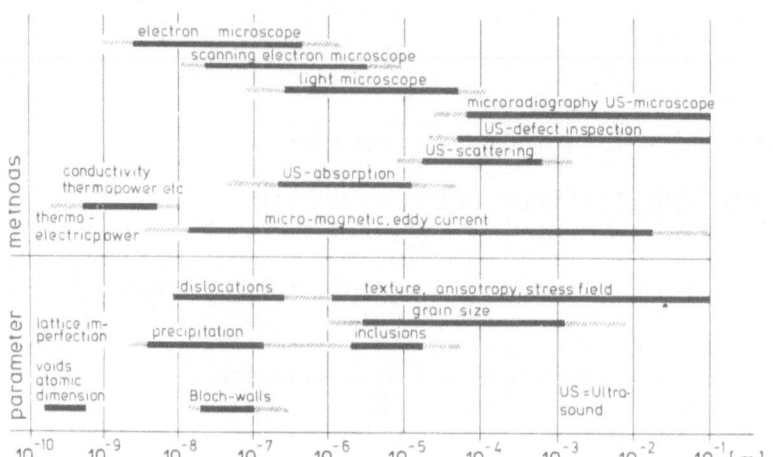

coefficient α_S depends like $\alpha_S \sim d^3 \cdot f^4$ on frequency f and grain
size d assuming Rayleigh scattering.[3]   Knowing these frequency
dependencies of the contributing attenuation mechanisms, the
ultrasonic attenuation coefficient α is separated into α_A and α_S
by measuring the ultrasonic attenuation at two different
frequencies.   From α_S the mean grain size is then evaluated
(two-frequency method[4]).

     To determine the attenuation coefficient α, ultrasonic
backscattering measurements are very appropriate.   Assuming
homogeneous material and neglecting multiple scattering, the
backscattered amplitude A_S decays as a function of sound path x
like $A_S \sim \exp(-\alpha x)$, from which α is easily derived (see Fig. 2).
In contrast to the usual way of determining α by analyzing
subsequent backwall echoes, the backscattering method offers some
important advantages: neither special geometries (like plane-
parallel specimens) nor complex soundfield corrections are
required.   Additionally, the local resolution achieved yields
information on local deviations from material homogeneity.

     Figure 1 shows a prototype built in our institute which
enables fast backscattering measurements, including automatic
evaluation of the measured curves.   This apparatus is described in
detail elsewhere.[5]   Here, only a brief description is given:
highly monochromatic pulses (variable pulse-length, Gaussian
envelope) are generated by means of a programmable frequency
synthesizer covering the frequencies from 2-30 MHz.   After
amplification (max. 250 W into 50 Ω), commercial narrow-band

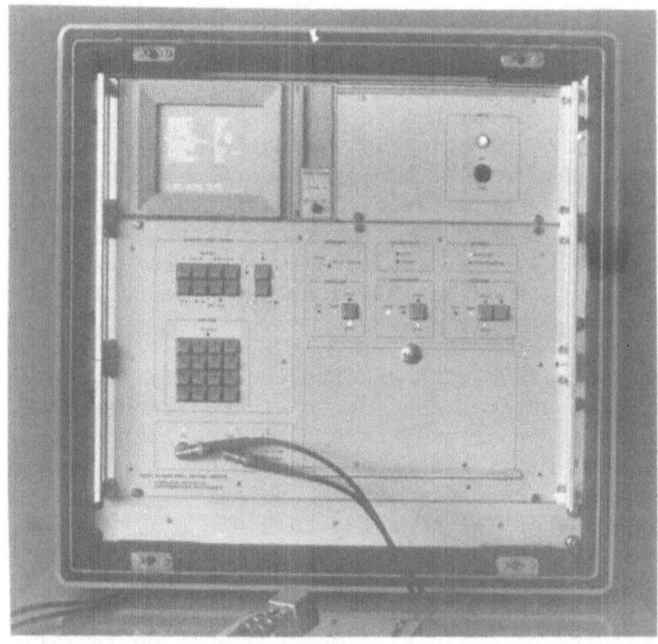

Fig. 1. Prototype for ultrasonic backscattering measurements.

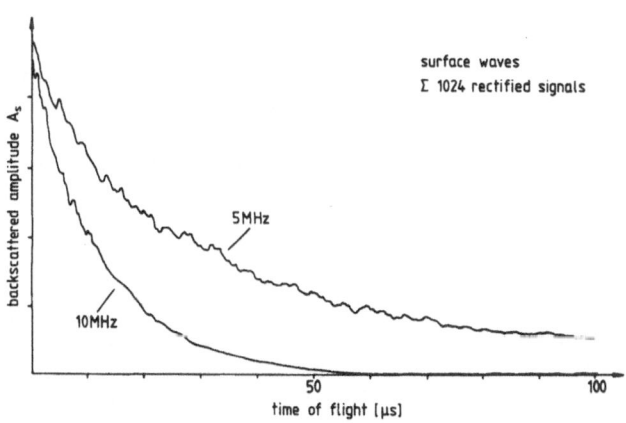

Fig. 2. Backscattering curves of surface waves at two different
         frequencies. Sample: ferritic steel with grain size 45 µm.

probes are used to generate ultrasound which is insonified either
in direct contact or in immersion technique. The backscattered
signals with amplitudes generally about 30-40 dB lower than usual
reflection echoes are received in pulse-echo technique and an
A-scan is obtained. Because of a pronounced interference pattern,
the backscattered signal has to be averaged in order to get decay
curves which can be numerically evaluated. This is done by
spatial averaging: during moving the probe slightly on the
specimen, which causes drastic changes in the interference pattern,
up to 1024 single A-scans are recorded, rectified, digitized,
summed and stored. As a result, a smoothed curve is obtained (see
Fig. 2). From its exponential decay the attenuation coefficient α
can be evaluated.

Besides operating with one frequency only, the apparatus
enables backscattering measurements at two frequencies
simultaneously in order to determine grain sizes by the two-
frequency method mentioned above. A grain size determination takes
about 5 seconds. The whole system is controlled by a microcomputer
which also evaluates the measured curves. All relevant settings
are programmable via keyboard. The results are displayed on a CRT
and can also be transferred to an external computer or xy-plotter.

Some results of grain size measurements on ferritic,
austenitic and also ferritic-pearlitic steels are shown in Fig. 3,
where the ultrasonic values d_{US} are compared with the metallographic values d_M. Besides shear waves, which are used for bulk
investigations, surface waves allow the investigation of the
surface region.[6,7] Working with frequencies between 2 and 30 MHz,
grain sizes in the range of ASTM 1 to ASTM 10 can be measured with
a typical error of ± half an ASTM class.

In addition, ultrasonic backscattering measurements offer some
other applications. Due to the high sensitivity on structural
changes, ultrasonic scattering combined with spatial averaging is
very appropriate to detect inhomogeneities in materials.
Transitions from the base material to inhomogeneous regions are
exhibited in backscattering curves by a deviation from exponential
decay caused by the corresponding change in acoustic impedances.
In this way inclusions, segregations, welding seams, etc. can be
detected in a simple manner, as shown in Fig. 4. At present, from
such measurements it is not possible to identify the type of an
observed inhomogeneity. However, the position, as well as the
extension of the inhomogeneous region, can be specified to some
extent (see Fig. 4). Qualitative characterizations of this kind
can be applied to a wide range of materials (including many types
of metals and ceramics), provided they show appreciable scattering.

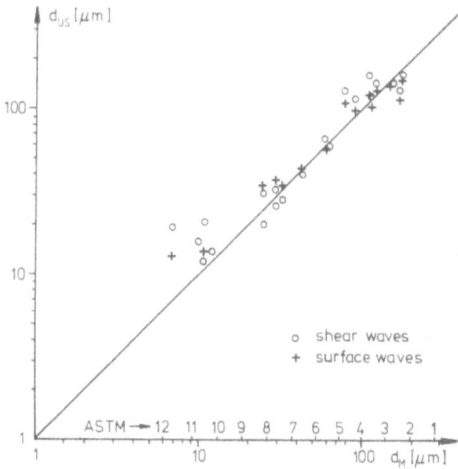

Fig. 3. Comparison between grain size determination by ultrasonic backscattering measurements (d_{US}) and by metallographic methods (d_M).

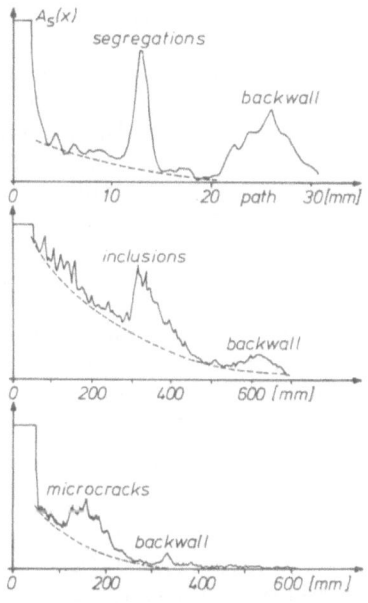

Fig. 4. Detection of inhomogeneities using ultrasonic backscattering; the strong deviations from exponential decay curve (dashed line) indicate inhomogeneous regions (6 MHz, shear waves).

Table 2. Ferromagnetic and Magnetoelastic ND Methods for
          Mechanical Stress Measurements

magnetic Barkhausen noise	irr	nss (+ ss)
acoustic Barkhausen noise	irr	ss
incremental permeability	rev, irr	nss, ss
dynamic magnetostriction	rev, irr	ss

MAGNETIC ND TECHNIQUES

The micromagnetic interactions with lattice imperfections, stresses and structural components depend upon the different magnetic reversals which can be classified in stress-sensitive and stress-insensitive processes. Certain 90° Bloch walls and all rotation processes are stress-sensitive (ss), whereas all 180° and the (100) 90° Bloch walls are stress-insensitive (nss). In the second class the residual stress field is essentially restricted to the Bloch wall dimension itself.[8]

In Figs. 5-7 some examples are presented how nd quantities-- coercivity H_c, incremental permeability μ_Δ, acoustic Barkhausen noise A--depend on hardness. In all illustrations the horizontal line region represents states which are tempered (T > 600°C). The index gives the annealing time in hours; U 13 is a weld-simulated state. All non-temperated states lie in the dot grid region; L means solution treatment and M60, e.g., 60% martensite and 40% bainite. The vertical line grid represents the plastic deformed region and the index gives the irreversible elongation.

The coercivity H_c in Fig. 5, measured by the magnetic Barkhausen noise or the incremental permeability, increases with the hardness values. If only the hardness or coercivity test is used, the test results are ambiguous. Therefore, one has to recognize the class--tempered, non-tempered or plastic deformed-- which is characterized by different microstructural conditions and Bloch wall arrangements.[11,12]

If the incremental permeability is additionally used to the H_c-test, the results become less ambiguous. The quantity $\Delta\mu_\Delta$ in Fig. 6 measures the broadening of the $\mu_\Delta(H)$-curve for a given μ_Δ-value and is very sensitive to the micro- and macro-stress state. The general information is quite similar to the H_c-results. With increasing annealing time the dislocation density decreases, as well as the arrangement of dislocation changes in dependence upon the lattice defect structure (atomic defects,

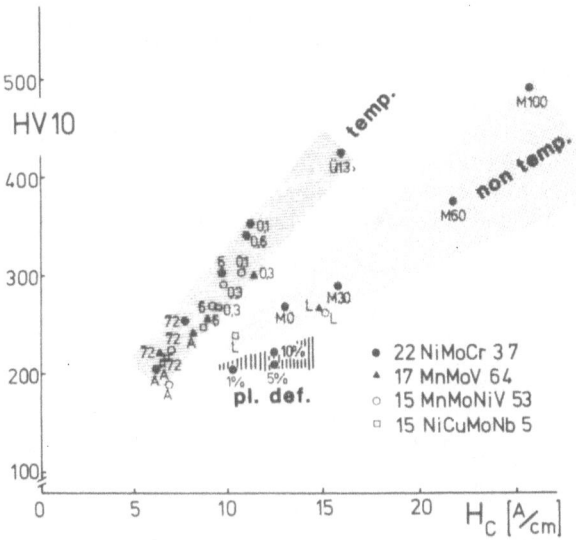

Fig. 5.   Hardness (HV10) and coercivity (H_c) for different
microstructural conditions and steel qualities.

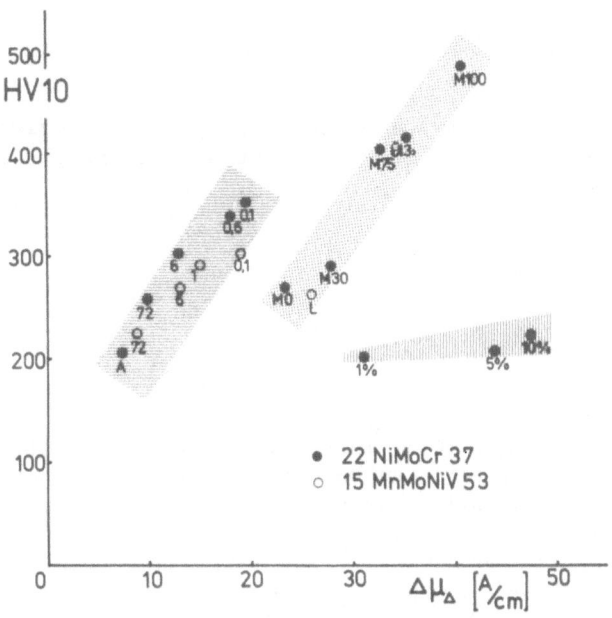

Fig. 6.   Hardness (HV10) and broadening of incremental
permeability ($\Delta\mu_\Delta$) for different microstructural
states and steel qualities.

precipitations). Additionally, the martensitic parts of the
macrostructure, the H_C and $\Delta\mu_\Delta$-values decrease. In the case of
plastically deformed states, one can see that the hardness test is
quite insensitive to separate the different states, whereas the
magnetic nd quantities, H_C, are sufficiently sensitive up to a
range ≤2%. Intermediate states, like the overheated sample U 13,
can lie in one or the other region, whatever the main interaction
between Bloch wall and microstructure components will be.

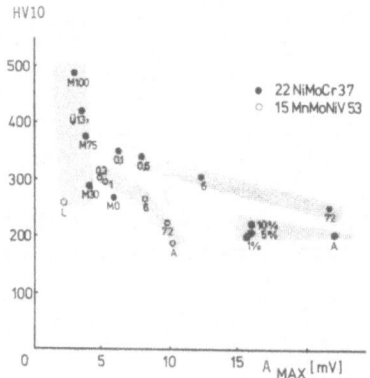

Fig. 7.   Hardness (HV10) and acoustic Barkhausen noise amplitude
          (A_{MAX}) for different microstructural states and steel
          qualities.

A quite different information about the introduced field
of microstructure states can be detected by the acoustic
Barkhausen noise method. The hardest states show the lowest noise
level (Fig. 7). This can be explained by microstresses and
martensite specific behavior. With increasing tempering also the
acoustic noise level increases. The reason for this behavior is
the augmentation in the 90° Bloch wall density and mobility.

If one correlates the yield point $R_{p0.2}$ or the upper shelf of
the notch impact strength (A_V), a direct correlation is obtained,
especially by using stress-sensitive magnetic nd quantities.

Fig. 8.   EMAG-analyzer.

The electromagnetic analyzer (EMAG-Analyzer E2302, Fig. 8) can be used for testing near-surface (≤ 1 mm) microstructure and stress states. The magnetic field strength (± 150 A/cm) and the frequency (20-100 Hz) can be selected. Quantities to be measured are the maximum of the magnetic Barkhausen noise M_{MAX} and the coercivity deduced from the maximum of the M(H)-curve.

First applications by the testing of low-pressure turbine blades, weld constructions and saw blades have been successful.

ACKNOWLEDGEMENT

This contribution is based on work performed with the support of the German Ministry for Research and Technology and the Eureopean Community for Carbon and Steel.

REFERENCES

1. W. Dahl, ed., "Grundlagen des Festigkeits- und Bruchverhaltens," Stahleisen, Düsseldorf (1974).

2. K. Goebbels, in: "Research Techniques in Nondestructive
    Testing, Vol. IV," R.S. Sharpe, ed., Academic Press,
    London (1980).
3. A. B. Ghatia, "Ultrasonic Absorption," Clarendon Press,
    Oxford (1967).
4. K. Goebbels, Materialprüf. 17, 231 (1975).
5. R. Neumann and K. Goebbels, in: "Proceedings of the 4th
    International Conference on Nondestructive Evaluation in
    Nuclear Industry," Deutsche Gesellschaft für Zerstörungs-
    freie Prüfung, Berlin (1981), pp. 337-346.
6. H. Willems and K. Goebbels, Materialprüf. 22, 356 (1980).
7. H. Willems and K. Goebbels, Metal Science 15, 549 (1981).
8. A. Seeger, eds., "Moderne Problems der Metallphysik,"
    Springer, Berlin (1974).
9. W. Theiner, B. Reimringer, H. J. Salzburger, K. Herz,
    D. Kuppler, P. Deimel, Neue Verfahrensansätze für die
    magnetische und magnetoelastische Gefügeprüfung, in:
    "Proceedings of the 4th International Conference on
    Non-Destructive Evaluation in Nuclear Industry," Lindau
    (1981).
10. W. A. Theiner, B. Reimringer, P. Deimel, D. Kuppler,
    Zerstörungsfreie Analyse der Gefüge von Druckbehälter-
    stählen mittels mikromagnetischer Verfahren, in:
    "8. MPA-Seminar," Staatliche Materialprüfungsanstalt,
    Stuttgart (1982).
11. K. Herz, D. Kuppler, P. Deimel, Microstructure and Bloch wall
    arrangement and Barkhausen noise in 22 NiMoCr 3 7 steel,
    J. of Magnetism and Magnetic Materials 24, 206 (1981).
12. P. Deimel, D. Kuppler, K. Herz, W. A. Theiner, Bloch wall
    arrangement and Barkhausen noise in steels 22 NiMoCr 3 7
    and 15 MnMoNiV 53, to be published in: J. of Magnetism and
    Magnetic Materials.

MICROSTRUCTURAL CHARACTERIZATION OF TITANIUM

BY ACOUSTIC MICROSCOPY

Donald E. Yuhas and Michael G. Oravecz

Sonoscan, Inc.
530 E. Green Street
Bensenville, Illinois 60106

ABSTRACT

The dependence of acoustic properties on microstructure can complicate the practice of non-destructive evaluation, but it can also suggest new areas of application. This influence of micro-structure on acoustic characteristics is particularly important when the wavelength approaches the size of the grains. In this investigation, 100 MHz acoustic waves are used to characterize titanium. Images are obtained on several different alloys (Ti 6Al-4V; Ti 3Al-8V-6Cr-4Mo-4Zr; Ti 5Al-2.5Sn), as well as gas contaminated Ti 6Al-4V welds. To augment the image data, average attenuation measurements and quantitative analysis of the image texture, (e.g. contrast) are obtained. Significant variations in average attenuation were found among the various alloys. The predominantly alpha phase Ti 6Al-4V alloy is found to have uniform acoustic images with average attenuation of 2.0 dB per millimeter. In contrast, the Ti 38644, a beta phase alloy, shows average attenuation of 14.9 dB per milli-meter. Variations in acoustic transmission over distances of a few hundred microns ranged from 5 to 28 dB leading to high contrast image texture. Comparison of contaminated and non-contaminated welds in Ti 6Al-4V leads to distinct differences in acoustic propagation at 100 MHz. These variations can be associated with hardness variations measured in the weld, even though gross differentiation on the basis of metallographic analysis is not observed. Contaminated welds were found to exhibit higher attenuation, a more coarse acoustic texture, and higher image contrast, relative to non-contaminated welds.

259

INTRODUCTION

Titanium alloys are finding increased usage as major structural
components. Accompanying this increased utilization, new alloys,
fabrication techniques, and testing methods need to be developed.
The motivation for this paper lies primarily with developing a nondestructive evaluation tool for detection of inferior or defective
welds in titanium.

Titanium is a unique material, with properties quite different
from more commonly used materials like steel. One particular property which becomes extremely important in the welding process is the
affinity of hot titanium for gases such as oxygen and nitrogen. The
presence of these gases during the welding process can cause considerable changes in microstructure as well as altered mechanical properties such as increased hardness and decreased fracture toughness.
As a result of different material properties, the nondestructive
evaluation of welded titanium represents a uniquely different problem
than that which is typically encountered in steel. Much of the nondestructive activity for welded steel is centered on detection of
cracks, slag inclusions, and lack of penetration. In welded titanium, it becomes equally important to detect zones which have become
brittle as the result of gas contamination. Since these zones, crack
precursors, have elastic properties that do not differ as radically
from the base material as do the properties of cracks, they are more
difficult to detect. However, these brittle zones are crack nucleation sites and need to be identified before a component is put into
service.

The detection of brittle zones (crack precursors) represents a
new challenge for nondestructive evaluation methods. Nondestructive
evaluation techniques, based on microstructural characterization, are
needed in order to detect crack precursors.

In this investigation, 100 MHz ultrasonic images were obtained
of different titanium microstructures. New methods for the quantitative analysis of these images, in terms of acoustic image texture
and average attenuation, were developed. The characterization
methods were applied to several different alloys as well as titanium
welds. The results show that microstructural variations can be
detected nondestructively. This represents an important initial
step in the nondestructive evaluation of welded titanium.

EXPERIMENT

A.   Specimens

Samples imaged in this investigation consist of a series of three
distinctly different titanium alloys as well as welded sections of
Titanium 6Al-4V.

The three alloys are Ti 3Al-8V-6Cr-4Mo-4Zr, Ti 6Al-4V, and Ti 5Al-2.5Sn. This choice of alloys represents a wide range in structure both in grain size, grain shape, and crystalline structure. Titanium, being allotropic, can exist in two crystalline forms, an alpha phase which is hexagonal close-packed, and a beta phase which is body centered cubic. The Ti 38644 is 100% beta phase while the Ti 5-2.5 is 100% alpha. The Ti 6-4 material contains less than 10% beta.

One hundred megahertz transmission acoustic images were obtained on the samples in the "as received" condition. The surface texture was similar in all three although the thickness varied, ranging from 1.3 mm for the Ti 38644 to 3 mm for both the Ti 6-4 and Ti 5-2.5 materials. Small sections of each sample were taken from the 2-inch square test coupons and subjected to standard optical metallographic analysis. Samples were polished and then etched with a 20% HF, 20% HNO_3 and 60% glycerol solution.

The metallography of Ti 38644 shows a largely beta titanium structure with etch pits that signify precipitation (aging) has initiated. The elongated grains of beta titanium suggest that the hot-rolled plate was subsequently cold rolled, 50%, and solution treated at a temperature at the lower extremity of the recrystallization range. The aging developed during cooling from the solution anneal temperature.

The Ti 5-2.5 microstructure shows a predominantly alpha titanium structure. The alloy was cold rolled after hot rolling and pickling. The cold rolling exceeds 70%. The plate was then annealed to give a partially recrystallized structure. The structure is dominated by elongated, ribbon-like crystals with a few interspersed equi-axed crystals.

The microstructure of the Ti 6-4 shows a microstructure which indicates an annealed product which is more than 90% alpha titanium with the rest in particles and stringers of residual beta. The divided aspect of the beta phase comes from the growth of the alpha-Ti as intersecting platelets ultimately isolating the residual beta phase.

Grain size estimates are difficult due to the wide variations in the shapes of the crystals found in the various alloys. In the moot crude sense, the Ti 38644 can be characterized as a coarse-grain in comparison to the Ti 6-4 and the Ti 5-2.5. Crystals in the Ti 38644 exceeded 50 microns, whereas the grains in the other two alloys were less than 15 microns.

Metallographic analysis was also performed on the welded Ti 6-4. The sample was a single piece 6.3 mm thick and 13 cm long, containing two welds. One of the welds, when viewed optically,

showed a dull grey appearance indicative of gas contamination. The
other weld was shiny in appearance with color similar to the base
metal.

Acoustic images were obtained at the center of the welded zone
by transmitting sound through the sample, perpendicular to the axes
of the weld. Metallography was done on transverse sections of the
weld. Results on the base metal, far removed from the welded zone,
are similar to those described earlier for the Ti 6-4. For both
welds, the fused zone penetrated to approximately 50% of the sample
thickness, and can be divided into two distinct zones, the "weld"
zone and a surrounding halo-like "haze" region. The weld zone corresponds to the material which has been melted and recrystallized
while the haze region corresponds to the heat affected zone. Comparing the microstructure of the contaminated and non-contaminated
welds, no gross differences were observed.

Rockwell hardness measurements were made at 1.0 mm intervals
along a line in the transverse section. These measurements extended
from the weld zone through the haze region and into the base metal.
In the non-contaminated weld, the weld, haze and base metal were
found to have the same Rockwell hardness values of 34.3 ±1. For the
contaminated weld, the average values in the haze plus base metal
region were found to be 33.9 ±1 with values in the central portion
of the weld in excess of 41.7.

B.   Equipment and Methods

The acoustic microscope used in this investigation is the Scanning Laser Acoustic Microscope. The key attribute of the SLAM technology is the production of real-time, high resolution acoustic
images. Both the speed and resolution arise from the focused, scanning laser beam used as a detector of the sound waves. A description
of the basic principles of operation has been presented previously[1].

Two acoustic imaging modes are commonly employed in the SLAM
technique, the acoustic amplitude mode and the acoustic interfero-
gram. The amplitude image shows the relative transmission level over
the field of view. In the amplitude mode, bright areas correspond to
zones in the sample with good acoustic transmission, whereas the
darker areas are more attenuating. In this study, quantitative
characterization data is derived from aplitude (attenuation) data
available with the SLAM. Use is made of the average attenuation as
well as structural information (variations in attenuation) available
from the SLAM images.

The imaging geometry and typical data are illustrated schematically in Figure 1. A sample is placed on the stage and insonified
with a plane wave. Sound is scattered, refracted and attenuated by

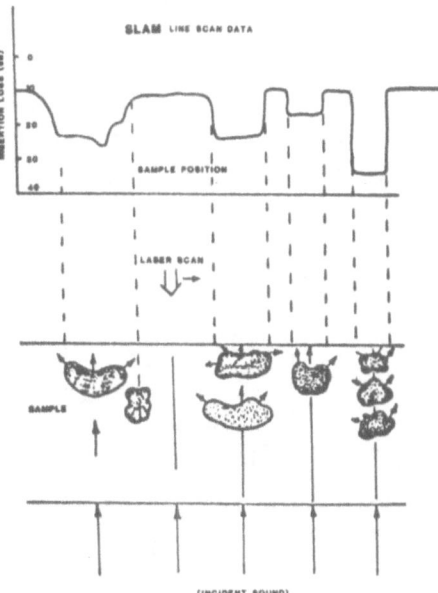

Fig. 1.   Schematic showing insonified sample and corresponding line
          scan data which has insertion loss plotted as a function
          of position.

the grain structure. The amplitude of the through transmitted sound
is detected by the scanning laser beam. The full raster scan of the
laser is used to display the image data in real-time (30 frames/sec).
The quantitative data is obtained by recording amplitude information
(peak detection) along a line. Average attenuation values are
obtained by either numerically averaging the line scan data or by
adjusting the gating circuitry in Figure 2 to integrate or average
over a predetermined window size. It is useful to view several
fields (2.3 x 3.0 millimeters) prior to selecting representative
zones for quantitative analysis. This insures that data is not
recorded in the vicinity of large flaws, e.g. cracks, and allows
identification of features which may arise from surface texture.

     Although the images provide a useful guide to selecting the
data, signal conditioning (e.g. contrast, brightness) and non-
linearity in the display and recording medium (film) make quantitative analysis from the micrographs difficult. Figure 2 shows
that the signals carrying the acoustic amplitude information are
taken directly from the receiver, gated, and then averaged or peak
detected. Thus, all quantitative data is recorded electronically
and not obtained from the micrographs.

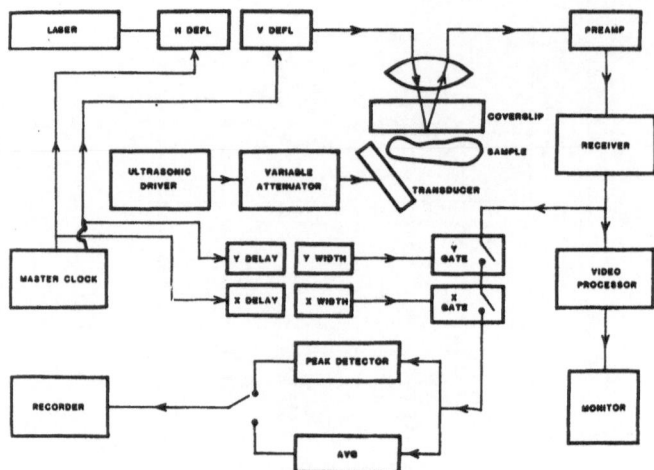

Fig. 2. Block diagram showing the gating scheme and variable
         attenuator required for quantitative line scan analysis.

     To convert the electronic signal levels from the receiver into
meaningful acoustic insertion loss data, a calibration procedure is
required. The relationship between the output voltage at the receiver and the acoustic amplitude impinging on the coverslip is
obtained using the variable attenuator shown in Figure 2. By monitoring the magnitude of the detected signal as the power to the
transducer is decreased (input attenuation), the relationship between detected voltage and acoustic amplitude at the detector plane
is established. Measurements show that the detected signal is linear
over a range of at least 30 dB.

     Two additional calibration procedures are required; one which
establishes a reference insertion loss level and another which
measures uniformity over the field of view. Reference level is
obtained by a substitution technique. The unknown sample is placed
between the transducer and the coverslip and the signal level at the
receiver recorded. The sample is then removed and the variable
attenuator adjusted to match the signal level found previously with
the sample in place. The amount of attenuation is equal to the
insertion loss of the sample relative to that of an equivalent thickness of water. In the strictest sense, the insertion loss values
measured in this way are relative and not absolute. However, the
attenuation in water is known (2.2 dB/mm)^2 and thus absolute attenuation values can be derived. Reproducible results with variation
of ±1.0 dB can be obtained with the substitution technique.
     Because one of the important characterization parameters in-
volves quantifying texture and contrast in the acoustic image, it

is important to establish a measure of acoustic field uniformity. This is most easily done empirically by producing line scan data on an acoustically homogeneous target. Figure 3 shows a line scan obtained on a 1.0 mm thick piece of crown glass. In an ideal system, the crown glass will produce a flat horizontal line. The deviation from flatness observed in this data provides a measure of the typical variations which are characteristic of our SLAM system and data recording methods. These variations will be superimposed on our characterization data.

The SLAM technique offers several advantages for materials characterization studies. First rapid, real-time full grey scale images of an object can be obtained. Thus, one can insure that areas selected for analysis do not contain specific flaws such as cracks. Features arising from surface features such as folds in the weld crown can be identified and eliminated from the analysis. In addition, large volumes of material can be imaged, thus assuring that the regions chosen for analysis are representative of the entire weld. Secondly, as applied here, the images are obtained in transmission using plane wave insonification; thus measurements of acoustic losses are meaningful in terms of acoustic attenuation and scattering. The information is much easier to interpret and quantify than the standard back reflection methods. Thirdly, the laser detection scheme approximates the idealized point receiver (23 microns); thus insuring that false readings arising from phase cancellation artifacts will not be present in the data. Lastly, high frequency, 100 MHz, insures increased sensitivity to microstructural changes.

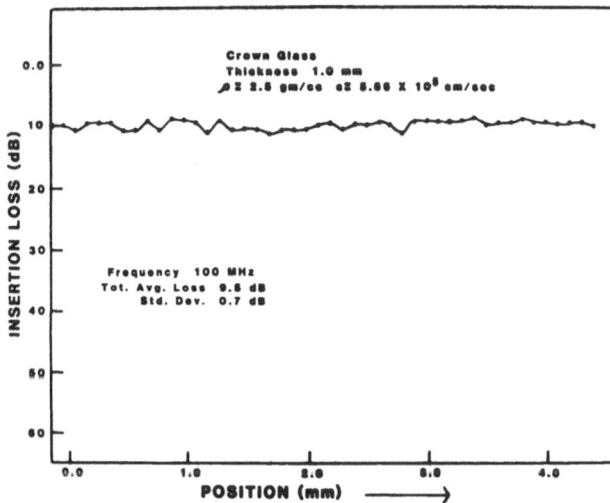

Fig. 3. Acoustic Line scan data obtained on an acoustically uniform glass sample. Curve empirically defines the acoustic field uniformity.

DATA PRESENTATION

A primary goal of this work is to explore the use of high frequency ultrasonic imaging methods as a means of characterizing welded
titanium. The basis of the method is detection of microstructural
changes related to the level of gas contamination in welded titanium.
As an aid to elucidating the nature of these changes, as well as
demonstrating the sensitivity of high frequency ultrasonic methods,
it is of interest to first explore several different alloys.

The 100 MHz acoustic micrographs obtained on two different titanium alloys, titanium 38644 and titanium 6-4, are compared in Figure 4. Qualitatively, the two alloys are distinctly different. The
titanium 6-4 shows uniform acoustic transmission, while the titanium
38644 has a highly structured appearance. Recall that the titanium
38644 has a primarily beta phase crystal structure, while the titanium 6-4 is primarily alpha phase. The grain size is also larger in
the titanium 38644.

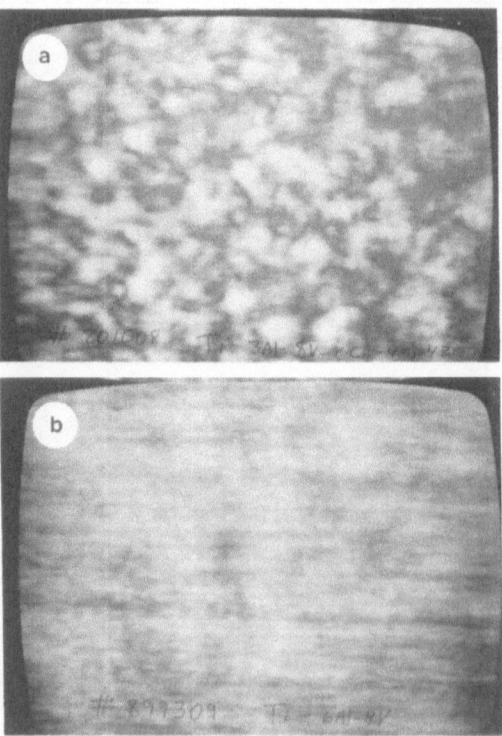

Fig. 4. Comparison of 100 MHz acoustic micrographs obtained on
         two titanium alloys, a) Ti 38644, and b) Ti 6-4.

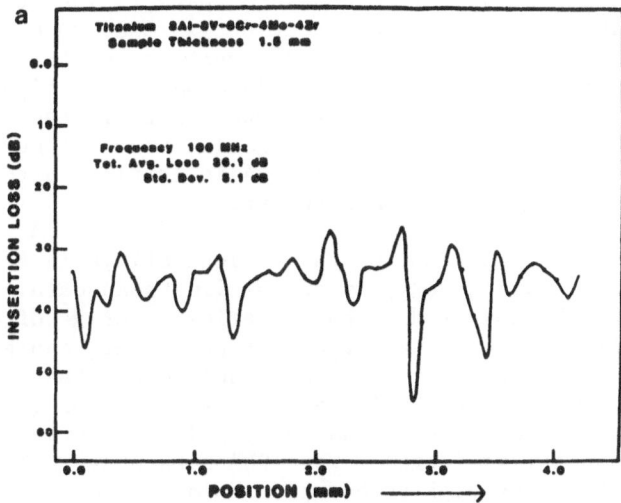

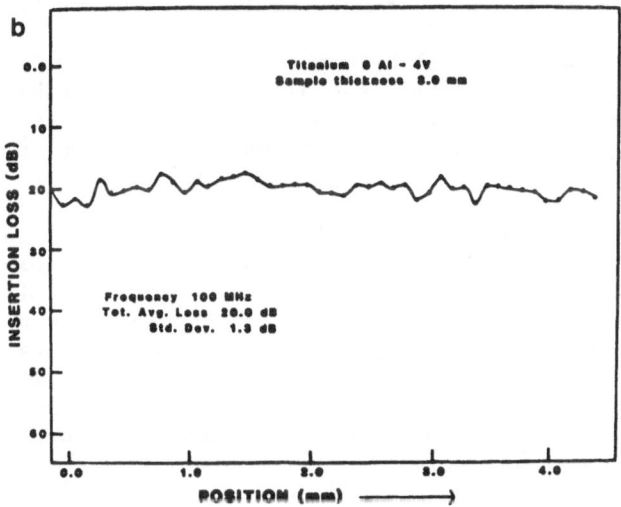

Fig. 5.   Comparison of acoustic line scan data obtained on two
           titanium alloys, a) Ti 38644, and b) Ti 6-4.

Figure 5 shows representative acoustic line scan data obtained on the two alloys depicted in Figure 4. The line scan data serves to quantify the grey levels on the micrographs. The 0.0 decibel point is defined as the signal level obtained if the sample is replaced with an equivalent thickness of water. The substitution method described in the previous section was used to obtain this value. The line scan data is used to quantify the differences in image structure which are qualitatively apparent in Figure 4. Clearly, there are substantial differences both in the average attenuation as well as the texture (deviation from the mean). The uniform acoustic transmission of Ti 6-4 is attributed to the finer grained, primarily alpha phase structure.

Quantitative comparison of the various alloys is most easily done by a more detailed analysis of the line scan data. The average attenuation can be computed by the numerical average of several lines. To quantify image texture, a variety of parameters can be used as a measure of image contrast. We have chosen to use the standard deviation. Figure 6 shows the (a) average attenuation (corrected for thickness and impedence loss) and the (b) image contrast for the three different alloys imaged. The 100% beta phase alloy distinguishes itself by its high attenuation relative to that of the two high alpha phase alloys. It is interesting to note that the differences are quite large. For example, a 4 mm thick sample of Ti 38644 would show 50 dB greater attenuation than a similar sample of Ti 6-4 at a frequency of 100 MHz. Figure 6b illustrates differentiation of the alloys on the basis of their contrast values. Clear separation between high alpha and high beta phase alloys is found. In addition, the two predominantly alpha phase alloys show measurably different contrast. Thus, it appears the image contrast as measured by the standard deviation represents a somewhat more sensitive monitor of microstructure.

Quantitative data comparing 100 MHz transmission characteristics of welded titanium 6-4 is shown in Figure 7. Attenuation and image contrast values are given for base metal, a non-contaminated weld and a contaminated weld. For the welded samples, the analysis is identical to that used for the alloy comparison (Figure 6). The acoustic data of the welded zones was taken along a line parallel to the weld axis, located at the weld center. Figure 7 shows that either attenuation or contrast can be used to distinguish contaminated from non-contaminated welds.

In addition to the quantitative acoustic data presented in Figure 7, Rockwell hardness measurements, 100 MHz acoustic micrographs, and optical micrographs of etched transverse sections of the weld were obtained. The acoustic micrographs could be used to qualitatively distinguish the "good" from the "bad" welds. Furthermore, qualitative differences between the "good" weld and the base metal are apparent. This distinction is not obvious in the attenuation and contrast value measurements presented in Figure 7. More complete analysis of the acoustic image texture must be done in order to enhance these differences. Metallurgical analysis of the etched transverse sections did not show any gross differences in the microstructure of the two welds. However, significant variations in the hardness values at the center of the weld were found. Values of Rc of 41.7 were found in the center of the contaminated weld, while the lower values of Rc 34 were recorded in the non-contaminated weld.

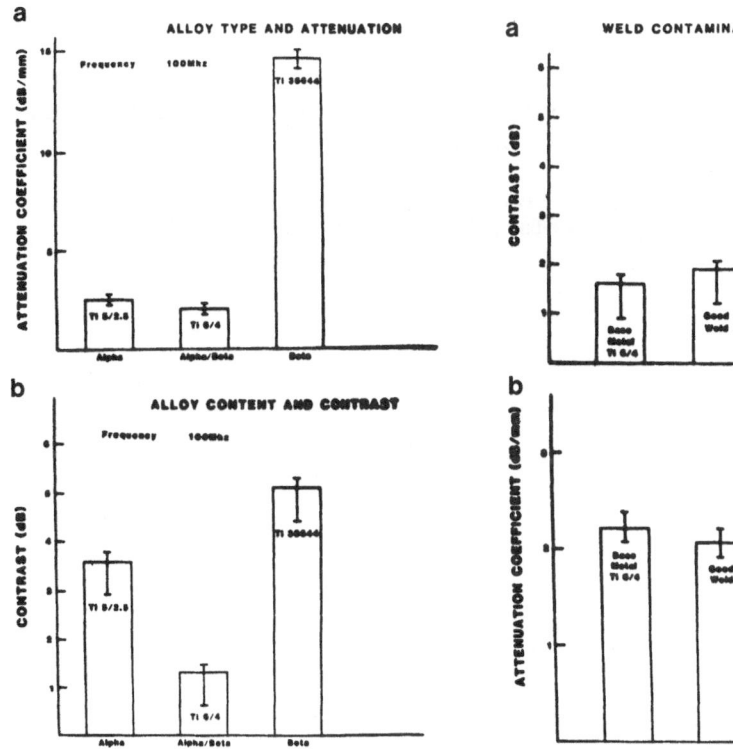

Fig. 6. Comparison of a) acoustic attenuation, and b) image contrast for three different alloy types. Substantial differences appear to be associated with the type of crystalline phases present.

Fig. 7. Comparison of acoustic (a) contrast, and (b) attenuation for base metal, contaminated (bad), non-contaminated (good) welds.

CONCLUSION

Analysis of high-frequency acoustic images can be used as a
sensitive monitor of microstructural variations. Results presented
here show large variations in average attenuation and image texture
in different alloys. Particularly noticeable is the increased attenuation associated with the high beta phase alloy. Differentiation
of gas contaminated and non-contaminated welds on the basis of
acoustic transmission characteristics has been achieved. The "good"
and "bad" welds are easily distinguished even in the absence of
gross microstructural differences as observed by conventional
metallography. More work on different alloys as well as welded
components of varying degrees of contamination is required to obtain
better understanding of the origin of the alteration in acoustic
properties.

REFERENCES

1. L.W. Kessler and D.E. Yuhas, Acoustic Microscopy 1979, Proc.
    IEEE, 67 (4) pp. 526-536 (1979).

2. J. Krautkramer and H. Krautkramer, Ultrasonic Testing of
    Materials, Springer-Verlag, New York, p. 242 (1977).

ULTRASONIC CHARACTERIZATION OF TITANIUM 6211 WELDMENTS

Sanford R. Buxbaum and Robert E. Green, Jr.

Materials Science and Engineering Department
The Johns Hopkins University
Baltimore, Maryland 21218

ABSTRACT

The use of ultrasonic testing for the inspection of thick section weldments in titanium alloys requires careful acoustic characterization of the materials and weldments in question. In the present work ultrasonic wave velocity measurements were performed on a series of test specimens of titanium alloy weldments and base plate material. Data provided by x-ray diffraction analysis and scanning electron microscopy were correlated with the ultrasonic results.

Dissolved oxygen contamination during GTA and GMA welding of titanium alloys can result in severe embrittlement of the weld region. In order to evaluate the feasibility of ultrasonic testing for quantitatively detecting the presence of interstitial gas contamination in weldments of Ti-6211, ultrasonic wave velocity and ultrasonic attenuation measurements were performed on a series of five specimens with nominal oxygen levels of 0.07, 0.14, 0.20, 0.24 and 0.29 percent by weight. Density measurements, in addition to the ultrasonic wave velocity data, enabled relative determination of elastic moduli. Variations in the ultrasonic data were correlated with results from scanning electron microscopy and hardness testing.

INTRODUCTION

Titanium alloys exhibit high strength-to-weight ratios and good corrosion resistance and are, therefore, desirable for use in structural applications. The safe, in-service use of titanium

271

alloy weldments in critical, load-bearing parts requires the development of reliable nondestructive methods for inspecting the mechanical integrity of the weld region. Unfortunately, current nondestructive evaluation technology has proven inadequate for such alloys and applications. Since one of the primary non-destructive inspection techniques is ultrasonics, it is expedient to determine the usefulness and limitations of ultrasonic inspection for the materials in question. This requires careful acoustical characterization of these materials in order that appropriate accept/reject criteria be established. Hence, the first stage of this research was undertaken with the following objectives:

(1) Acoustically characterize weldments of Ti-6211 plate through ultrasonic wave velocity measurements.
(2) Calculate elastic moduli for base plate and weld metal from measured ultrasonic wave velocities and measured densities.
(3) Correlate ultrasonic wave velocity measurements with microstructural variations as observed through metallographic analysis and x-ray diffraction analysis.

The mechanical properties of titanium alloys are known to be sensitive to oxygen content. Above a certain threshold concentration oxygen appears to have a severe embrittling effect on titanium alloys. Since gas tungsten-arc (GTA) and gas metal-arc (GMA) welding techniques are susceptible to dissolved gas contamination, it is of specific interest to investigate methods for nondestructively determining the oxygen content of a Ti-6211 welded joint. The objective of the second stage of this research was to evaluate the feasibility of ultrasonic testing for detecting quantitatively the presence of interstitial gas contamination in weldments of Ti-6211 by:

(1) Performing ultrasonic wave velocity and ultrasonic attenuation measurements on a series of five specimens with varying oxygen contents.
(2) Comparing variations in ultrasonic data with results of scanning electron microscopy and hardness testing.

BACKGROUND

Pure titanium has a hexagonal close-packed structure (alpha phase) at room temperature, transforms to a body-centered cubic structure (beta phase) at 883°C, and melts at 1668°C. In this investigation the alloy used was titanium 6211, which has a nominal composition of 6 weight percent aluminum, 2 weight percent columbium, 1 weight percent molybdenum, and 1 weight percent tantalum with the balance ideally being pure titanium. This alloy of structural importance is classed as a near alpha alloy, because it contains primarily the alpha phase with small amounts of beta phase material interspersed.

The welded specimens evaluated in this work were joined either by a gas tungsten-arc (GTA) or a gas metal-arc (GMA) process. Gas tungsten-arc welding is a fusion welding process in which metals are joined by melting them with an electric arc which is generated between a nonconsumable tungsten electrode and the sections of metal to be welded. An inert gas or gas mixture shields the arc and the weld puddle during the welding process. Pressure may or may not be applied to the joint and filler metal may or may not be added. Plate section GTA welds in titanium and titanium alloys typically require the addition of filler metal. In gas metal-arc welding the metals are joined by melting them with an electric arc between a continuous, consumable filler metal and the sections to be welded. As in GTA welding, an inert gas or gas mixture shields the arc and the weld puddle during the welding process.[1] In both GMA and GTA welding the filler metal used was Ti-6211. Contaminants such as oxygen, nitrogen, hydrogen, and carbon can be introduced into the weld by incomplete gas shielding and/or slight contaminant gas impurities in the inert cover gas mixture.

Ultrasonic wave propagation analysis of a material is a nondestructive evaluation technique that provides information about the elastic properties and absorption characteristics of the material in which the wave propagates. Absolute measurements of shear and longitudinal wave speed can be used to calculate useful material parameters, such as the effective Young's modulus and the effective shear modulus. The term "effective modulus" refers to the fact that the modulus is calculated for a material that is assumed to be linear elastic, homogeneous, and isotropic, which is a fair approximation for some fine-grained polycrystalline materials. The energy loss, or ultrasonic attenuation of an elastic wave propagating in a solid, may be divided into contributions from geometrical and intrinsic effects. Of interest here are the intrinsic effects which include scattering of the ultrasonic wave by inhomogeneities, conversion of sound energy to heat as a result of elastic deformation, interaction with thermal phonons, and dislocation damping.[2]

Since the concentration of an alloying element, oxygen in this case, affects the same set of variables that determine ultrasonic wave propagation characteristics, it was expected that ultrasonic techniques would be suitable for determining oxygen

concentration in contaminated Ti-6211 specimens. It has been
shown that foreign solute atoms invariably change elastic
moduli.[3-6] Additionally, solute atoms can act as pinning points
for dislocations and would, therefore, be expected to influence
attenuation measurements. Some previous studies of the effects
of interstitial gases on elastic wave propagation also indicate
the feasibility of this method for determination of dissolved gas
concentration. Hsu and Conrad[6] looked at the effect of oxygen on
the elastic properties of titanium-oxygen alloys through ultrasonic
wave velocity measurements performed on specimens with oxygen
content in the range of 0.04 weight percent to 2 weight percent
and controlled impurities. It was found that both density and
longitudinal wave velocity increased with increasing amount of
oxygen. Since the relative change in the wave velocity was greater
than that in the density, the dynamic elastic modulus was increased
by the oxygen in solid solution. Ultrasonic attenuation was also
observed to increase with increasing oxygen content. The authors
concluded that it was likely that the interstitial oxygen atoms in
the titanium lattice actually increased the binding forces between
atoms.

EXPERIMENTAL PROCEDURES

     The material used in the present investigation, Ti-6211, was
examined using various destructive and nondestructive techniques.
These techniques included scanning electron microscopy, ultrasonic
wave velocity and attenuation measurements, density measurements,
Rockwell "C" hardness measurements and x-ray diffraction analysis.

Metallography

     Metallographic specimens were taken from sections cut from
the weld region, the base metal, and from the oxygen contaminated
Ti samples. The specimens were ground on successively finer
silicon carbide papers down to 600 grit, and then polished on
lapidary wheels using 15 micron and 0.05 micron compounds. A
two-step etching technique was used to reveal microscopic detail.
In the first step, the polished specimen was briefly swabbed with
a 2 ml HF, 98 ml water solution. This etched the alloy and
stained the alpha phase. The second step, swabbing with a 1 ml HF,
2 ml HNO_3, 97 ml water solution, removed the stain, leaving a dark
field of alpha phase material in which the beta phase appeared as
finely dispersed white lines when viewed through a scanning
electron microscope (SEM).

Ultrasonic Measurements

     The specimens used for ultrasonic examination were fabricated

from welded specimens and oxygen contaminated specimens provided
by the Naval Research Laboratory in Washington, D.C.   These
plate-section weldments required many welding passes and were not
suitable for ultrasonic inspection with contact transducers in
the as-received condition due to surface irregularities resulting
from welding.   The oxygen contaminated specimens were also
unsuitable in the as-received condition for contact ultrasonic
inspection due to poor surface finish.   Flat parallel faces were
machined on each specimen prior to performing both ultrasonic
measurements and hardness measurements.   A high degree of
parallelism of specimen faces was required to minimize diffraction
errors caused by divergence of the ultrasonic pulse as it
propagated back and forth through the specimen.

The ultrasonic wave velocity and attenuation measurement
system used in the present research is shown schematically in
Fig. 1.   Conventional pulse-echo overlap techniques as described
by Chung, Silversmith, and Chick[7] were used to measure the
ultrasonic wave velocities.   Attenuation measurements were made
with an automatic attenuation recorder, which includes a time gate
permitting selection of any two echoes from the received wave
train, an automatic gain control to stabilize the amplitude of the
first echo, and circuitry to obtain the logarithm of the ratio of
two selected echo amplitudes and display the result in deciBels.

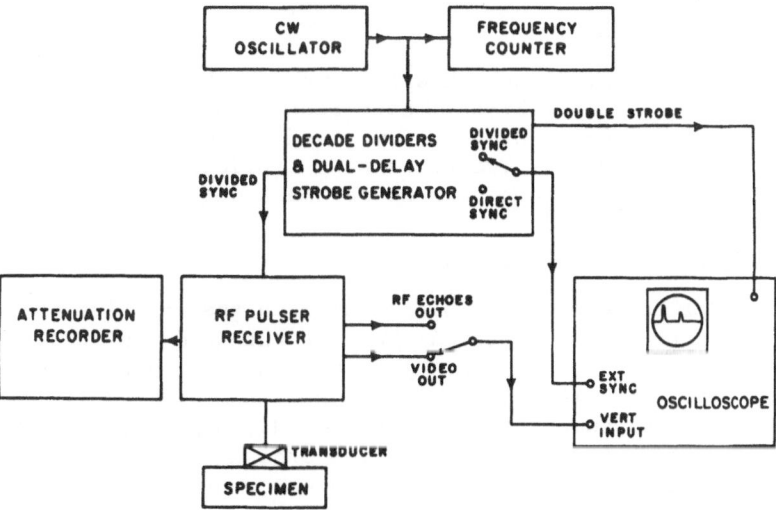

Fig. 1.   Ultrasonic wave velocity and attenuation measuring system.

Commercial, longitudinal wave, ceramic transducers along with appropriate couplants were used in this research. An Aerotech couplant (a light oil) was used for the wave velocity measurements. A 2 mm thick elastomer coating (developed by Martin Marietta Laboratories in Catonsville, MD) was applied to a 2.25 MHz, longitudinal wave, commercial transducer and was used to acoustically couple the transducer to the oxygen contaiminated specimens for the attenuation measurements. The viscoelastic properties of this coating make it an ideal couplant for attenuation measurements. When subjected to low frequency stress cycles (low rates of deformation), the elastomer is flexible and can be forced down on a sample for a reproducible acoustical couple. When subjected to high frequency stress cycles (high rates of deformation) the coating behaves like a rigid body. The ultrasonic pulses effectively propagated through a rigid acoustical coupling media. This elastomer coating provided more reproducible attenuation measurements than was possible with conventional couplants. The thickness of the coating made it unsuitable for accurate velocity measurements.

Hardness Measurements

Rockwell "C" hardness measurements were made on the oxygen contaminated Ti-6211 specimens in accordance with ASTM Standard E18. A diamond tipped "Brale" indentor with a 150 kg load was used. Since Ti-6211 continued to exhibit plastic flow after the application of the major load, the dial indicator continued to move after the operating lever stopped. For this reason the operating lever was brought to its latched position at an elapsed time of 30 seconds between application and removal of load.

Density Measurements

Cubes nominally 1 inch on a side were machined from each of the oxygen contaminated Ti-6211 specimens. The volume was measured to an accuracy of 0.001 cm^3, and the mass was measured to an accuracy of 0.001 g. Two cubes were fabricated from each of the five oxygen contaminated specimens for mass density determination.

X-Ray Diffraction Analysis

In order to investigate differences in the relative amounts of alpha and beta phase material between the weld region and the base metal and changes in the relative amounts of alpha and beta phase material with increasing oxygen content in Ti-6211, x-ray diffractometer measurements were performed. A proportional counter was used to record intensities and angular positions of the diffracted beam filtered at the detector by a thin foil of nickel. The x-ray tube had a copper target (Cu K_α radiation, λ = 1.54 Å) and was generally operated at 44.5 kV and 33 mA. The diffractometer was equipped with a beam slit of 3° MR, a soller slit of medium resolution, and a detector slit of 0.2°. This produced a rectangular beam of primary radiation. The target angle was 4° with a scan rate of 4° per minute. Full scale for the detector was 5000 counts per second (100 cps = 1 milliroentgen/hr). The samples were initially scanned through 2θ angles of 10° to 100°; after appropriate peaks were chosen for closer examination, the samples were only scanned through angles of 33° to 45°.

RESULTS AND DISCUSSION

Metallography

Metallographic analysis revealed significant microstructural differences between the base metal and the weld region in the Ti-6211 alloy. At low magnifications the microstructures observed in the base plate metal are broken up and show texturing due to rolling. Prior beta phase grain boundaries are decorated by the alpha phase, because the grain boundaries are the first regions to tranform during cooling at higher temperatures. Narrow platelet formations characterize the microstructure of the alpha phase; the remaining beta phase is interspersed between the acicular alpha platelets.

The weld region exhibits a very different microstructure from the base plate metal. Acicular alpha and alpha prime (martensitic-type structure) has formed on preferred planes in the prior beta crystals. The alpha platelets are narrower, straighter, and more sharply defined in the weld region than in the base metal. This is probably a result of the faster cooling rates present during welding. The boundaries of the large, equiaxed prior beta grains are decorated by alpha metal as in the base plate. Figure 2 illustrates the microstructural differences between weld metal and base metal.

Metallographic analysis performed on the five oxygen contaminated specimens showed distinct changes in microstructure with increasing oxygen content (Fig. 3). As the oxygen content increased, the alpha platelets became wider; hence, the separation of the layers of beta increased. Additionally, the beta layers thickened until they began to break up and become more globular in appearance. This is especially evident in the SEM photomicrograph of specimen E, the most heavily contaminated specimen. It should be noted that the microstructures of the oxygen contaminated plates are not representative of those observed in the weld region of welded Ti-6211 specimens.

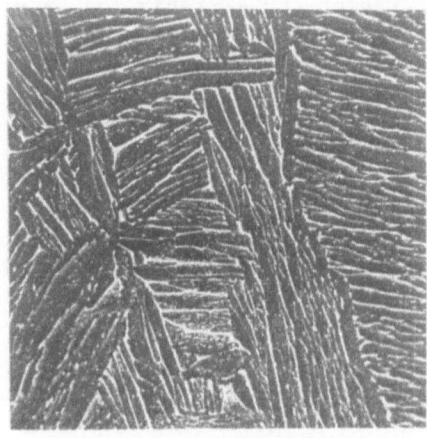 

Base Metal   Weld Region
Mag = 1000X   Mag = 1000X
────── 10 μm   ────── 10 μm

Fig. 2.   Comparison of microstructures of base metal and weld
          region.

The oxygen was introduced into the molten Ti-6211 samples in
the form of TiO_2 powder, and the samples were subsequently beta
processed (heated and rolled in the beta phase field at 925°C in a
controlled, inert atmosphere).   The samples were then air-cooled.
Chemical analysis of the specimens was performed by the RMI Company
of Niles, Ohio, producers of the plate samples.   The results of
this analysis are presented in Table I.

Density Measurements

The results of density measurements performed on this series
of five oxygen contaminated specimens are presented in Fig. 4.
Note that the mass density (ρ) increases with increasing oxygen
content, because the oxygen enters the titanium lattice
interstitially.   These results are consistent with the results of
Hsu and Conrad.[6]

Ultrasonic Wave Velocity Measurements

Longitudinal wave velocity was measured at various locations
along several welded specimens.   Data for such a scan are shown in
Fig. 5.   The sound wave travel path corresponds to the vertical
path up and down the paper.   Wave velocity was lowest in the center
of the weld where the ultrasound primarily passed through weld
metal.   The velocity difference from weld metal to base metal was
about one percent.   Peaks in the velocity versus distance curve

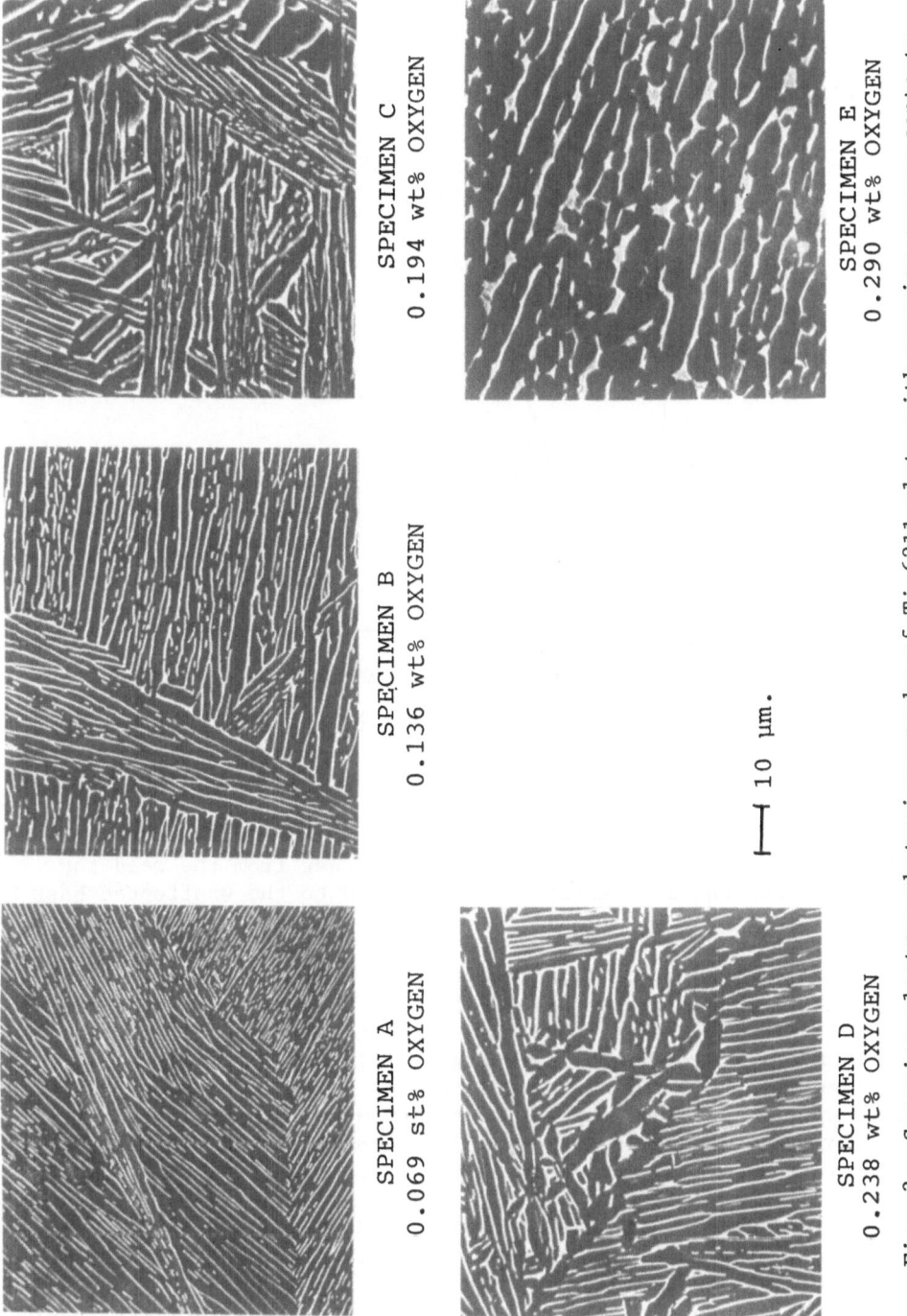

SPECIMEN A
0.069 st% OXYGEN

SPECIMEN B
0.136 wt% OXYGEN

SPECIMEN C
0.194 wt% OXYGEN

SPECIMEN D
0.238 wt% OXYGEN

SPECIMEN E
0.290 wt% OXYGEN

10 μm.

Fig. 3. Scanning electron photomicrographs of Ti-6211 plate with varying oxygen contents.

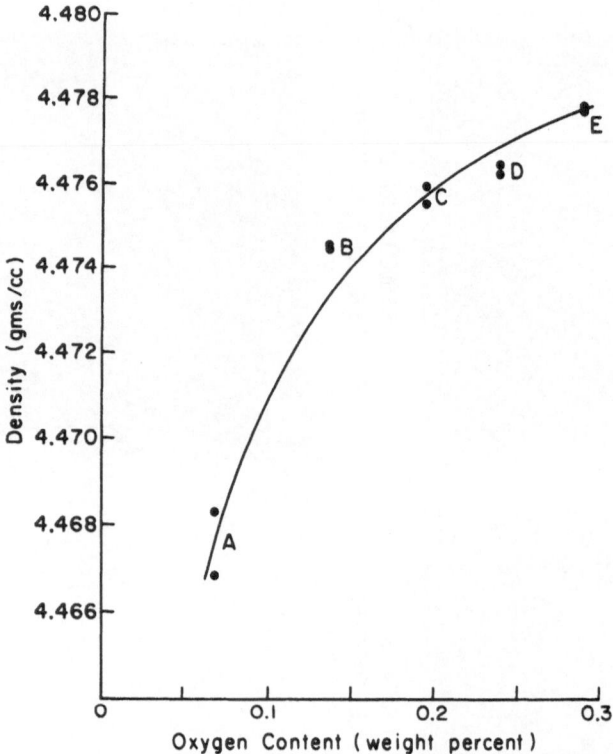

Fig. 4.   Mass density vs. oxygen content.

occurred just outside the weld zone. Further from the weld the
velocity decreased slightly as it returned to the unaffected base
plate value.

Longitudinal wave velocity (V_L) was found to decrease with
increasing oxygen content (see Fig. 6). Specimen C gave anomalous
velocity results; it is suspected that this was due to improper
thermomechanical preparation. Figure 7 is a plot of of the
average product $\rho V_L{}^2$ (normalized) versus oxygen content for the
five specimens. The behavior of the quantity $\rho V_L{}^2$ is indicative
of the behavior of the elastic moduli, so Fig. 7 shows that the
elastic moduli tend to decrease with increasing oxygen content.

Discussion of the observation of a decrease in the
longitudinal wave velocity requires consideration of the material
properties that affect V_L in both the oxygen contaminated Ti-6211
specimens and the welded specimens. In 1949 Zener[3] theoretically
demonstrated that the residual stress introduced by slight

Table I. Specimen Composition

SPECIMEN	COMPOSITION (weight percent)									
	Ti	Al	Cb	Mo	Ta	C	N	Fe	O
A	90.3	6.0	1.95	0.7	0.88	0.02	0.010	0.05	0.069
B	89.9	6.0	2.09	0.9	0.97	0.02	0.006	0.03	0.136
C	90.2	5.9	1.90	0.8	0.94	0.02	0.005	0.03	0.194
D	90.1	5.8	2.05	0.8	0.99	0.02	0.006	0.03	0.238
E	89.8	5.9	2.16	0.7	1.06	0.03	0.008	0.03	0.290

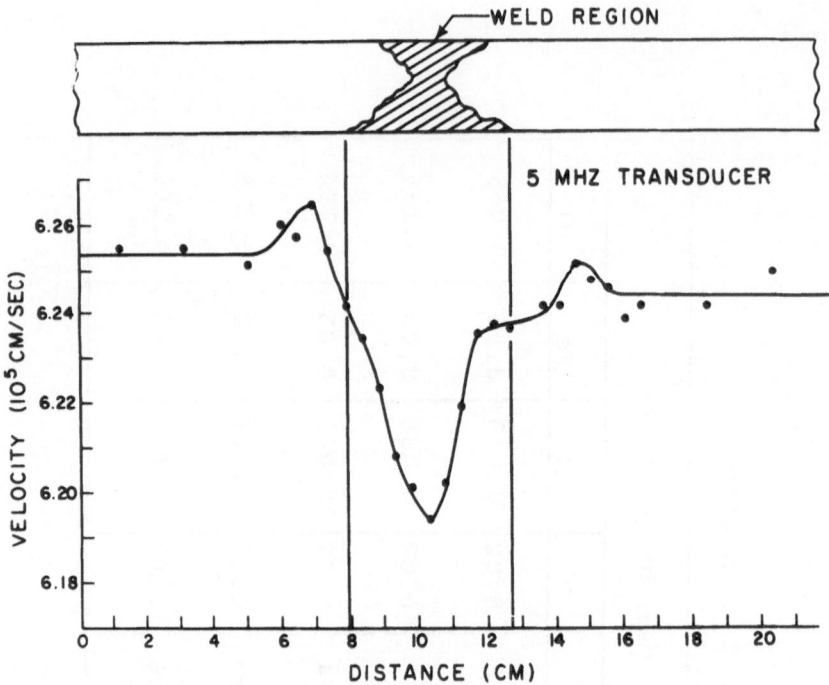

Fig. 5. Longitudinal wave velocity vs. transducer location along
         a welded Ti-6211 specimen.

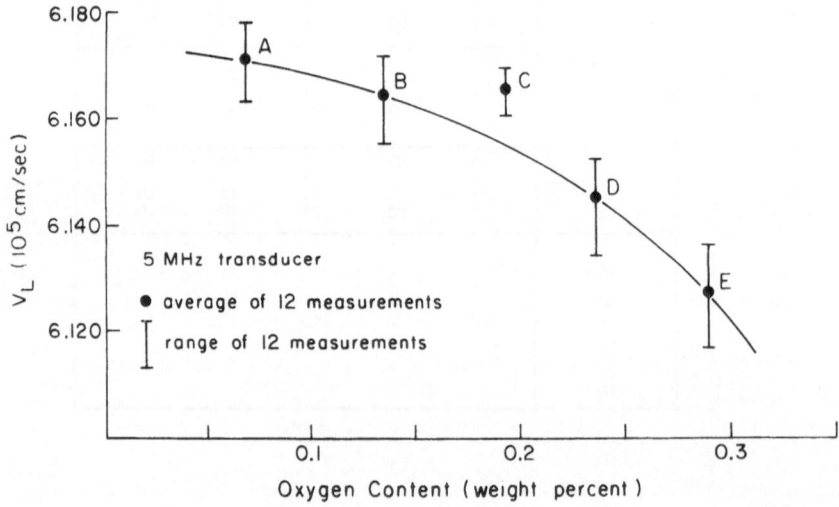

Fig. 6. Longitudinal wave velocity vs. oxygen content.

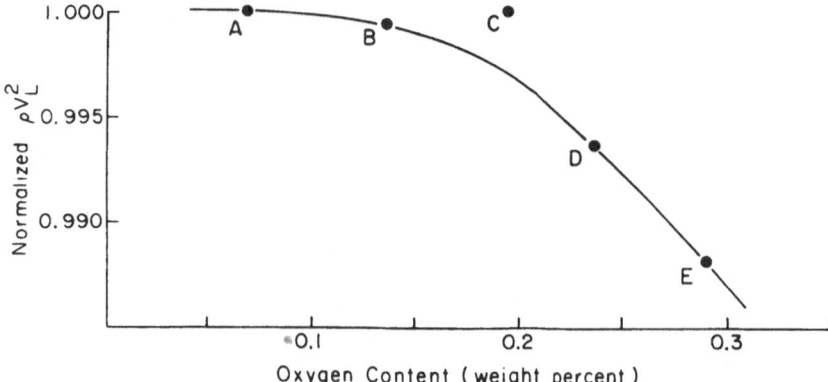

Fig. 7. Normalized ρV_L^2 vs. oxygen content.

interstitial alloying in any metal necessarily reduces the elastic moduli. However, his theory was based on purely geometric considerations and did not consider the importance of the complex Ti-O electronic interaction. Additionally, Zener's conclusions were in conflict with the experimental results of Hsu and Conrad[6] who showed that in 100% alpha titanium, longitudinal wave velocity and, therefore, the elastic moduli, increased with increasing oxygen content. Hence, it is doubtful that the theoretical arguments of Zener[3] account for the results obtained here. In fact, if the only parameter changing was the amount of oxygen in the alpha-phase lattice of the Ti-6211 alloy, following Hsu and Conrad[6] we would expect V_L to increase with oxygen content rather than decrease.

Changes in the relative amounts of the alpha and beta phase could also affect the wave velocity in both the welded specimens and the oxygen contaminated specimens. As mentioned before, the Ti-6211 alloy is a near alpha alloy containing primarily alpha phase (HCP) metal with some beta phase (BCC) metal. The sound wave velocity should be lower in the beta phase, because it has a more open crystal structure than the alpha phase. Hence, the decreased sound velocity observed in the weld region could possibly be caused by an increased amount of beta phase metal present in the weld region. X-ray diffractometer measurements were performed to test this hypothesis.

Initial diffractometer scans indicated that the BCC (110) and the HCP (101) peaks could be used to qualitatively assess changes in the relative amounts of alpha phase and beta phase material. Subsequent scans demonstrated that the weld region of the specimen tested contained more beta phase metal than the base

plate. These results, along with consideration of the thermal
history of the weld zone, support the hypothesis presented above.
The rapid cooling rates present during welding should result in
more retained, high temperature beta phase metal, hence, lower
sound wave velocities.

If as a result of the thermomechanical processing the oxygen
contaminated specimens received, increasing oxygen content somehow
resulted in more retained beta phase material, then this effect
could account for the longitudinal wave velocity behavior
observed in Fig. 6. However, both x-ray diffraction analysis and
quantitative metallographic analysis failed to support this
hypothesis and in fact indicated that, as would be expected, the
amount of beta phase actually tended to decrease with increasing
oxygen content.

Other factors that could influence V_L would be differences in
texture[8] and residual stress introduced by rolling the plate
specimens. For example, the higher the oxygen content, the
greater the resistance to rolling deformation would be, which in
turn would cause larger residual compresssive stresses and the
decreasing wave velocity behavior observed in Fig. 6. The
contribution of residual stress can be studied by performing a
stress relieving heat treatment on the specimens (assuming this
heat treatment does not affect the texture and microstructure) and
then repeating the ultrasonic wave velocity measurements. This
set of experiments is currently underway.

Ultrasonic Attenuation Measurements

Attenuation measurements were performed on the oxygen
contaminated specimens to evaluate the usefulness of attenuation
for quantitatively detecting the presence of oxygen. The results,
shown in Fig. 8, demonstrate the inherent scatter present in
attenuation measurements using contact transducers due to the
sensitivity of this technique. The elastomer coating used to
couple the transducer to the specimen provided a fairly
reproducible transducer specimen coupling and helped decrease the
data scatter. Between 24 and 32 attenuation measurements were
performed on each specimen; the average and range of these
measurements are presented in Fig. 8. Statistical reduction of
the attenuation data (see Fig. 9) indicates that ultrasonic
attenuation may have potential for quantitatively detecting the
presence of oxygen in Ti-6211. It is not known whether the
decrease in attenuation with increasing oxygen content is due to
the effect of the oxygen on the alpha titanium lattice, the change
in the amount of retained beta phase, or the microstructural
variations induced by oxygen and thermal history.

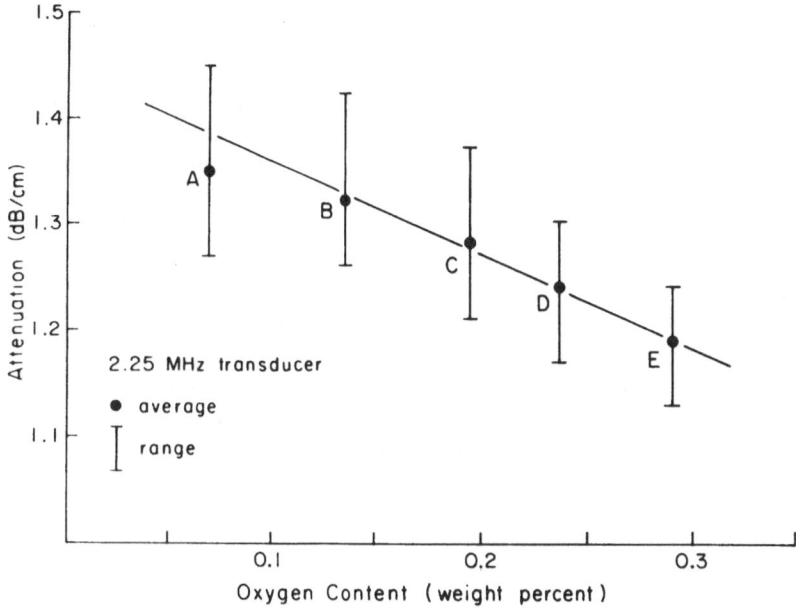

Fig. 8.   Longitudinal wave attenuation vs. oxygen content.

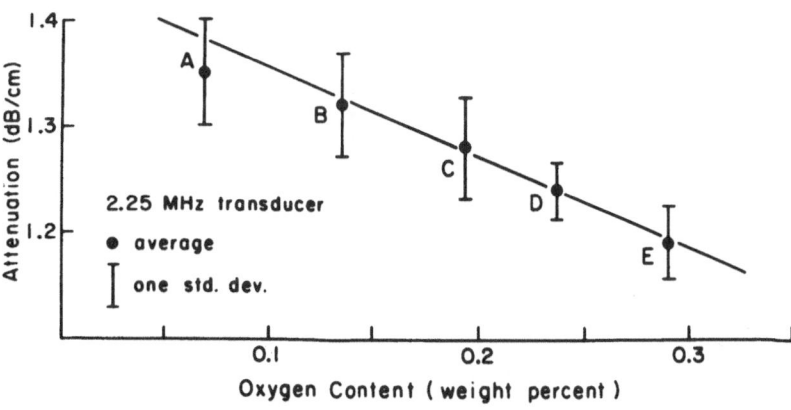

Fig. 9.   Longitudinal wave attenuation vs. oxygen content.

Hardness Measurements

Rockwell "C" hardness, a measure of material response to plastic deformation, was found to increase with increasing oxygen content (see Fig. 10). The increase was probably due to the solid solution strengthening provided by the interstitial oxygen, which partitions preferentially into the alpha phase.[9-11] This increase is consistent with results reported in the literature. An alternative explanation could involve the changing distribution of alpha phase and beta phase material as the oxygen content changes.

CONCLUSIONS

Analysis of the data collected in this research has underscored some of the material effects, such as high temperature phase retention, microstructural alteration, texture, and residual stress that influence the ultrasonic testing of the Ti-6211 alloy and metal alloys in general. Both ultrasonic wave velocity and

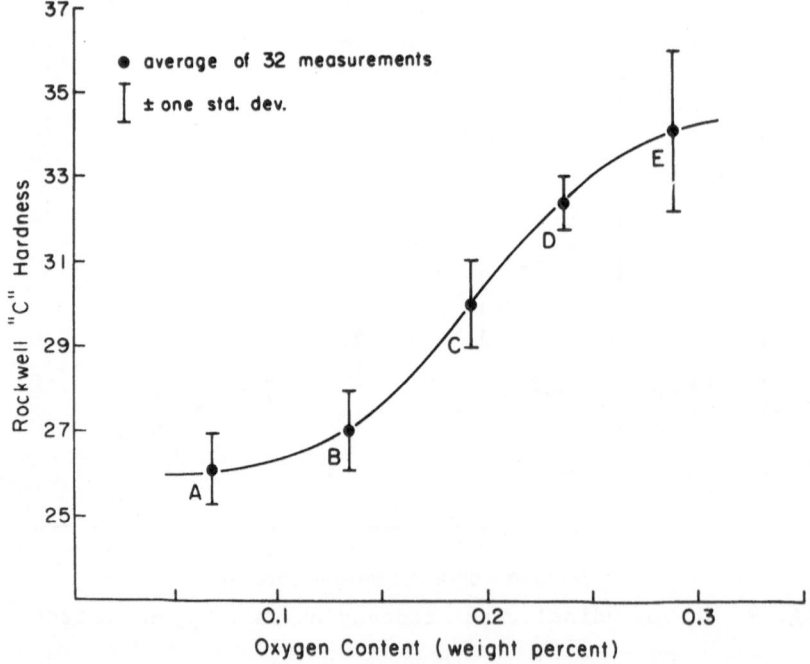

Fig. 10. Rockwell "C" hardness vs. oxygen content.

ultrasonic attenuation were found to decrease with increasing
oxygen content and the ultrasonic wave velocity was observed
to be lower in the weld region than in the base metal as a
result of these material effects. Results of scanning electron
microscopy, x-ray diffraction analysis, and hardness testing
helped explain the variations in ultrasonic data observed. This
work also indicated that ultrasonic attenuation shows promise as a
nondestructive tool for quantitatively evaluating the amount of
oxygen present in a Ti-6211 sample.

ACKNOWLEDGMENTS

The authors wish to thank Miss Angela Guarda for her technical
assistance; Dr. B. B. Djordjevic of Martin Marietta Laboratories in
Catonsville, Maryland, for providing the elastomer coating which
served as a couplant for the attenuation measurements; Ms. Romilda
Schack for her patience and skill in typing this manuscript;
Mr. Henry Chaskelis of the Naval Research Laboratory, Dr. Bruce
MacDonald of the Office of Naval Research and Dr. Hans Vanderveldt
of the Naval Sea Systems Command for supporting this work.

REFERENCES

1. "Glossary of Metallurgical Terms and Engineering Tables,"
American Society for Metals, Ohio (1979).
2. R. E. Green Jr., "Treatise on Materials Science and
Technology," Vol. 3, Academic Press, New York (1973).
3. C. Zener, Relation Between Residual Strain Energy and
Elastic Moduli, Acta Cryst 2:163 (1979).
4. J. N. Pratt, W. J. Bratina, and B. Chalmers, Internal
Friction in Titanium and Titanium-Oxygen Alloys,
Acta Met. 2:204 (1954).
5. C. F. Ying and R. Truell, The Effect of Hydrogen on
Ultrasonic Attenuation and Velocity Measurements in
Titanium, Acta Met. 2:374 (1954).
6. N. Hsu and H. Conrad, Ultrasonic Wave Velocity
Measurements on Titanium Oxygen Alloys, Scripta Met.
5:905 (1971).
7. D. H. Chung, D. J. Silversmith, and B. B. Chick, A
Modified Ultrasonic Pulse-Echo-Overlap Method for
Determining Sound Velocities and Attenuation of
Solids, Rev. Sci. Inst. 40:718 (1969).
8. E. G. Henneke II and R. E. Green Jr., Compilation of
Elastic Wave Modes in Hexagonal Metals, J. Appl.
Phys. 40:3626 (1969).
9. R. I. Jaffee and I. E. Campbell, Alloys of Titanium with
Carbon, Oxygen, and Nitrogen, Trans. AIME 188:1261
(1950).

10.   R. I. Jaffee and I. E. Campbell, The Effect of O, N, and
      H on Iodide Refined Titanium, Met. Trans. 185:646
      (1949).

11.   W. L. Finlay and J. A. Snyder, "Effects of Three
      Interstitial Solutes (N, O, C) on the Mechanical
      Properties of High-Purity Alpha Ti, Trans. AIME
      188:277 (1950).

MEASUREMENT OF NEAR-SURFACE ULTRASONIC

ABSORPTION BY THERMO-EMISSIVITY [+]

Jean-Pierre Monchalin [*]

Physical Metallurgy Research Laboratories
CANMET Energy, Mines and Resources, Canada
568 Booth Street
Ottawa, Ontario, K1A 0G1, Canada

and

Jean F. Bussière

Industrial Materials Research Institute
National Research Council of Canada
750 Bel-Air Street
Montreal, Quebec, H4C 2K3 Canada

It is generally recognized that ultrasonic absorption, i.e. the relaxation of mechanical energy into heat is related to mechanical properties for several materials. In particular, in polycrystalline metals and within the frequency range generally used in ultrasonic testing, absorption is mainly caused by thermoelastic losses, intergranularthermal currents, magnetic domain wall motion, the relaxation of interstitials and the interaction with dislocations.[1,2] Among these different mechanisms, the last three can be related (to a variable extent) to the strength of the metal and its state of stress or fatigue.

Our work has been mostly directed to the development of a technique to measure directly ultrasonic absorption in polycrystalline

[*] Present address: Industrial Materials Research Institute
                    750 Bel-Air Street, Montreal, Que. H4C 2K3

[+] Work performed in the laboratories of the Industrial Materials
    Research Institute

metals. Such a technique would be interesting since it is rather difficult to deduce absorption from conventional pulse-echo or pulse transmission ultrasonic techniques. These techniques measure attenuation, i.e. the combination of absorption and scattering. In order to determine absorption, one has to subtract the scattering contribution from attenuation data. This operation requires a previous knowledge of the variation of scattering with ultrasonic frequency, which, in practice, is rather difficult since the scattering contribution is very sensitive to grain size distribution[3] (which is unknown unless precise metallographic examination is performed).

Several detection schemes consisting basically in measuring the heat produced by ultrasonic absorption have been considered. We present below one which enables to measure the rather weak absorption of metals (by opposition to materials such as plastics), at the surface and near the surface. The technique is based on the measurement of the temperature variations at the surface of the metal specimen caused by absorption of an amplitude modulated ultrasonic beam. These variations are monitored by an infrared detector which views the part of the surface on which impinges the beam. In the present technique, the ultrasonic source is slowly amplitude modulated (typically between 0.1 Hz to 1 KHz), so that the temperature modulation observed is related to the absorption within a certain thermal depth below the surface. Mathematically, this depth is characterized by the thermal diffusion length μ_s given by[4]

$$\mu_s = \left[K_s / \pi \rho_s C_s f \right]^{\frac{1}{2}} \qquad (1)$$

where K_s, μ_s, C_s are the thermal conductivity, the density and the specific heat of the material, respectively, and f is the frequency of modulation. μ_s is typically 5.4 mm in aluminum and 2 mm in mild steel at 1 Hz. μ_s is the characteristic damping length of the thermal waves inside the sample (this parameter is analogous to the skin depth parameter which characterizes the damping of electromagnetic waves in metals). Heat produced several μ_s away from the surface is therefore not detected. Since absorption in metals is weak, it is important to ensure that the heat produced by the transducer and by the coupling fluid is not detected by locating the transducer several μ_s away from the viewed area. This condition being satisfied, several configurations for ultrasonic excitation and detection may be considered.

For a parallel or nearly parallel sample of sufficient thickness ($\gg \mu_s$), ultrasound may be produced by a transducer (compression or shear) bonded on the opposite side of the viewed surface. We have performed a few experiments with parallel steel samples and have observed, as expected, a very large increase of the detected infrared signal when the ultrasonic frequency is tuned to a resonance of the sample. From the Q of this resonance, it is possible to deduce attenuation:[5]

$$Q^{-1} = \Delta f_u / f_u = v\, \alpha_{at} / \pi f_u \qquad (2)$$

where Δf_u is the full width at half maximum of the resonance ($\frac{1}{2}$ maximum instead of $1/\sqrt{2}$ of maximum since one detects a power effect), f_u is the ultrasonic frequency, v is the ultrasonic velocity and α_{at} is the attenuation coefficient (α_{at} is defined as the coefficient for the ultrasonic stress or strain, the coefficient for ultrasonic power or intensity are twice). Using this technique, we were able to obtain values of α_{at} in agreement with measurements of exponential decrease of the echos produced in a conventional pulse-echo method.[6]

When the ultrasonic beam is at perpendicular incidence to the surface and covers a much larger area than that seen by the detector (uniform excitation), it can be readily shown, by solving unidimensional heat diffusion equations, that the rms value and the phase of the surface temperature modulation τ is given by the following equations:

$$\tau_{rms} = \frac{\pi\, \alpha\, Z\, f_u^2\, U^2}{2\sqrt{2}\, f\, \rho_s C_s} \qquad (3)$$

phase of τ = phase of applied
ultrasonic stress - $\pi/2$ (4)

where Z is the acoustic impedance of the specimen assumed to be uniform (for longitudinal or shear waves), U is the maximum amplitude of the ultrasonic surface displacement, α is the absorption coefficient (for the stress or strain) and the other parameters have been defined previously (Eq. 3 is slightly approximated: we have used the fact that $\alpha \ll 1/\mu_s$ and we have neglected a thermal factor which differs from 1 by less than 10^{-3}). These formulas can also be derived by substituting the appropriate expression for the modulated heat source in expressions used in photoacoustic or photothermal (optical) spectroscopy[7]: in the present case, the modulated heat source is produced by ultrasound instead of being caused by the absorption of light beam.

The voltage we apply to the ultrasonic transducer is typical of the voltages generally used in ultrasonic NDE techniques (< 200V). The corresponding ultrasonic intensity is generally less than 100 W/cm^2 and the ultrasonic strains range from less than 10^{-5} to a few 10^{-5} at maximum. For such these intensities and powers, the surface temperature modulation is small and always much less than 1°C. The modulated radiation flux, F, collected by a broadband infrared detector is given by the following equation:[8]

$$F_{rms} = 4\varepsilon \; K \; T^3 \; E \; \tau_{rms} \qquad (5)$$

where ε is the surface emissivity, K is the Stefan-Boltzman constant, T is the temperature of the sample and E is the étendue (or through-put) of the viewing geometry between the sample and the detector element. When using pyroelectric detectors (or better a Golay detector) at a few mm away from the surface and a viewing area of about 1 cm diameter, a detection limit of less than $10^{-5}/\varepsilon$ $^{\circ}C/Hz^{\frac{1}{2}}$ is obtained with narrow band synchronous detection. It was found that this sensitivity enables to do more than mere detection and is sufficient for performing absorption measurements at the power levels mentioned above. Eqs 3 and 5 show that, in order to determine α, it is also necessary to measure the surface emissivity, the étendue of the collecting geometry (actually the product of these 2 para-meters) and the amplitude of the surface displacement. The product of the emissivity and étendue can be measured by comparison to a sample of known emissivity whereas the amplitude of the surface dis-placement can be determined with a Michelson interferometer.[9] We are pursuing work to incorporate the calibration feature and the Michelson into a single unit which will work as an ultrasonic absorp-tion meter.

As an additional remark, we mention that the technique has the capability of depth profiling by varying the modulation frequency. In particular, if the absorption, instead of being uniform, was lo-calized at the surface, the phase lag would be equal to $-\pi/4$. This was verified by spreading a thin but highly absorbing layer of oil or grease at the surface of the sample.

The use of the technique is not limited to samples excited near normal incidence from the opposite side or to bulk waves. We present below an example of use with an inclined shear wave. The geometry of the set-up is depicted in the insert of fig. 1. Ultrasound (shear wave) is produced by a lithium niobate shear transducer (2.5 MHz) mounted on an aluminum wedge (the use of a plastic wedge is generally proscribed because of excessive heating in the wedge). The sample is a 30 cm long, 2.5 cm thick piece of mild steel and the infrared detector scans the side opposite to the transducer (the same side could have been scanned as well). The magnitude of the infrared signal (fig. 1) shows the 2 bumps corresponding to positions where the beam bounces off the surface (these places are in the far field of the transducer). From this recording, it should be possible to deduce attenuation after taking into account a possible effect of the beam spread and after emissivity and étendue calibration. Mea-surement of absorption can be performed with the additional use of the interferometer, as mentioned previously. Also seen in fig. 1,

is a plot of the phase variation, which agrees at least qualitatively, with what is expected: as the heat source moves deeper, a
phase lag results and a phase lead in the opposite case. We note
in particular that the phase is about the same in the space between
the two rebounds, where the signal comes from some scattered ultrasonic energy, and at the middle of the first rebound. This should
be expected since in both cases the detector sees a nearly uniform
heat source.

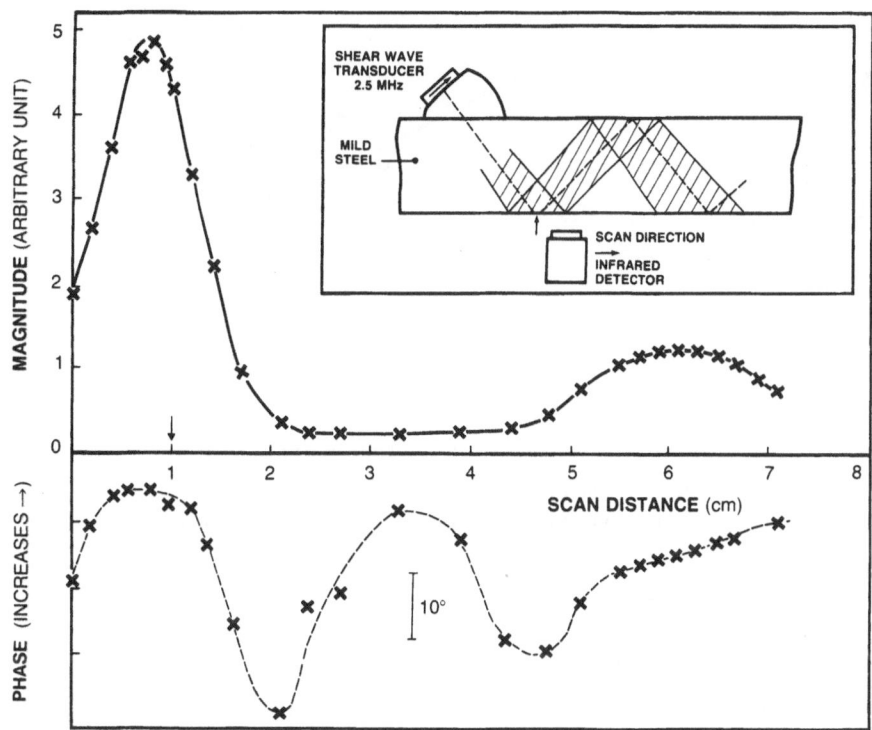

Fig. 1 Plot of the magnitude and the phase obtained
        from a scanned infrared detector, the experi-
        mental set-up being depicted in the insert.
        The arrow in the insert and on the abscissa
        axis marks approximately the center of the
        first rebound on the beam.

        It is also possible to measure the absorption of surface waves.
The experimental set-up is depicted, in this case, by the insert
of fig. 2. The sample is the same as before and the Rayleigh wave
is generated by a lithium niobate shear transducer mounted on a
brass wedge. The variation of the signal magnitude when the detec-

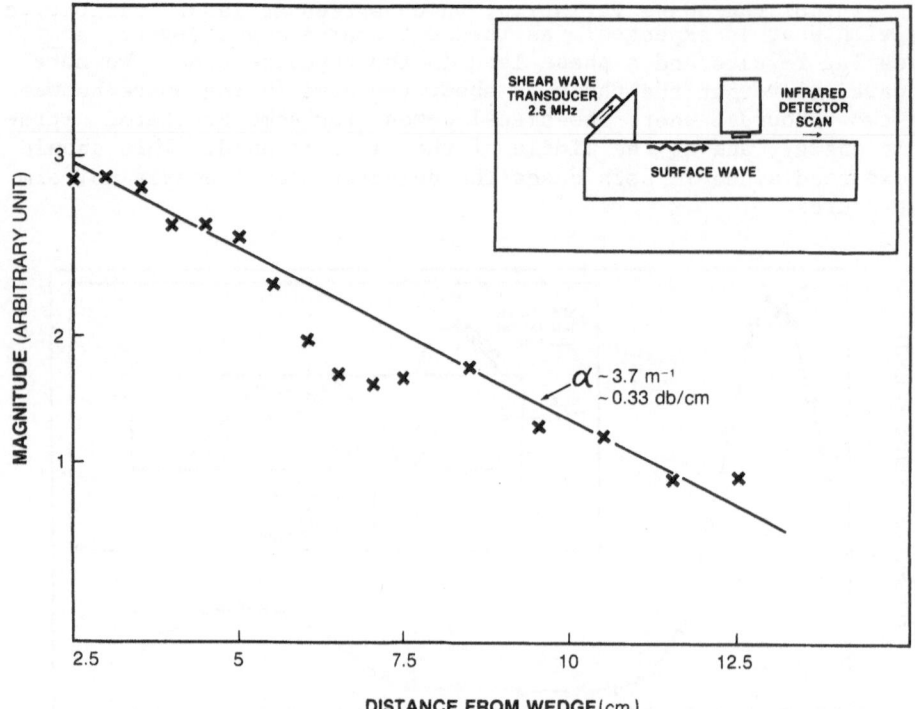

Fig. 2   Logarithmic plot of the magnitude of infrared
         signal versus distance for surface wave exci-
         tation. The experimental set-up is shown in
         the insert.

tor is scanned is plotted in fig. 2 using a logarithmic scale.
From the slope of the plot, the attenuation can be deduced. The
deviation from a straight line observed in fig. 2 is attributed to
local changes of surface emissivity. We have verified that when
excited at a higher ultrasonic frequency (resulting into a shallower
penetration of the surface wave) and in the case where the thermal
diffusion length is longer than the wavelength, a phase lead occurs,
as expected.

   Using the technique we have described, we are currently pursuing several experiments where absorption is changed by application of a static stress or a magnetic field. As an example of this
type of work, we present below a measurement of the absorption of
a Metglas ribbon (as-quenched, 2505 SC from Allied Chemical) placed
in a magnetic field. The experimental set-up is sketched in the
insert of fig. 3. A shear transducer mounted on a wedge is used

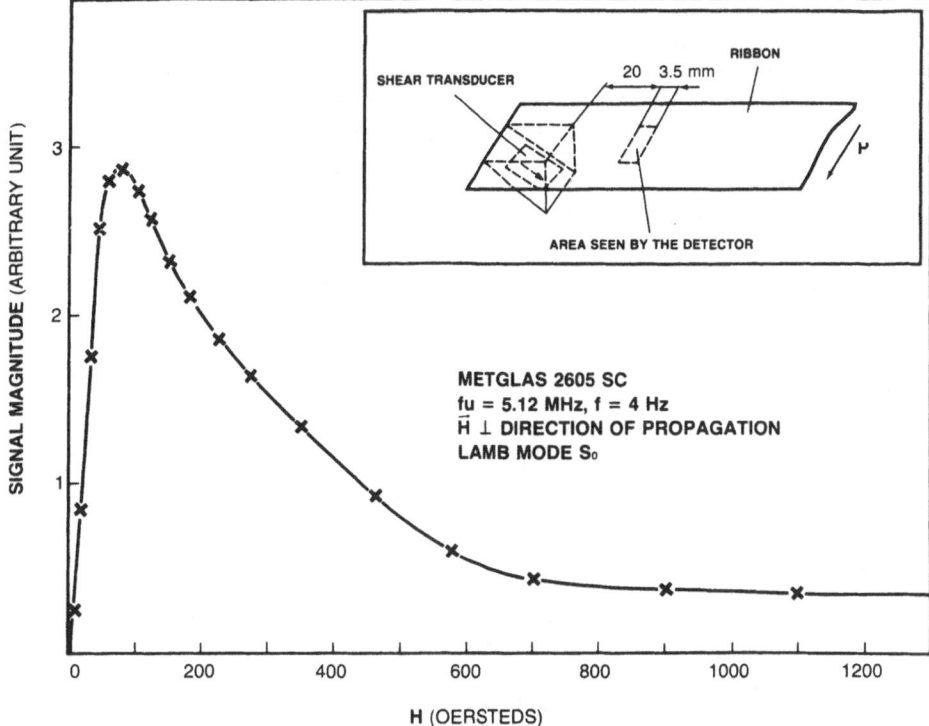

Fig. 3   Plot the infrared signal produced by a symetric
         Lamb wave S_O at 5 MHz launched along a Metglas
         ribbon (2605 SC from Allied Chemical) versus
         magnetic field. The experimental set-up is shown
         in the insert.

to launch a symetric Lamb wave (mode S_O) along the ribbon. The
infrared detector views a well defined area about 2 cm away from
the wedge. The experimental curve is shown in fig. 3. Because the
magnetic domains in these Metglas ribbons are very mobile, there is
virtually no propagation at zero field and the infrared signal is
zero as observed. The experimental results can be explained by
noting that the signal varies as $\alpha \exp(-2 \alpha z)$ where z is the distance from the emitting wedge (the ribbon is very long and there is
no reflection from its end) and by knowing that α decreases from a
very large value to some residual value (caused by absorption of
origin other than magnetic). This explains why the experimental
signal goes through a maximum (for $\alpha = 1/2z$) and reaches an asymto-
tic value at high field. In order to determine α as a function of
the field, the function $2\alpha z \exp(-2\alpha z)$ was tabulated for several
values of $2\alpha z$ and then normalized to the same maximum value as the

experimental plot. By using this tabulation, the value of α corresponding to each experimental point was found. The result is plotted
in fig. 4. We should note that this technique gives absolute values
of absorption coefficients, which are otherwise difficult to obtain
since good parallelism between the emitting transducer and the re-
flecting end is required. We are currently exploring the extention
of this method which does not require a measurement of the surface
displacement, to bulk magnetic polycrystalline samples (steel,
nickel).

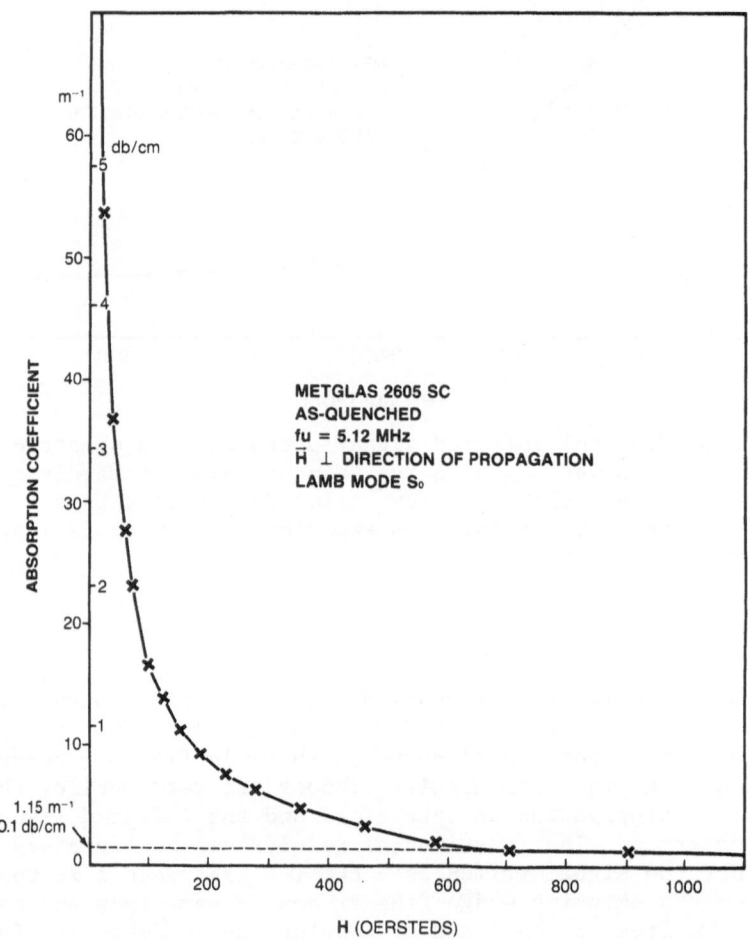

Fig. 4 Plot the absorption coefficient of a Metglas
         ribbon 2605 SC versus applied magnetic field
         (symetric Lamb wave S_O at 5 MHz).

In conclusion, we have devised a technique based on thermo-emissivity which enables the measurement of surface and near-surface ultrasonic absorption. We have presented several uses of the technique and discussed the realization of a versatile absorption meter.

REFERENCES

1. A.B. Bhatia, "Ultrasonic Absorption", Clarendon Press, Oxford (1967).
2. A.E. Lord, Jr. and D.N. Beshers, Acta Metallurgica, 14: 1659 (1966).
3. R.L. Smith, Ultrasonics, 20: 211 (1982).
4. H.S. Carslaw and J.C. Jaeger, "Conduction of heat in solids", 2nd edition, Oxford Press, 1959, chap. 2.6.
5. A.S. Nowick, in "Physical Acoustics", vol XIII, edited by W.P. Mason, Academic Press, New-York, 1977, chap. 1, p. 6.
6. R. Truell, C. Elbaum and B.B. Chick, "Ultrasonic Methods in Solid State Physics", Academic Press, New-York, 1969, chap. 2.
7. A. Rosencwaig, "Photoacoustics and Photoacoustic Spectroscopy", John Wiley, New-York, 1980, chap. 9.2 and 9.3.
8. "The infrared handbook", edited by W.L. Wolfe and G.J. Zissis, Office of Naval Research, Department of the Navy, Arlington, Va, 1978, chap. 1 and 20.
9. C.H. Palmer and R.E. Green, Jr., Appl. Opt. 16: 2333 (1977).

ULTRASONIC CHARACTERIZATION

OF AMORPHOUS METALS, POLYMERS, AND COMPOSITES

ULTRASONIC MATERIALS CHARACTERIZATION

OF MELT SPUN METALLIC RIBBONS

C. Lee Friant and Moshe Rosen

Materials Science and Engineering Department
The Johns Hopkins University
Baltimore, MD 21218

INTRODUCTION

The rapid cooling rates obtainable by melt spinning techniques have enabled the production of nonequilibrium metallic glasses and microcrystalline alloys with enhanced solubility of solute elements. The recent and intensive development of rapidly solidified metal alloys has been accompanied by an extensive body of research concerning characterization of both material properties and preparation processes. Structural, magnetic, electrical and tensile properties have been investigated thus far. However, accurate characterization of elastic properties has been difficult because the final product achieved by melt spinning is usually a ribbon or foil with a thickness on the order of 50 μm or less. This geometry precludes the use of bulk ultrasonic waves for characterization of the elastic properties of these materials. Instead, guided waves must be utilized for the determination of elastic properties. Unlike bulk waves, the velocity of guided waves usually shows dispersion; velocity is a function of extrinsic factors, such as specimen thickness and ultrasound frequency. Hence, determination of elastic properties of ribbons from sound wave velocities cannot be achieved by direct substitution into appropriate equations. Rather, the effect of these extrinsic factors on velocity must be characterized before elastic moduli can be calculated. In the present study this characterization and subsequent isotropic moduli calculation were performed for ribbon or foil samples of Metglas, Pd-Cu-Si, Cu-Zr and commercial purity aluminum using conventional ultrasonic velocity measuring apparatus. The assumption of isotropy is valid for the amorphous or microcrystalline product obtained by melt spinning.

THEORY OF GUIDED WAVES-ELASTIC MODULI DETERMINATION

Guided waves are the type of waves which propagate in bounded elastic media, such as plates, sheets, foils, ribbons, cylinders, rods and wires. For some of these geometries, the exact solutions to the equations of motion can be derived from classical elasticity theory.[1] These solutions must satisfy specific boundary conditions. For example, the two plane parallel surfaces of plates, sheets, foils and ribbons are assumed to be stress free. Similarly, in the case of cylinders, rods and wires, the single-curved surface is assumed to be stress free. The primary types of guided waves that exist in solids fall into three categories: longitudinal, flexural and shear (or torsional). The longitudinal and flexural waves that travel in plate-type materials are often referred to as symmetric and anti-symmetric Lamb waves, respectively.

Several characteristics of guided waves distinguish them from bulk waves. In particular, these waves show dispersion of wave velocity with frequency and plate thickness or rod diameter. Whenever dispersion exists, the group velocity deviates from the phase velocity. Also, higher order modes of each type of wave exist, each exhibiting its own unique dispersion relation. Regions of high dispersion have also been shown to correlate with regions of high attenuation. Due to the multiplicity of modes for each type of wave, coupling of modes is often observed at certain values of thickness (or diameter) to wavelength ratio. A typical theoretical phase and group velocity dispersion curve for the first-order longitudinal (symmetric) Lamb wave is shown in Fig. 1. The velocity is plotted versus thickness (t) to wavelength (λ) ratio to facilitate comparison of experimental measurements on specimens of various thicknesses at a number of frequencies. At the zero value for the thickness to wavelength ratio, the phase and group velocity converge to the value, V_L, which is related to the isotropic Young's modulus (E), density (ρ) and Poisson's ratio (ν) by the equation that appears at the right of the figure. Underneath is the equation for the longitudinal extensional velocity, V_E, which applies to cylinders, wires and rods. As the thickness to wavelength ratio increases, the phase velocity decreases monotonically from V_L to V_R while the group velocity passes through a minimum. To the right the Rayleigh or surface wave velocity, V_R, is given in terms of the bulk shear wave velocity, V_S, for a Poisson's ratio, ν, of 0.29. In Fig. 1 an equation is also given for the velocity of a bulk shear wave, V_S; this is identical to the equation for the velocity of the fundamental horizontally polarized shear wave, V_{SH} (0), the only nondispersive guided wave. The SH (0) mode is useful for determining the shear modulus directly because no other elastic moduli appear in the equation. The only limitation on this wave

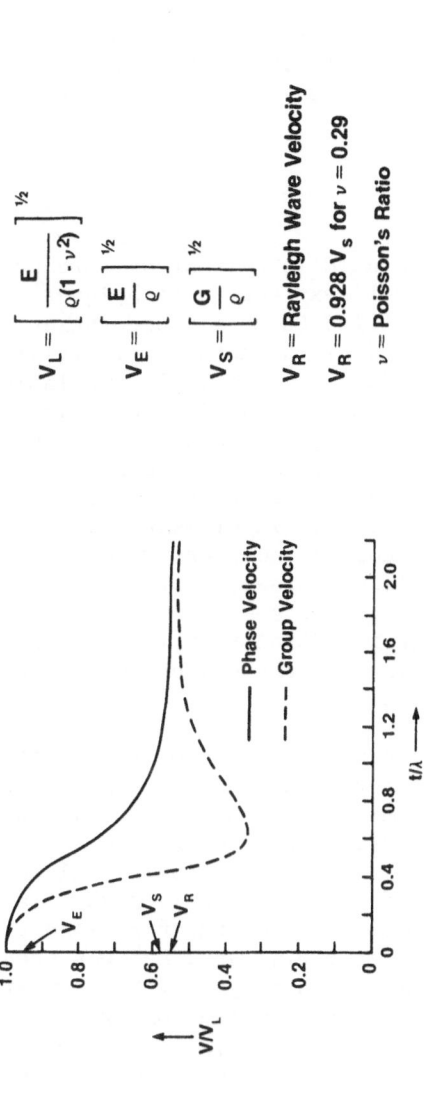

$$V_L = \left[\frac{E}{\varrho(1-\nu^2)} \right]^{\frac{1}{2}}$$

$$V_E = \left[\frac{E}{\varrho} \right]^{\frac{1}{2}}$$

$$V_S = \left[\frac{G}{\varrho} \right]^{\frac{1}{2}}$$

V_R = Rayleigh Wave Velocity

$V_R = 0.928\ V_s$ for $\nu = 0.29$

ν = Poisson's Ratio

Fig. 1. Theoretical phase and group velocity dispersion curves for an infinite sheet or plate ($\nu = 0.29$, after Kolsky[2]) and equations for V_L, V_E and V_S in terms of elastic moduli and density.

is that the plate, sheet, foil or ribbon must be several
wavelengths in width for the wave to be truly nondispersive.

Measurement of both the fundamental SH-type velocity V_{SH} (0)
and the velocity of the first-order longitudinal Lamb wave at zero
frequency (V_L) for any ribbon, foil or sheet specimen is sufficient
for the complete determination of all the isotropic elastic moduli
of the material if the mass density is known. A summary of the
equations required to calculate the more important of these moduli
is presented in Table 1.

EXPERIMENTAL TECHNIQUE

The ribbon specimens used in this study were produced by
either melt-spinning or melt-extraction techniques. Metglas
2605-SC was provided by Allied Chemical Corporation. Pd-Cu-Si and
Cu-Zr samples were supplied by the Metallurgy Division of the
National Bureau of Standards. Commercially available aluminum
foil and sheet of greater than 99% purity were also used. In
Table 2 the atomic composition, ribbon cross-sectional dimensions
and mass density are listed for each specimen. All specimens were
at least 4 cm long. In addition to these ribbon and foil specimens
an amorphous Pd-Cu-Si rod 3.5 mm in diameter and 2.1 cm long was
available for bulk longitudinal and shear wave velocity measurements. Unlike most metallic glasses, amorphous Pd-Cu-Si can be
produced in relatively large cross-sections. Metglas 2605-SC and
Cu-Zr were only available in ribbon form. Mass density values for
ribbons were obtained using a submerged balance technique similar
to that described by Davis.[3]

ln this study the group velocities of the first-order
longitudinal Lamb wave and the fundamental horizontally polarized
shear wave (SH (0)) were measured. Conventional longitudinal and
shear polarized piezoelectric transducers were used to generate
and receive these two types of waves. Commercial longitudinal
transducers with nominal frequencies of 1, 2.25, 5 and 10 MHz were
used for generation and detection of the first-order longitudinal
Lamb wave. AC-cut quartz operating at 5 MHz was used to generate
and receive the SH (0) wave. This type of transducer required
impedance matching networks whereas the commercial longitudinal
transducers did not. For both wave types a viscous silicone fluid
was used as an acoustic couplant between the transducer and the
ribbon specimen. A two-transducer pitch-catch arrangement was used
for the time-of-flight measurements instead of the single-
transducer pulse-echo method due to the absence of at least two
echoes of appreciable amplitude.

The orientations of the generating and receiving transducers
relative to the ribbon specimen proved to be an extremely important

Table 1. Equations for Isotropic Elastic Moduli in Terms of Measured Guided Wave Velocities

$$G = \rho V_S^2 \qquad\qquad E = 4G\left[1 - \frac{V_S^2}{V_L^2}\right]$$

$$\nu = 1 - 2\left(\frac{V_S}{V_L}\right)^2 \qquad\qquad k = \frac{E}{3(1-2\nu)}$$

where G = Shear Modulus
ρ = Mass Density
E = Young's Modulus
ν = Poisson's Ratio
k = Bulk Modulus
V_S = V_{SH} (0); Velocity of fundamental horizontally polarized shear wave
V_L = Velocity of first-order longitudinal Lamb wave at zero frequency

Table 2. Chemical Composition, Size and Density of Ribbon and Foil Specimens

Specimen	Atomic Composition (%)	Average Ribbon Thickness (μm)	Average Ribbon Width (mm)	Mass Density (g/cm^3)
Amorphous Pd-Cu-Si	$Pd_{77.5} Cu_6 Si_{16.5}$	53	1.48	10.52
Amorphous Cu-Zr	$Cu_{50} Zr_{50}$	51	1.68	7.33
Metglas 2605 · SC	$Fe_{81} B_{13.5} Si_{3.5} C_2$	41	⩽ 51	7.3
Aluminum Foil	> 99% AL			
Regular		25	100	2.7
Heavy Duty		50	100	

consideration. The transducer-specimen configurations used in this study are based on those reported by Gelles[4] in his work on films, foils, thin plates, whiskers, fibers and fine wires. In particular, two transducer positions for each type of wave were found to yield the largest received signal amplitudes. The

longitudinal Lamb wave transducer-specimen configuration is shown
in Fig. 2 and the SH (0) wave transducer-specimen configuration is
shown. in Fig. 3. In both cases the "a" configuration was used for
samples that could not support their own weight but could be easily
bent around a sharp corner. The "b" configuration was used for
stiffer samples that could support their own weight but could not
be easily bent around a sharp corner. For generation and detection
of the SH (0) type wave, the transducers were positioned in such a
manner as to orient the shear particle displacements parallel to
the width direction in the plane of the ribbon or sheet specimen.

A schematic diagram of the experimental apparatus used for
velocity measurements is shown in Fig. 4. The apparatus is
essentially equivalent to that used for the pulse-echo overlap
method.[5] The continuous wave (cw) oscillator provided a variable
frequency sync signal which has fed into the decade-divider and
dual-delay strobe generator. The decade-divider divided the
frequency of the sync signal by selected powers of 10. This
divided sync signal was used to trigger the rf pulse modulator and
the monitoring oscilloscope. The dual delay strobe generator
provided strobed intensification of the attenuated rf pulse and a
selected portion of the signal received by the second transducer.
The rf pulse was fed into an attenuator so that it did not overload
the oscilloscope's vertical input. When the "divided" sync was
used to trigger the oscilloscope sweep, it was possible to view
and tune the signals received by the second transducer on one
vertical input channel and the attenuated rf pulse on another. In
the "direct" mode, the oscilloscope is swept at a frequency whose
period is equal to exactly the time difference between the rf pulse
sent to the first transducer and the received signal detector by
the second transducer. In practice this sweep is obtained by first
estimating the travel time between signals, calculating the
reciprocal, then adjusting the cw oscillator to this frequency
value. In the "direct" mode the strobed signals appear superimposed on the oscilloscope screen. At this point fine frequency
adjustment is made until a cycle-to-cycle match of these signals
is obtained. The time difference used to determine the velocity is
exactly equal to the reciprocal of the sync frequency that provides
the correct overlap of signals.

There is a correction that must be made to time-of-flight
measurements for the pitch-catch technique that is not necessary
for pulse-echo overlap method. This correction arises from the
sum of the time delays introduced by the two transducers, impedance
matching networks (if used) and receiving amplifier. In order to
find the actual travel time through the specimen, the appropriate
time delay must be subtracted from the total travel time measured.
The major contribution to the time delay was that due to the
transducers. Commercial transducers contain both piezoelectric
elements and matched wear plates, both of which increase in

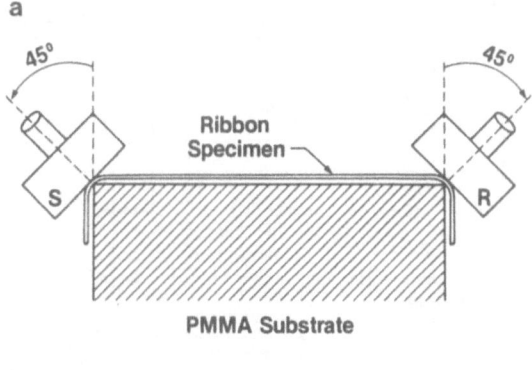

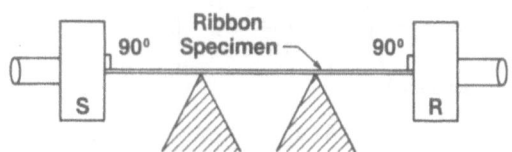

Fig. 2.   Sending (S) and receiving (R) transducer configurations relative to ribbon position for longitudinal Lamb wave generation and detection.

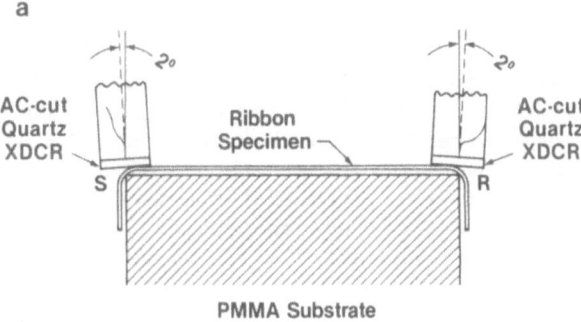

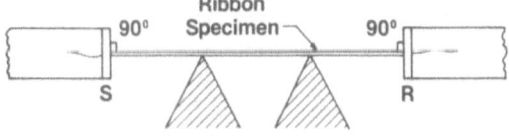

Fig. 3.   Sending (S) and receiving (R) transducer configurations relative to ribbon position for SH-type shear wave generation and detection.

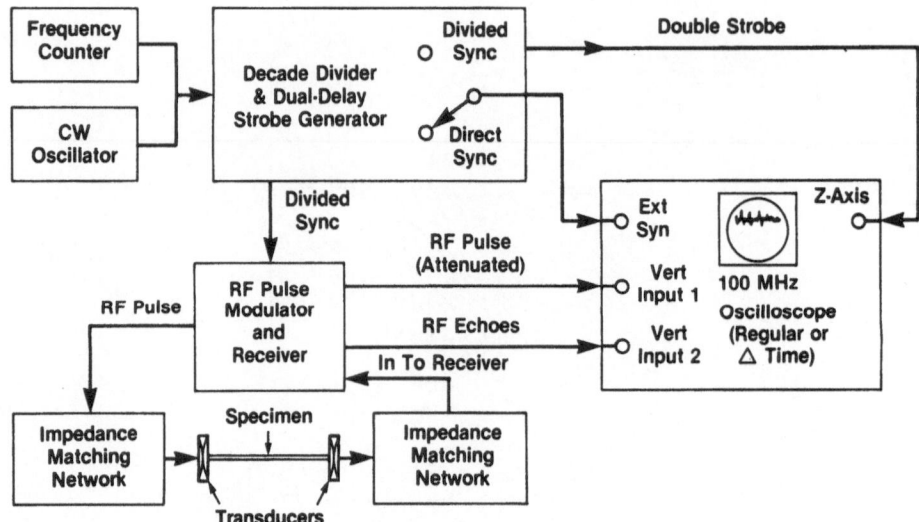

Fig. 4.   Experimental apparatus for measuring the sound velocity
          in melt spun ribbons.

thickness as the nominal frequency decreases. Hence, the lower
the frequency, the greater the time delay introduced by the
transducers. Using a pair of 1 MHz commercial longitudinal
transducers, the total delay time was measured to be 1.4 μsec. Not
correcting for a delay of this magnitude can lead to a Lamb wave
velocity error as high as 17% for a specimen 4 cm in length.

In addition to time-of-flight measurements for each type of
wave, it is necessary for velocity calculation to make an accurate
determination of specimen length. The length of each ribbon or
sheet specimen was made with a vernier caliper that had a precision
of ±0.02 mm. It is estimated that the velocity measured by the
technique described above has an accuracy of 1 to 2% depending on
ultrasonic frequency and attenuation value of the material.

RESULTS AND DISCUSSION

Commercial purity aluminum was selected as a reference material
because it was readily available in bulk, plate, sheet and foil
geometries. Velocity measurements were taken on various thickness
specimens at several frequencies to check the agreement between
experimentally measured values of group velocity and the theoretically derived dispersion curve. In Fig. 5 the shear velocity
divided by the group velocity of the first longitudinal mode is
plotted versus the thickness-frequency product divided by the
shear wave velocity, V_S. The nondispersive shear wave velocity
was used to normalize the thickness-frequency product because the

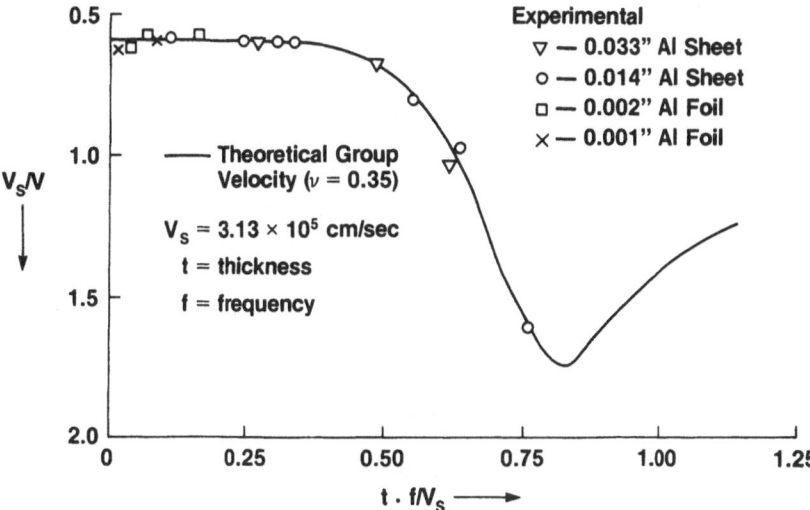

Fig. 5. Group velocity of first-order longitudinal Lamb wave in
        aluminum vs thickness-frequency product (theoretical
        curve after Meeker and Meitzler[1]).

phase velocity of the first longitudinal mode was not measured,
and the group velocity does not give an accurate value for the
ultrasonic wavelength. The theoretical curve was due to Meeker
and Meitzler[1] and was calculated for a Poisson's ratio of 0.35.

The theoretical group velocity displays a minimum at
$t \cdot f/V_S = 0.80$, where t is the ribbon thickness. Experimental data
points taken on aluminum foil and sheet at various frequencies fell
very near the theoretical curve. Agreement to better than 1% was
obtained between the group velocity predicted by theory and that
measured by experiment for commercial purity aluminum. Hence, it
could be assumed that accurate experimental group velocity
determinations could be made on materials for which no theoretical
velocity dependence was known.

Figure 6 shows the relationship between group velocity of the
first longitudinal mode and frequency for Metglas sheet 5 cm in
width, 41 μm in thickness and 10 cm in length. In this case, the
group velocity was plotted versus frequency because all samples
were the same thickness. For reference, 10 MHz corresponds to
$t \cdot f/V_S = 0.14$; this is well within the linear portion of the
dispersion curve in Fig. 5. It is necessary to determine, V_L, the
velocity of the first-order longitudinal Lamb wave at zero

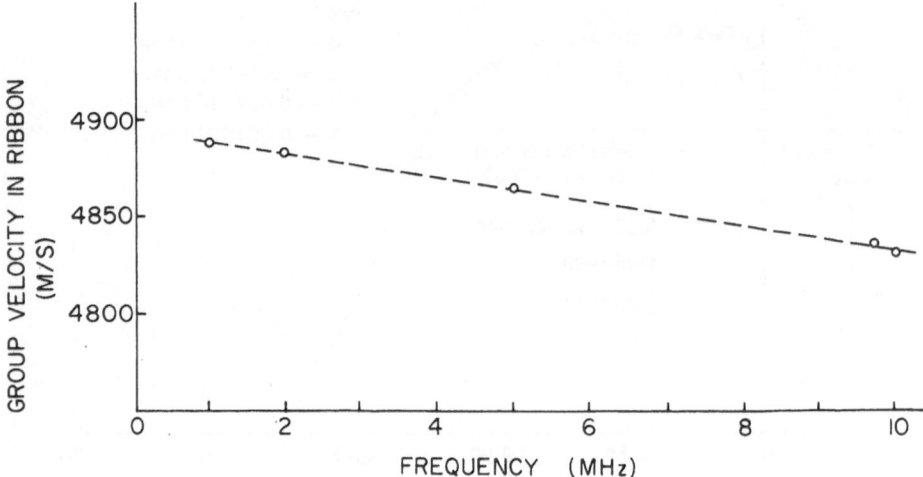

Fig. 6. Group velocity of first-order longitudinal Lamb wave vs
         frequency for Metglas ribbon.

frequency so that the elastic moduli of the ribbon, foil or sheet
can be calculated using the equations given in Table 1. In Fig. 6
a dotted line is drawn through the velocity data points taken at
various frequencies and is extended in the direction of zero
frequency. For Metglas, the line intersects the vertical axis at
a value of 4895 m/sec, which is the Lamb velocity V_L. The SH (0)
wave velocity of Metglas was also measured (2860 m/sec at 5 MHz).
Ideally, the SH (0) mode is nondispersive for a constant thickness.
In practice, it was found that the ribbon must be several
wavelengths wide to prevent dispersion due to edge effects.

     A rather different velocity-frequency behavior was exhibited
by amorphous Pd-Cu-Si ribbon, as can be seen in Fig. 7. A
prominent dip in the group velocity was observed that looks similar
to the dip in Fig. 5. On closer inspection the dip occurred at
approximately 2.5 MHz which corresponds to a t·f/V_S ratio of 0.07.
Since this value definitely lies within the linear region of
Fig. 5, some factor other than ribbon thickness must be responsible.

     It was shown earlier (Figs. 1 and 5) that dispersion took
place in a region where the thickness was on the same order as the
acoustic wavelength. At high values of thickness-wavelength ratio,
the velocity approached that of a Rayleigh wave and at low values
it approached V_L, the value for a thickness-wavelength ratio of
zero. It is also possible for dispersion to occur due to the
width of the ribbon.[6] In this case the dispersion due to ribbon
width is superimposed on that due to ribbon thickness. This

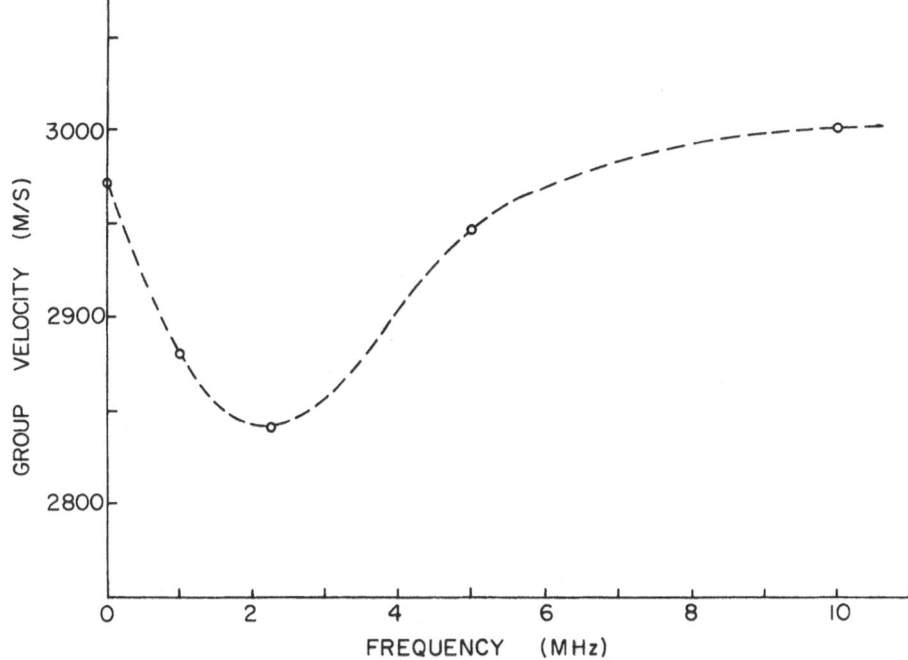

Fig. 7. Group velocity of first-order longitudinal Lamb wave vs
frequency for amorphous Pd-Cu-Si ribbon.

yields a very complicated dispersion relation in which the
character of the wave is initially extensional (rod-type), then
with increasing frequency changes to Lamb (plate-type) and finally
to Rayleigh (surface-type). If the specimen has a square or
round cross-section, the wave remains extensional in character
until the frequency is high enough for the transition to a
Rayleigh wave to occur. The width-to-thickness ratio for the
Pd-Cu-Si ribbons was 28:1. In the region of the group velocity
minimum at 2.5 MHz, the acoustic wavelength became comparable to
the ribbon width. A group velocity dip due to ribbon thickness
was not observed for Pd-Cu-Si because it is estimated that the
acoustic wavelength becomes comparable to the thickness at 50 MHz
which was above the 11 MHz limit of this investigation. No
dispersion due to width was observed in the cases of aluminum
(Fig. 5) and Metglas (Fig. 6) because the width-to-thickness
ratios were greater than 1000:1. A group velocity dependence
similar to that of Pd-Cu-Si was observed for amorphous Cu-Zr
ribbon which had almost identical thickness and width dimensions.

A group velocity minimum due to width can, as in the case of
Pd-Cu-Si and Cu-Zr ribbon, obscure the low-frequency value for the
velocity of the first-order longitudinal Lamb wave, V_L. For these
specimen geometries the ribbon width is small enough for the

longitudinal extensional wave velocity, V_E, to be determined at the
low-frequency end of the dispersion curve. In this regime the
acoustic wavelength is much greater than both the thickness and the
width of the ribbon specimen. The low-frequency extensional wave
velocities were measured by Rosen et al.[7] and Chang et al.[8] for
Pd-Cu-Si and Cu-Zr, respectively. Their technique utilized a
Nd-Yag laser to generate ultrasonic waves and a piezoelectric
transducer to receive them. This technique has been shown to be
very useful for the measurement of the change in Young's modulus
of amorphous Pd-Cu-Si and Cu-Zr ribbons with time during
crystallization heat treatment.

The 2.1 cm by 3.5 mm diameter cylinder of Pd-Cu-Si was large
enough for the propagation of bulk longitudinal and shear
ultrasonic waves. The pulse-echo overlap technique,[5] which is more
accurate than the pitch-catch method, was used to measure the
velocity in both cases. The longitudinal wave velocity was
4495 m/sec and the shear wave velocity was 1787 m/sec. These
values are in very good agreement with those reported by other
researchers for the same material.[9]

Using the equations listed in Table 1, values for the
Young's, shear and bulk moduli and Poisson's ratio were calculated
for Pd-Cu-Si, Cu-Zr, Metglas and commercial purity aluminum. In
addition, values for the velocity of the bulk longitudinal wave,
V_ℓ, and the longitudinal-type extensional wave, V_E, were calculated
for Metglass from V_L and V_S. For Pd-Cu-Si and aluminum, V_ℓ, V_S and
V_E were obtained from direct measurements. Finally, for Cu-Zr, V_E
and V_S were determined by direct measurements and V_ℓ was calculated.
Table 3 summarizes the results obtained for elastic moduli,
Poisson's ratios and sound wave velocities.

CONCLUSIONS

The present investigation showed that it is possible to
determine all the elastic moduli of melt spun metallic ribbons
50 µm or less in thickness. This was achieved by measurement of
the wave velocities V_L, V_E, and V_S along with the mass density of
the material and insertion of these values into proper equations.

A method for extrapolating the velocity of the first-order
longitudinal Lamb wave, V_L, from measurement of the group
velocity at several frequencies was utilized successfully for
ribbons and foils which were many acoustic wavelengths in width.

The low-frequency Lamb wave velocity, V_L, could not be
determined for narrow ribbons whose width was on the same order as
the acoustic wavelength. Instead, the longitudinal extensional
wave velocity, V_E, must be used for elastic moduli determination.

Table 3. Summary of Experimental and Calculated Values for Sound Velocities and Isotropic Elastic Moduli of Ribbon and Foil Specimens

Alloy	V_E(cm/s)	E(dynes/cm^2)	V_S(cm/s)	G(dynes/cm^2)	V_l(cm/s)	K(dynes/cm^2)	ν
Pd·Cu·Si	2.97×10^5	9.45×10^{11}	1.787×10^5	3.36×10^{11}	4.495×10^5	16.78×10^{11}	0.41
Cu·Zr	3.30×10^5	7.98×10^{11}	1.98×10^5	2.87×10^{11}	$4.67 \times 10^{5*}$	12.15×10^{11}	0.39
Metglas 2605·SC	4.64×10^5	15.72×10^{11}	2.86×10^5	5.97×10^{11}	$5.56 \times 10^{5*}$	14.56×10^{11}	0.32
Aluminum Foil	5.12×10^5	7.09×10^{11}	3.13×10^5	2.65×10^{11}	6.36×10^5	7.19×10^{11}	0.34

*Longitudinal velocities calculated from experimentally observed extensional and shear velocities.

Conventional ultrasonic velocity measurement techniques and apparatus can be adapted to make accurate group velocity measurements of guided waves that propagate in ribbon and foil geometries.

ACKNOWLEDGEMENTS

This research was supported by the Defense Advanced Research Projects Agency (DARPA) under DARPA Order No. 4275. The authors wish to gratefully acknowledge Dr. Robert Mehrabian, Dr. Louis Testardi, and Dr. H.N.G. Wadley of the National Bureau of Standards (Gaithersburg, Maryland), and Dr. Emanuel Horowitz, Director of the Center for Materials Research, The Johns Hopkins University, for their encouragement and support. The technical assistance of F. Biacaniello of the National Bureau of Standards, S. Fink and B. Elkind of Johns Hopkins is appreciated.

REFERENCES

1.   T. R. Meeker and A. H. Meitzler, Guided Wave Propagation, in: "Physical Acoustics," Vol. I, Part A, W.P. Mason, ed., Academic Press, New York (1964).
2.   H. Kolsky, "Stress Waves in Solids," Dover Publications, New York (1963).
3.   R. S. Davis, Determination of Silicon Density to High Precision Using a Submersible Servo-Controlled Balance, Metrologia. 18, 193 (1982).
4.   I. L. Gelles, Ultrasonic-Pulse Propagation Through Films, Foils, Fibers, and Whiskers, J. Acoust. Soc. Am. 40, 138 (1966).

5. J. E. May Jr., Precise Measurement of Time Delay, 1958 IRE
    Natl. Conv. Record 6(2), 134 (1958).
6. R. W. Morse, Dispersion of Compressional Waves in Isotropic
    Rods of Rectangular Cross Section, J. Acoust. Soc. Am. 20,
    833 (1948).
7. M. Rosen, H. N. G. Wadley, and R. Mehrabian, Crystallization
    Kinetics Study of Amorphous Pd-Cu-Si by Ultrasonic
    Measurements, Scripta Met. 15, 1231 (1981).
8. J. C. Chang, F. Nadeau, M. Rosen and R. Mehrabian,
    Crystallization Kinetics Study by Ultrasonic and
    Microhardness Measurements, Scripta Met. 16, 1073 (1982).
9. L. A. Davis, Mechanics of Metallic Glasses, in: "Rapidly
    Quenched Metals, Sec. I," N.J. Grant and B.C. Giessen,
    eds., MIT Press, Cambridge (1976).

APPLICATION OF ULTRASONIC TECHNIQUE

TO CURE CHARACTERIZATION OF EPOXIES

H. Thomas Hahn

Washington University

St. Louis, MO 63130

INTRODUCTION

During cure a polymer changes from viscous liquid to visco-elastic rubber, and finally, to a glassy state. On the microscopic level, polymer molecules grow into longer chains with branches and also crosslinking occurs [1,2].

The knowledge of property changes during cure is essential to the establishment of an optimum process cycle. Let alone the prevention of under or over cure, the information can be used to determine and modify the flow field during a molding operation. When fiber reinforcement is involved, the flow field controls the fiber orientation and may result in an undesired fiber distribution.

Voids can form during cure because of the presence of volatiles. In composite laminates, the voids mostly are found as interlaminar defects between constituent plies. These voids can be suppressed by applying pressure at a proper time. If the pressure is applied too early, much resin will be squeezed out, resulting in a resin-starved composite. On the other hand, a late application of pressure may not remove voids because of the increased viscosity of the resin. Thus, one must know the change of viscosity to establish a proper pressure window.

Residual stresses are induced when there is a nonuniformity in cure or temperature distribution, and also when there are geometric constraints [3]. The residual stresses in neat resins may be removed by annealing since their elastic change with temperature is uniform. However, the residual stresses in composite laminates are caused by the expansional mismatch between the matrix

resin and the fibers [4]. Since these stresses are usually low at
elevated temperatures, annealing does not work well. Thus, an
accurate analysis of residual stresses becomes more important for
composite materials.

There are many techniques available for monitoring cure. All
these techniques are based on the change of properties of one type
or another. Some of the properties frequently used are electrical
resistance, dielectric loss tangent, wave absorbancy, complex
modulus, and heat generation [2,5,6]. Since the design of a process
cycle depends mostly on the thermomechanical properties, the last
two methods are of prime interest.

The cure of most epoxies is exothermic and generates heat. The
time rate of heat generation is measured in differential scanning
calorimetry (DSC). The completion of cure is then indicated by
the lack of heat generation. The heat generated to the present
time is compared with the total heat of reaction to determine the
degree of cure. This method, however, does not provide any information on mechanical behavior.

The mechanical properties frequently measured are the complex
moduli in shear and the viscosity. Both the storage modulus and
the shear viscosity increase with cure, the type of increase being
sigmoidal for the former and exponential for the latter. The loss
modulus initially increases to a peak and then decreases. The
measurements of the complex moduli are usually taken in a low frequency range of up to 30 Hz.

In the ultrasonic characterization of cure, wave speed and
attenuation are monitored [7,8,9]. Both the longitudinal and shear
waves have been used. However, no attempt has been made to correlate
the wave characteristics to the material property parameters.

This paper describes an ultrasonic characterization of cure of
two epoxy systems by using the longitudinal wave. A simple constitutive model is proposed to explain the experimental data on the
wave speed and attenuation.

EXPERIMENTAL PROCEDURE

Two epoxy systems were used in this investigation: Epon 828/Z
(80/20) and Epon 815/V140 (60/40) where the numbers inside the paren-
theses denote the weight ratios. Epon 828, Epon 815 and curing agent
Z are made by Shell Chemical Company whereas curing agent V140 is a
product of Henkel Corporation. Epon 828 is a diglycidyl ether of
bisphenol A and Epon 815 is a diluted version of the same resin.

Epoxy was mixed with the appropriate curing agent and degassed in vacuum. It was then poured ~4 mm deep into an aluminum container ~58 mm in diameter. A hole was drilled in the bottom of the container for direct contact between the epoxy and the transducer. A vacuum grease was applied on the face of the transducer to prevent adhesion.

The wave speed and attenuation were measured by using a Sonoray 303B ultrasonic flaw detector at a frequency of 10 MHz. An Alpha series 10 MHz transducer was used in the pulse-echo mode.

Three different cure temperatures were employed: room temperature (RT), 50°C and 65°C. The cure at the elevated temperatures was achieved in an air circulating oven. A thermocouple was embedded in the epoxy to monitor the temperature.

RESULTS AND DISCUSSION

An acoustic planar wave traveling at a speed c through a homogenous medium is described by

$$A = A_o e^{-\alpha x} e^{i\omega(t - x/c)} \tag{1}$$

Here ω is the frequency, t the time, x the position coordinate, and α is the attenuation coefficient.

The wave speed c after the start of cure was normalized with respect to the initial wave speed c_L, i.e., the wave speed in liquid state at RT right after mixing. Thus, the relative wave speed is defined as c/c_L.

The relative attenuation (RA) is defined as the ratio of A_L/A expressed in dB, i.e.,

$$RA = 20 \log(A_L/A) \tag{2}$$

Here A_L is again the initial amplitude of the echo. Combining Eqs. (1) and (2), we obtain

$$\alpha - \alpha_L = \frac{1}{20 \log e} \frac{RA}{h} \tag{3}$$

where h is twice the depth of the specimen and e is the Naperian base. Thus, the change of the attenuation coefficient can be assessed by measuring RA.

Figure 1 shows the variations of c/c_L, RA and temperature with cure time for Epon 815. Although the reaction is exothermic, it is

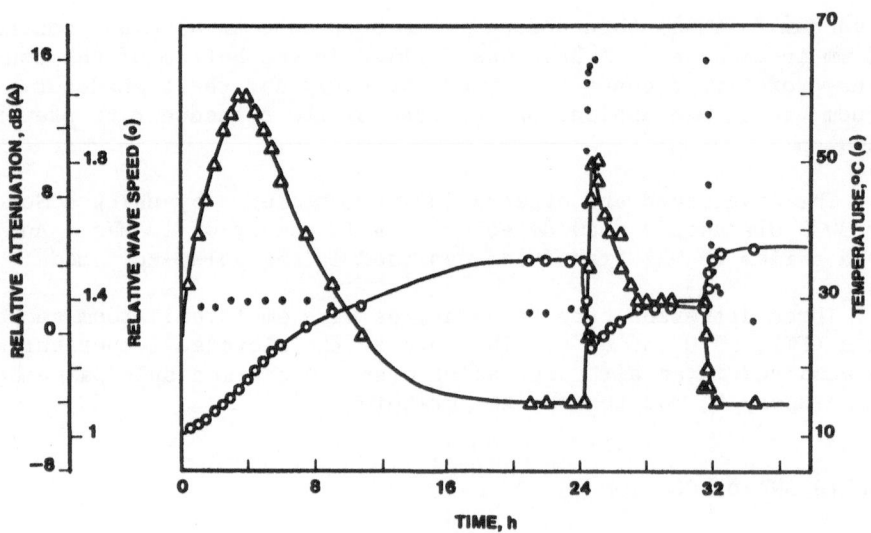

Fig. 1.   Relative wave speed, relative attenuation and temperature
during cure at room temperature, Epon 815/V140.

too slow to give rise to any appreciable temperature increase. The
wave speed increases monotonically as the epoxy hardens. The relative attenuation also increases, but after reaching a peak, it drops
down to a value lower than the initial value.

To ascertain if the stabilization of the wave speed and attenuation is an indication of complete cure, the specimen was put into
an oven at 65°C. The resulting changes in Fig. 1 indicate that the
cure was not complete. After the temperature cycle, the cure was
confirmed to be complete by applying another temperature cycle,
although not shown in the figure. It is to be noted that the characteristics of the transducer also change at elevated temperature and
these changes are included in the measurements.

The same type of changes are observed for Epon 828 at RT, Fig. 2.
Similarly to Epon 815, cure is not complete even after 24 h at RT.
When the specimen is brought back to RT, the attenuation unexpectedly
jumps up and then exponentially decreases. Such behavior is not
observed for Epon 815 in Fig. 1.

At 50°C, the cure is much faster for both epoxies, Figs. 3 and 4.
After a combined temperature history at 50°C and RT, a complete
cure is realized as evidenced by the absence of unstable changes during the reheating. The changes of attenuation for Epon 828 during
the temperature changes are quite different from those for Epon 815.
The differences appear at all three temperatures.

The initial drop in attenuation is the result of the epoxy

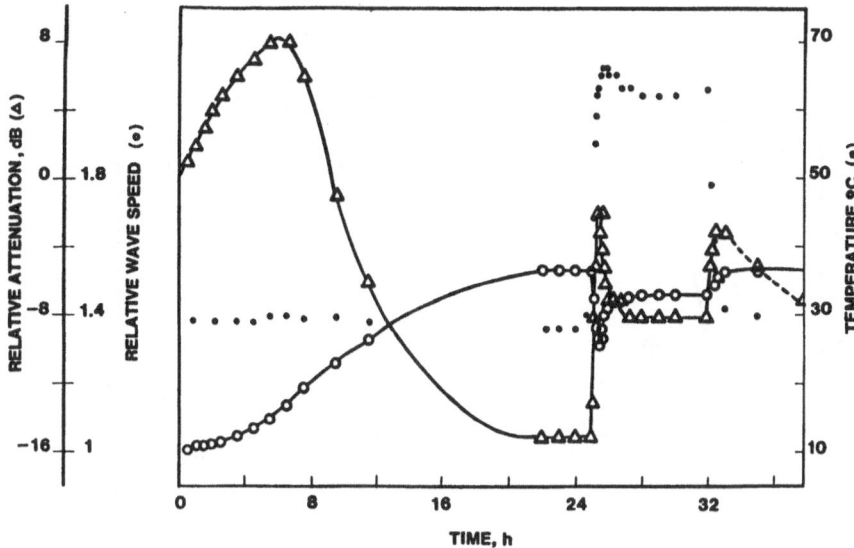

Fig. 2. Relative wave speed, relative attenuation and temperature during cure at room temperature, Epon 828/Z.

losing its viscosity as the temperature increases. There are two competing mechanisms affecting the viscosity. A higher temperature reduces viscosity but accelerates cure. Advancing cure decreases viscosity. The final viscosity and hence the attenuation depend on the temperature and the degree of cure.

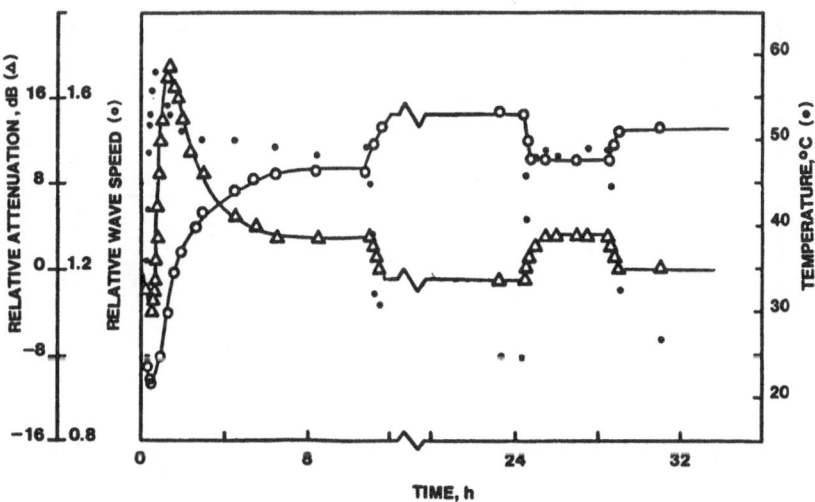

Fig. 3. Relative wave speed, relative attenuation and temperature during cure at 50°C, Epon 815/V140.

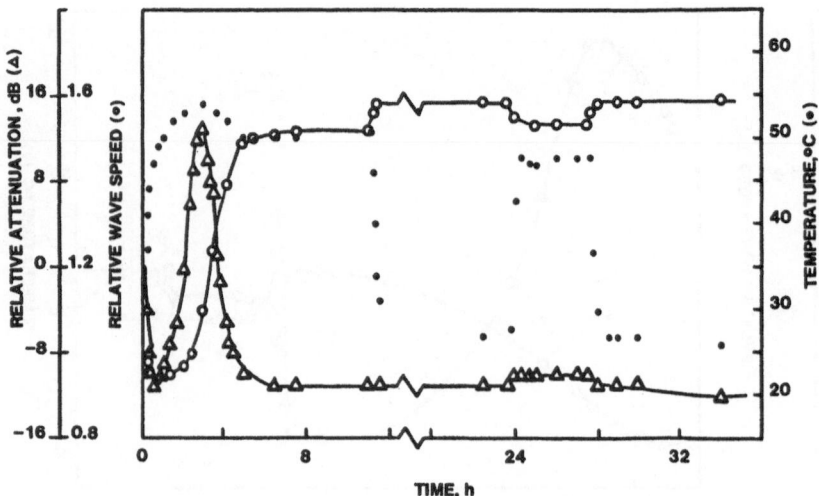

Fig. 4.   Relative wave speed, relative attenuation and temperature
          during cure at 50°C, Epon 828/Z.

     The changes at 65°C are similar to those at 50°C, Figs. 5 and 6.
Again, the change of attenuation is as expected for Epon 815 but not
quite so for Epon 828.   That is, an increase in attenuation is
observed with decreasing temperature for the latter epoxy.

     At the elevated temperatures, an accelerated reaction can be
seen from the epoxy temperature rising above the oven temperature.
The epoxy temperature peaks out about the same time as the attenuation does.

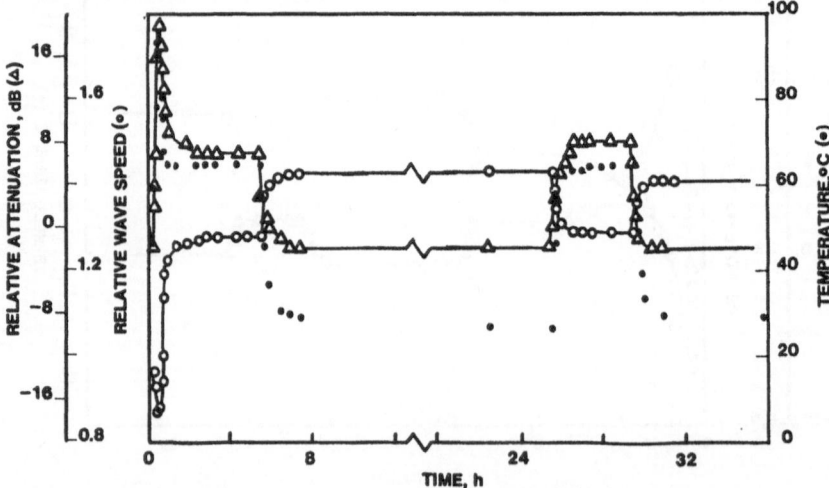

Fig. 5.   Relative wave speed, relative attenuation and temperature
          during cure at 65°C, Epon 815/V140.

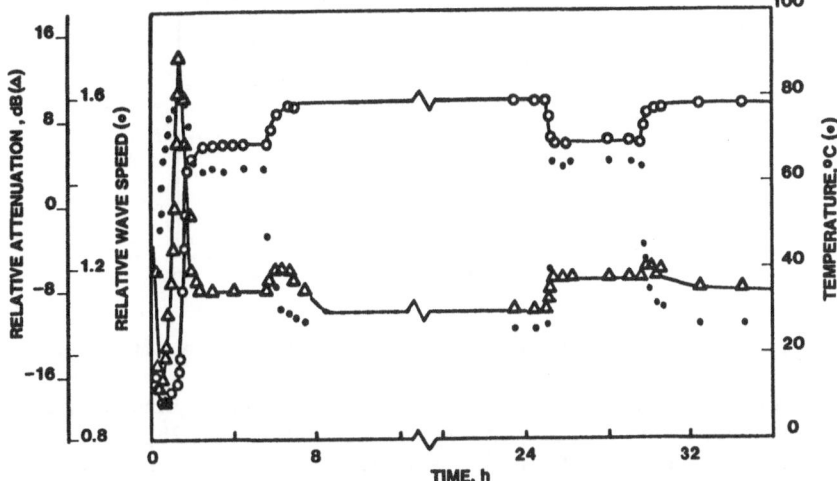

Fig. 6. Relative wave speed, relative attenuation and temperature during cure at 65°C, Epon 828/Z.

The times to peak attenuation were measured and are plotted against the cure temperatures in Fig. 7. The figure is similar to the time-temperature-transformation diagram proposed in [2]. Since there is only one attenuation peak, the cure temperatures chosen in the present study are believed to be below the respective gel glass transition temperature of the epoxies.

The variation of wave speed with temperature after complete cure is shown in Fig. 8. For Epon 828, the wave speed does not change much over the temperature range studied. However, the wave speed in Epon 815 starts to drop around 65°C. The glass transition

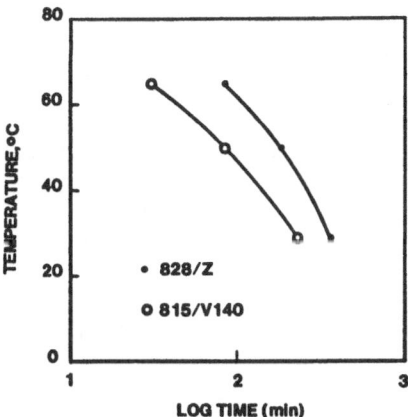

Fig. 7. Change of time to peak attenuation with cure temperature.

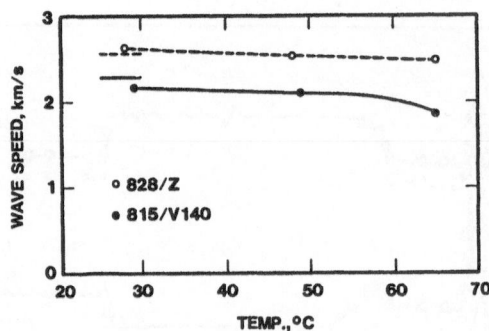

Fig. 8. Change of wave speed with temperature after cure.

temperatures inferred from the separate thermal expansion measurements are ~60°C and ~115°C for Epon 815 and Epon 828, respectively.
Thus, the softening of Epon 815 around 60°C is expected to reduce
the corresponding wave speed.

The wave speeds calculated from the elastic moduli separately
determined by tensile testing are shown as two lines between 25 and
30°C. They are quite close to the measured speeds.

CONSTITUTIVE MODELING

Since the wave propagation involves an infinitesimal distur-
bance, the epoxy can be regarded as a linear viscoelastic material.
Introducing a complex longitudinal stiffness C_{11}^{*}

$$C_{11}^{*} = C_{11}' + iC_{11}''$$ (4)

we obtain the wave speed c as [10]

$$\rho c^2 = 2C_{11}' \frac{1 + \tan^2\delta}{1 + (1+\tan^2\delta)^{1/2}}$$ (5)

Here $\tan \delta$ is the ratio of the storage stiffness C_{11}' to the loss
stiffness C_{11}'', i.e.,

$$\tan \delta = C_{11}''/C_{11}'$$ (6)

The attenuation coefficient α is related to the complex stiffness by

$$\alpha = \frac{\omega}{c} \tan \frac{\delta}{2}$$ (7)

If the loss stiffness is much smaller than the storage stiffness, i.e. tan $\delta \ll 1$, Eqs. (5) and (7) reduce respectively to

$$c = (C_{11}'/\rho)^{1/2} \tag{8}$$

$$\alpha = \frac{1}{2} \frac{\omega}{c} \tan \delta = \frac{1}{2} \omega \rho^{1/2} C_{11}''/C_{11}'^{3/2} \tag{9}$$

Equations (8) and (9) indicate that the changes in C_{11}' and C_{11}'' will be similar to those in c and α, respectively.

To illustrate the change of C_{11}^* with cure, it is more convenient to introduce the complex bulk modulus K^* and the complex shear modulus G^* such that

$$C_{11}^* = K^* + \frac{4}{3} G^* \tag{10}$$

We now proceed to express the complex moduli in terms of the elastic moduli and viscosities.

In liquid state, the epoxy has a bulk modulus but no shear modulus. The viscosity changes from a low to a very high value as cure proceeds. Therefore, the bulk behavior requires a parallel combination of a Maxwell element and a spring element while the shear behavior requires a Maxwell element only. Referring to Fig. 9, we can express the components of the complex moduli in terms of the elastic constants and viscosities:

$$K' = \frac{k_f + k_s \omega^2 \tau_k^2}{1 + \omega^2 \tau_k^2} \quad , \quad K'' = (k_s - k_f) \frac{\omega \tau_k}{1 + \omega^2 \tau_k^2} \tag{11}$$

$$G' = g_s \frac{\omega^2 \tau_g^2}{1 + \omega^2 \tau_k^2} \quad , \quad G'' = g_s \frac{\omega \tau_g}{1 + \omega^2 \tau_g^2} \tag{12}$$

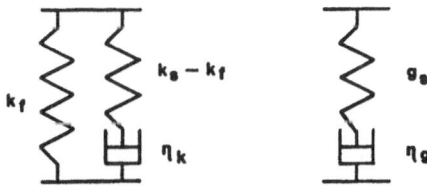

Fig. 9.   Models for bulk modulus and shear modulus.

Here the relaxation times τ_k and τ_g are defined by

$$\tau_k = \frac{\eta_k}{k_s - k_f} \quad , \quad \tau_g = \frac{\eta_g}{g_s} \tag{13}$$

If we further assume that

$$\tau = \tau_k = \tau_g \tag{14}$$

the storage and loss stiffnesses are given by

$$C_{11}' = \frac{k_f + (k_s + 4g_s/3)\omega^2\tau^2}{1+\omega^2\tau^2} \tag{15}$$

$$C_{11}'' = \frac{(k_s - k_f + 4g_s/3)\omega\tau}{1+\omega^2\tau^2} \tag{16}$$

The changes of C_{11}' and C_{11}'' with τ are thus similar to those of the wave speed and the relative attenuation, respectively, with time. In the limiting cases of $\tau = 0$ and ∞, C_{11}' reduces to k_f and $(k_s+4g_s/3)$, respectively. The peak of C_{11}'' occurs at $\tau = 1/\omega$ and is equal to $(k_s+4g_s/3)/2$.

The change of the complex modulus with time can be determined if the reaction between the relaxation time and the real time is known. To this end, we note [11] that the shear viscosity η_g depends on the degree of cure F by

$$\eta_g = \eta_{go} \exp(\frac{E_\eta}{RT}) \exp(B_1 F) \tag{17}$$

where η_{go} and B_1 are constants, E_η the activation energy for viscosity, R the universal gas constant, and T is temperature.

The degree of cure is defined as the ratio of the heat generated to the total heat of reaction [12]. The reaction kinetics is then expressed by an equation

$$\frac{dF}{dt} = B_2 \exp(-\frac{E_F}{RT}) f(1-F) \tag{18}$$

where B_2 is the constant and E_F is the activation energy for cure reaction. The function f depends on the order of kinetics involved [2].

The foregoing equations describe the changes of wave speed and attenuation during cure in terms of more basic variables. In the absence of any data for Eqs. (17) and (18), no quantative correlation between the theory and the ultrasonic data is made in the present paper. Yet the theory provides a framework in which property constants can be determined from the ultrasonic measurements.

Before cure, the shear viscosity is much smaller than the shear modulus of the cured epoxy, i.e., $\tau \ll 1$. Using Eqs. (8) and (15), we can calculate the bulk moduli from the wave speeds. It is interesting to note that the liquid bulk modulus is about half the solid bulk modulus:

k_f = 2.09 GPa, k_s = 4.76 GPa for Epon 815

k_f = 3.34 GPa, k_s = 6.65 GPa for Epon 828

The liquid densities used in the calculations are 1059 and 1125 kg/m^3 for Epon 815 and Epon 828, respectively. The solid bulk moduli were obtained from tensile tests.

CONCLUSIONS

Ultrasonic wave speed and attenuation are very sensitive to the degree of cure, and can be used to monitor cure. The measurements can be used to develop a constitutive model which is needed to optimize the cure cycle. A simple model consisting of springs and dashpots has been proposed in the present paper. The model identifies appropriate material constants and provides a general framework in which various cure monitoring techniques can be combined.

ACKNOWLEDGEMENTS

This paper is based on work supported by the National Science Foundation under Grant No. MEA-8110777 with Clifford J. Astill as Program Director. An appreciation is extended to J. Choi for the experimental work.

REFERENCES

1. P. J. Flory, "Principles of Polymer Chemistry," Cornell University Press, Ithaca (1953).
2. J. B. Enns and J. K. Gillham, The time-temperature-transformation (TTT) cure diagram: modeling the cure behavior of thermosets, ACS, Div. Org., Coatings, & Plastics Chem., Preprints, 48:575 (1982).

3. M. Levitsky and B. W. Shaffer, Thermal stresses in chemically hardening elastic media with application to the molding process, J. Appl. Mech. 96:647 (1974).

4. H. T. Hahn, Residual stresses in polymer matrix composite laminates, J. Composite Materials 10:266 (1976).

5. J. F. Carpenter, Instrumental techniques for developing epoxy cure cycles, MCAIR 76-003, McDonnell Douglas, (1976).

6. C. D. Han and K. W. Lem, Rheology-curing kinetics-molecular weight relationships of thermosetting polyester resin, ACS, Div. Org. Coatings & Plastics Chem., Preprints 48:581 (1982).

7. G. A. Sofer, A. G. H. Dietz and E. A. Hauser, Cure of phenol-formaldehyde resin, Ind. Eng. Chem. 45:2743 (1953).

8. A. G. Dietz, E. A. Hauser, F. J. McGarry and G. A. Sofer, Ultrasonic waves as a measure of cure, Ind. Eng. Chem. 48:75 (1956).

9. A. M. Lindrose, Ultrasonic wave and moduli changes in a curing epoxy resin, Exp. Mech. 18:227 (1978).

10. R. M. Christensen, "Theory of Viscoelasticity: An Introducrion," Academic Press (1971).

11. A. M. Stolin, A. G. Merzhanov and A. Y. Malkin, Nonisothermal phenomena in polymer engineering and science: a review. Part II: Non-isothermal phenomena in polymer deformation, Polym. Eng. Sci. 19:1074 (1979).

12. S. Sourour and M. R. Kamal, Differential scanning calorimetry of epoxy cure: Isothermal cure kinetics, Thermochemica Acta 14:41 (1976).

ULTRASONIC MEASUREMENTS OF ELASTIC MODULI OF THERMALLY

CYCLED METAL MATRIX COMPOSITE PRECURSOR WIRES

Robert W. Reed

Applied Research Laboratory
The Pennsylvania State University
University Park, Pennsylvania

and

Albert L. Bertram
Naval Surface Weapons Center
Silver Spring, Maryland

INTRODUCTION

Metal matrix composite structures are often made by the
consolidation of precursor wires typically on the order of 1 mm
in diameter. The elastic and mechanical properties of the consolidated parts depend critically on the elastic and mechanical
properties of the precursor wires. The consolidation process
involves hot pressing at elevated temperatures. The final end-
item products may also be subjected to elevated temperatures in
their normal use cycles.

We have developed an ultrasonic method that permits the
continuous measurement of the ultrasonic attenuation and velocity
of longitudinal and/or torsional waves as the precursor wire is
manufactured[1]. The longitudinal and torsional velocities are
convertible into Young's and the shear moduli. Since very precise
and repeatable measurements of these velocities are readily made,
it was decided to use this apparatus to measure the modulus
changes induced by the thermal cycling of the precursor wires. As
a point of calibration, the velocities were also measured in a set
of untreated wires manufactured to contain a range of fiber volume
fractions. Our work to date has been concentrated on graphite

(Gr)/Aluminum (Al) metal matrix precursor wires. The apparatus
has been shown to be compatible with other metal matrix composite
systems.

ACOUSTIC MEASUREMENT METHODS

    Electromagnetic acoustic transducers (EMATs) have been used
for the generation of both sound modes employed. This permits
high speed scanning of the precursor wires since no fluid acoustic
couplant is required with these noncontacting transducers. In
addition, access to the ends of the wire is unnecessary since the
driver and receiver transducers require access only to the sides
of the wire.

    The precursor wires suffer from the characteristic of having
diameters which are irregular (including non-circular) with
typical variations of 10%. Hence, the transducer coils must have
sufficiently large inside diameters to permit the largest diameter
precursor wire to pass through unhindered. As a result, the lift
off distance of the wire to the transducer windings varies
substantially as the wire travels through the transducer array.
Typically, the individual receiver output voltage levels may show
variations of ± 10 dB or more as the lift-off distance changes
with the wire motion. Since the typical attenuation of the
ultrasound is less than 1 dB/cm, the lift off variations can
easily obscure the signal changes of interest.

    We have developed apparatus which completely accounts for the
lift off variations and permits accurate attenuation and velocity
measurements. Figure 1 schematically shows the experimental
arrangement. Two receiver EMATs are placed between two transmitter EMATs. The distance Δx between the two receiver EMATs is
the gauge distance for the velocity and attenuation measurements.
The transmitter EMATs are bidirectional, sending acoustic pulses
in both directions along the wire. Let V_{ij} be the peak voltage
at receiver EMAT j when transmitter EMAT i is excited by a narrow
dc pulse of ~ 100 amps peak current. Simple algebra can be used
to show that the attenuation, α, in the wire is then given by

$$\alpha = \frac{1}{2} \frac{1}{\Delta x} \left[\log_{10} \left(\frac{V_{11}}{V_{12}} \right) - \log \frac{V_{21}}{V_{22}} \right] \tag{1}$$

independently of the EMAT lift off distances and conversion
efficiency differences. This result is exact for a stationary
measurement. In our case, the wire is continuously pulled through
the transducer array at speeds of ~ 1 cm/sec. Since the acoustic
travel times between receiver transducers are typically ~ 10 to
40 μsec, the stationary approximation has been found to hold quite
well for the present case.

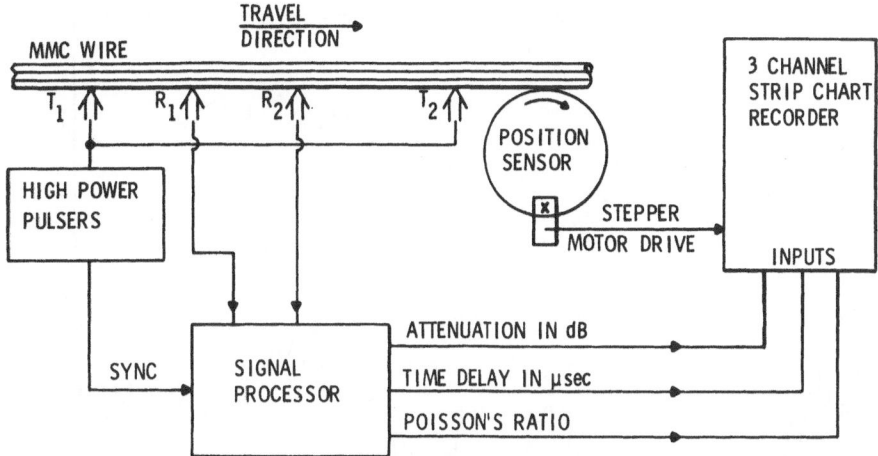

Figure 1.   Schematic representation of ultrasonic apparatus. T_1
and T_2 are transmitter EMATs, R_1 and R_2 are receiver
EMATs.

A four-channel gated peak detector of novel design has been
constructed to implement Eq. 1. The circuitry provides for the
automatic gain control of the signal voltages to permit stable
triggering, at a precise phase location, of a time interval
measuring circuit. Hence, the precise time delay of the signal
propagating between any two channels is measured. Typically,
10 ns resolution for 1 mHz single cycle pulses is achieved on a
single shot basis. With signal averaging this resolution can be
readily improved to better than 1 ns. The attenuation output has
a 40 dB dynamic range with a resolution of ~ 0.25 dB. The
attenuation and time interval measurements can follow signal level
changes of up to 3 dB per repetition cycle. Analog outputs for
both attenuation and velocity are available for recording on x-y
or strip chart recorders. By careful calibration of the gauge
length, Δx , between receiver transducers the attenuation in
dB/cm or velocity of sound in cm/sec can be computed from the
measured attenuation and time interval values.

VELOCITY AND ATTENUATION MEASUREMENTS IN GR/AL MMC PRECURSOR WIRES

Figure 2 shows representative attenuation, α , and time delay
data, Δt , for VSB-32 GR/6061 Al metal matrix composite wire of
good quality (i.e., fully infiltrated). The finest details in the
Δt plot are entirely repeatable, although the exact cause of the
structure in the curve is not fully understood. The smallest
details of the α curve are random noise whereas the larger details
are repeatable, i.e., changes of $\geqslant 0.025$ dB/cm are repeatable. A
comparable plot for simple Al alloy wire would show almost
completely flat Δt and α curves which are devoid of any of the

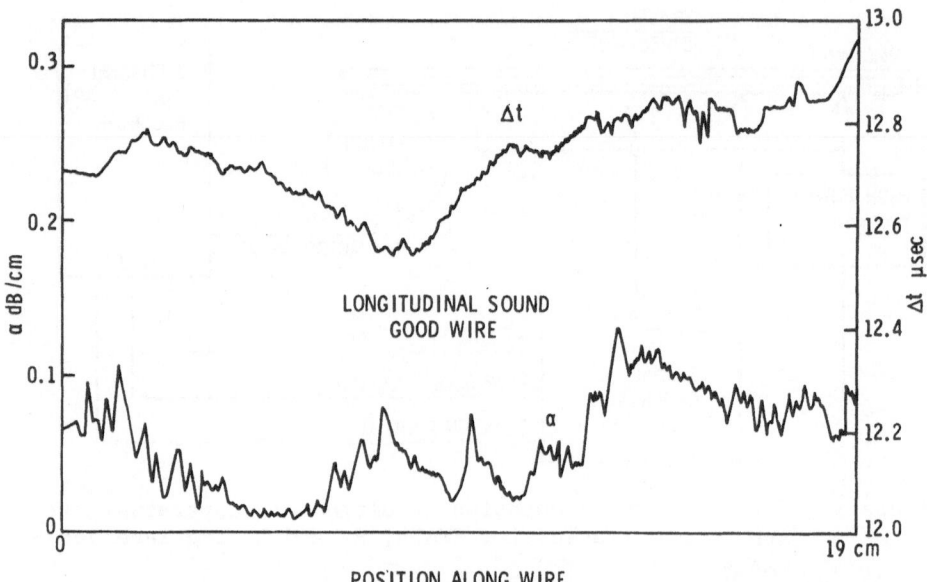

Figure 2. Representative longitudinal attenuation and time-delay
          measurements in good quality Gr/Al precursor wires.
          The sound frequency is ~ 2 MHz.

details seen in this figure. We have previously shown that α is a
good measure of the quality of the infiltration of the Al into the
Gr fiber tow[1].

    The effects of the fiber moduli can be readily seen in the
Δt measurements. Figure 3 gives an example of longitudinal and
torsional velocities measured in Gr/Al of ~ 45% fiber volume. The
VSB-32 (Union Carbide designation) fiber has nominal longitudinal
modulus of 345 GPa, whereas the VS0054 has a nominal modulus of
627 GPa. The shear moduli for both fibers are approximately
10.3 GPa[2,3]. It is apparent from the figure that the longitudinal
velocities, for a 2 MHz frequency, differ as expected for the
different moduli, whereas the torsional velocities, which depend
only on the shear modulus for a torsional frequency of 200 kHz,
are essentially the same for the two wires.

    Figure 4 shows the average velocities measured in a group of
wires with calibrated fiber volume fractions. All the fibers are
type VSB-32 Gr, and the matrix is 6061 Al. The data are compared
with velocity values calculated using Hashin's theory[4] for
calculation of the shear modulus, μ , of a composite wire, and
with values calculated from a composite Young's modulus, E ,
derived from a rule of mixtures calculation. The velocities are
obtained from the moduli by

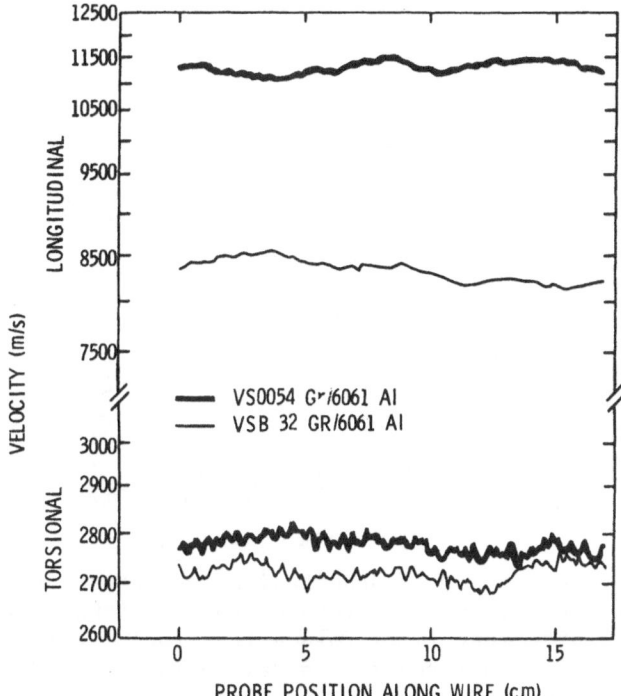

Figure 3.   Comparison of the sound velocities for precursor wires
            containing VS0054 and VSB-32 graphite fibers. The
            torsional frequency is ~ 200 kHz and the longitudinal
            frequency is ~ 2 MHz.

$$V_L = (E/\rho)^{1/2} \qquad\qquad\qquad (2)$$

and

$$V_T = (\mu/\rho)^{1/2} \qquad\qquad\qquad (3)$$

where ρ is the composite mass density computed by a rule of
mixtures calculation. It is seen from this figure that the
measured longitudinal velocity, V_L , is in generally good
agreement with the predicted values. Except for six maverick
points (the wires for which show good V_L values), the torsional
data are in fairly good agreement with the predicted values.
(It is noted that the longitudinal properties, in particular
tensile strength, are generally better behaved than the shear
properties of Gr/Al composites.) Of particular importance to our
later discussions is the fact that the maverick torsional data
points are comparable to velocities of graphite alone which has a
significantly smaller shear modulus than that of 6061 Al alone.
The implication here is that this may be due to very poor interfacial bond strengths at the fiber/matrix interface.

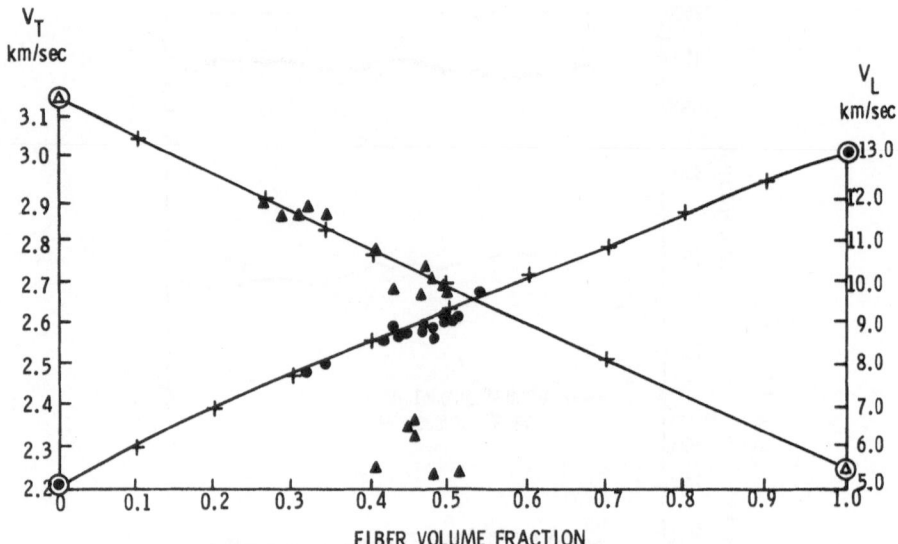

Figure 4. Comparison of longitudinal and torsional sound
           velocities with theoretical predictions for
           different fiber volume fractions. ▲ indicates
           torsional velocity data. ● indicates
           longitudinal velocity data.

HEAT TREATMENT RESULTS

The Gr/Al precursor wires were heat treated in a tube furnace
for which the temperature profiles are shown in Fig. 5. All of
the velocity measurements were made at room temperature after heat
treatment periods of a specified duration. After each velocity
scan, the wire was reinserted in the furnace for additional
treatment. Except for the case of extreme heat treatment by long
exposure to temperatures near the Al alloy melting point, the
longitudinal velocities show very little change with accumulated
heat treatment. (Tensile strength measurements after a completed
heat treatment cycle also show no significant degradation.)

The torsional velocities show considerable reduction with
heat treatment. Figure 6 shows typical torsional time delay for a
Gr/Al wire heat treated at 595°C for an accumulated time of
25 hours. The time delay is seen to continually increase with
additional heat treatment, i.e., the torsional velocity is slowing
down. Note also that the wire was inserted into the furnace in
such a way that the wire for position 0 to ~ 10 cm was quickly
pushed through the hot furnace but was otherwise untreated. This
led to a jump reduction in the velocity in this slightly heat
treated region with little further deterioration on each
additional insertion through the furnace. Since the central

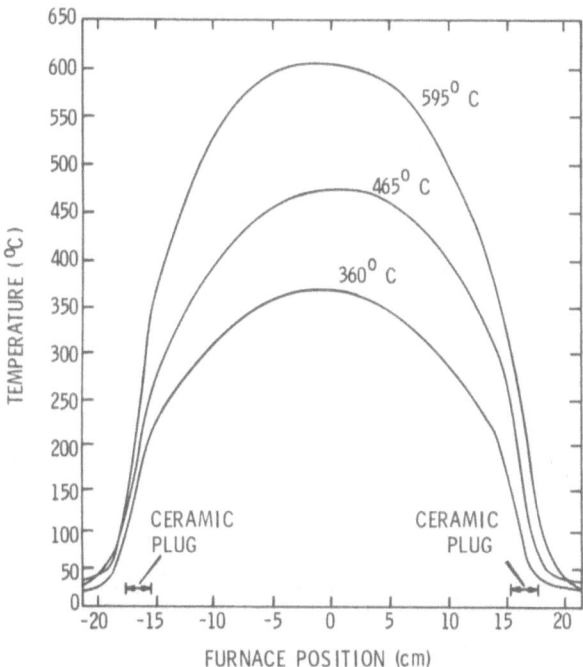

Figure 5.   Temperature profiles of the tube furnace.

10 cm portion of the furnace was uniformly heated to within
± 10°C, data from the wire located in this region was averaged to
obtain a velocity representation of heat treatment at the central
furnace temperature.

Figure 7 shows the velocity changes for three different wires
heat treated at 595°C. These wires include both VSB-32 and VS0054
fibers in a 6061 Al matrix. The initial torsional velocities are
seen to be in agreement with the calculated values of Fig. 4 for
wires with 45 to 50% graphite fiber. The final values, after
40 hours at temperature, show about a 25% reduction in torsional
velocity (~ 2,000 m/s which is below that for pure graphite alone,
see Fig. 4), which implies a destruction of the interfacial bond
between the Gr and Al. This information seems to imply that somehow differences in the wire production process led to the maverick
points in Fig. 4. (Five of the six bad wires in Fig. 4 were from
a different manufacturing source than all but one of the others in
in the figure.) There appear to be no differences in the
torsional velocity behavior of wires made from either of the
graphite fiber types.

Wires heat treated at 465°C show similar behavior to those
discussed above. Figure 8 shows data for wires exposed at 465°C.
Only the time scale of the deteriorization in velocity is changed.

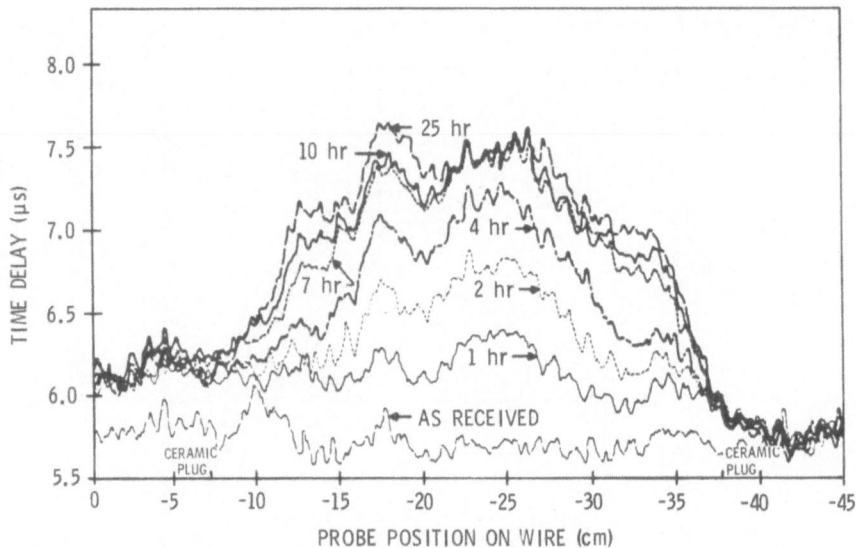

Figure 6.   Composite of torsional time delays for a wire heat
treated at 595°C.

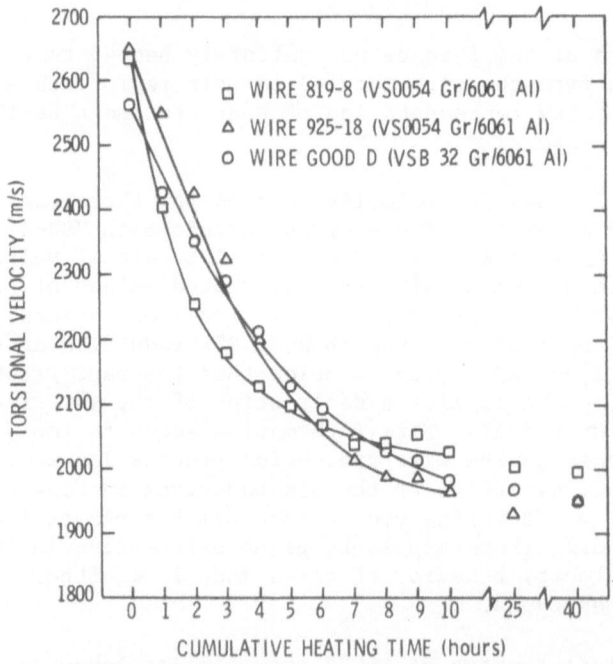

Figure 7.   Torsional velocity variation as a function of heat
treatment time at 595°C.

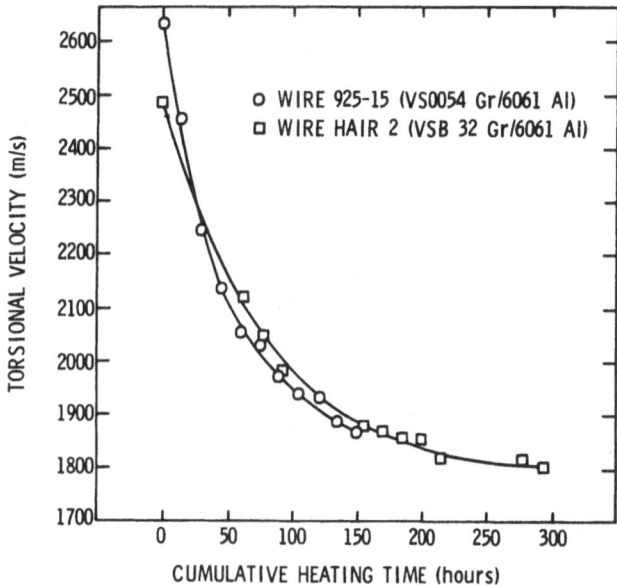

Figure 8. Torsional velocity variation as a function of heat
treatment time at 465°C.

Apparently the presumed destruction of the Gr/Al interface takes
place more slowly at this lower temperature. The lower "final"
velocity found at this temperature is not presently understood.

     Several wires were examined from a lot sold as production
material. The wires were first examined and found to be separated
into two groups - those with nominal torsional velocities and
those with abnormally low values, similar to the "maverick" values
in Fig. 4. These are termed as "good" and "bad" wires
respectively. The longitudinal velocities and attenuations of the
two wire kinds were the same within the normal variations seen.
Samples of each wire type were heat treated at 595°C for
successive one hour periods. The results are shown in Fig. 9.
It is seen that the bad wires show substantially more rapid
velocity reductions in the early treatment time intervals. This
result could be quite significant since precursor wires used to
make end products undergo significant heat treatment while they
are hot pressed in the consolidation process.

CONCLUSIONS

     Thermally cycled Gr/Al precursor wires were found to show
significant reductions in their shear moduli. The results may be
significant in view of the hot processing method used to consol-
idate precursor wires into end product materials. In addition,
the predicted thermal history for various end product applications

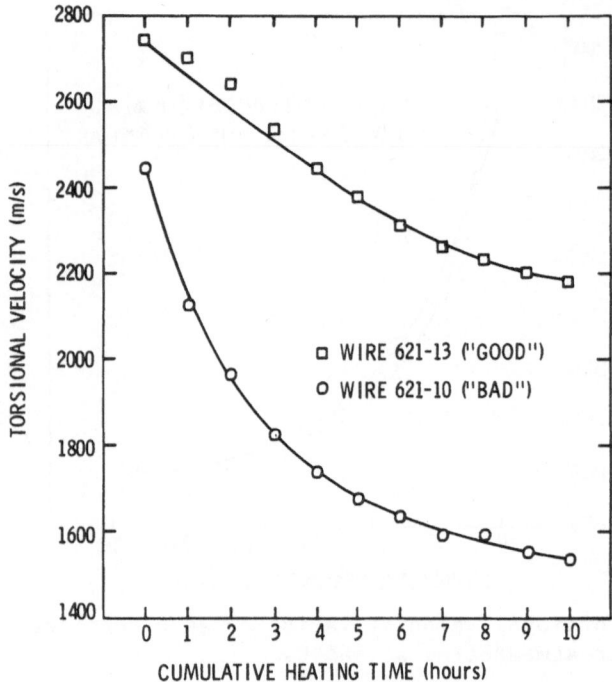

Figure 9. Differences in torsional velocity degradation for
          "good" and "bad" precursor wires heat treated at
          595°C.

should be considered in light of the present results. Additional
measurements involving more carefully controlled environments
in the tube furnace, and on consolidated plates are clearly
implied.

REFERENCES

1. Reed, R. W., "Quantitative Attenuation and Velocity
    Measurements in Metal Matrix Precursor Wires," Proceedings
    of the AF/DARPA Review of Progress in Quantitative NDE,
    La Jolla, CA, 1-6 August 1982.
2. Frost, H. M., Prout, J. H., & Reed, R. W., "Torsional
    Velocity Measurements in Wire, with Application to Metal
    Matrix Composites," J. Appl. Phys. 53:4218 (1982).
3. Ramnath, V. & Buesking, K. W., "Evaluation of Potential
    Practical Applications for Very High Modulus Carbon Fibers,"
    Materials Sciences Corp., MSC TFR 1303/1409, Spring House, PA,
    April 1982.
4. Hashin, Z., J. Appl. Mech. 46:543 (1979).

ACOUSTO-ELASTIC EFFECT

OF STRESS IN ALUMINUM AND STEEL

EFFECTS OF ROLLED PLATE THICKNESS ON ANISOTROPY,

WITH APPLICATION TO ACOUSTIC STRESS MEASUREMENT

Richard B. Mignogna*, Alfred V. Clark
Bhakta B. Rath, and Carl L. Vold

Naval Research Laboratory
Washington, DC 20375

ABSTRACT

Common structural materials which are normally idealized as isotropic for most engineering purposes must be considered anisotropic for purposes of acousto-elastic stress measurement. The degree of anisotropy affects the number of moduli, as well as their departure from the usual isotropic values. For instance, rolled plates made of the same alloy, with the same temper designation, may have different moduli for different amounts of rolling. It is the purpose of this paper to determine whether the acousto-elastic constants are identical for commercial rolled plates of the same alloy and temper but different thicknesses.

INTRODUCTION

Early theories developed for acousto-elastic stress measurement[1] treat polycrystalline materials as isotropic. However, in the presence of stress, materials behave as if they were anisotropic. For states of plane stress, pure-mode SH-waves can be propagated through the material if the waves are polarized along the principal stress axes. The phase velocities of waves polarized along these axes will be unequal unless the principal stresses are equal. The normalized difference in phase velocity is known as the acoustic birefringence. For an isotropic material the initial (unstressed) birefringence is zero.

However, it has been recognized by Rollins [2,3], Sullivan and Papadakis[4], Mahadevan[5] and Crecraft[6], to name a few, that most polycrystalline materials have some degree of initial anisotropy or texture (preferred grain orientation) due to fabrication, such as

---

* NRC-NRL Research Associate

that found in rolled plate. The polycrystalline aggregates that result from various fabrication procedures have certain material symmetry directions, much like anisotropic single crystals. The initial birefringence is the normalized difference in phase velocity of SH-waves polarized along these symmetry directions. The initial birefringence caused by texture may, in fact, be greater than the birefringence due to stresses as high as the yield stress for some materials.

In some cases it is possible to subtract the initial birefringence from the total birefringence, so that the remainder is due solely to stress. This was successful in the cases of local uniaxial or biaxial stress states, with the stress axes oriented parallel to directions of material symmetry.[6,7] We will discuss this in some detail later in this paper.

A theory for wave propagation through slightly anisotropic stressed materials was recently developed by Okada.[8] Okada's theory considers materials which exhibit orthotropic symmetry; i.e., the material has three two-fold orthogonal axes. His theory results in three acousto-elastic constants necessary to characterize the material for acoustic stress measurement. However, his approach was to use an optical analog with a second order index of refraction tensor.

Previously, another more rigorous theory had been developed by Iwashimizu and Kubomura[9] using the Christoffel equation for wave propagation in anisotropic solids. However, they considered the effects of anisotropy in the second order elastic moduli and ignored it in the third order elastic moduli resulting in only one acousto-elastic constant.

Clark and Mignogna[10] have recently shown that the results of Okada's theory can be rigorously derived from the Christoffel equation and that the results of the Iwashimizu and Kubomura theory are a special case of the Okada theory.

Using the results of Okada's theory we have been able to map out the shear-stress field around a simulated crack in an ASTM E561 modified compact tensile (MCT) specimen of 6.3mm (0.25 in.) thick 2024-T351 rolled aluminum plate.[11] The results agreed very well with the theoretical shear-stress field as derived for this type specimen by Sanford[12]. The theoretical and experimental shear-stress fields are shown together in Fig. 1 for comparison. The stress intensity factor calculated from the measurements also agreed very well with the theoretical stress intensity factor, 28.6 MPa-$\sqrt{m}$ (26.0 ksi-$\sqrt{in}$) for the experimental value and 28.3 MPa-$\sqrt{m}$ (25.7 ksi-$\sqrt{in}$) for the theoretical value.

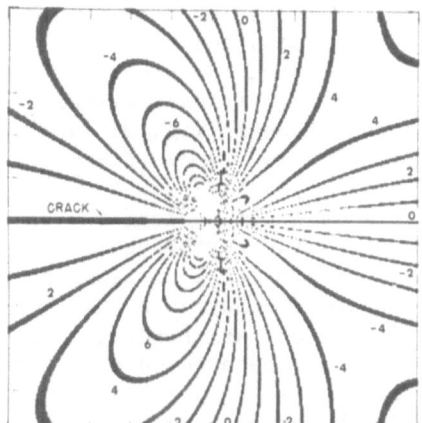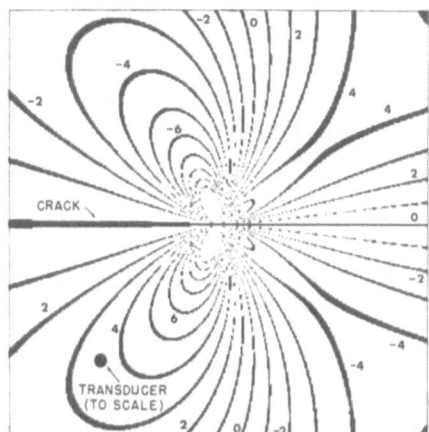

Fig.1 Theoretical shear stress contours (left) and experimental
       contours (right) within a 51 mm x 51 mm (2 in. x 2 in.) region
       of the MCT specimen centered at the crack tip. Theoretical
       contours were generated from the optical birefringence data of
       Sanford [13]. Experimental countours were generated from
       acousto-elastic data. The size of the transducer used in the
       experiment is also shown. Contours are plotted for σ_{xy} = (N $\pm$
       0.08) ksi, N=0, 1, 2.. To obtain stress in MPa, multiply by
       6.9.

       Although excellent agreement has been shown between theory and
experiment, the question addressed in this paper is whether the
acousto-elastic constants measured for one thickness of 2024-T351
rolled aluminum plate can be used for a thicker plate of 2024-T351
aluminum. It is realized that the anisotropy in polycrystalline
materials is controlled by both the deformation and heat treatment
history of the material. Since the fabrication history for plates
of the same material but of various thicknesses is somewhat
different, some degree of difference would be expected in the
amount or type of anisotropy produced. Thus, a possible difference
in the measured acousto-elastic constants may be found.

MATERIAL AND THEORETICAL CONSIDERATIONS

       Two thicknesses of 2024-T351 rolled aluminum plate were used
in this work. The plate thicknesses were 6.3 mm (0.25 in.) and
19.0 mm (0.75 in.) and will be henceforth referred to as the thin
and thick plates, respectively. Both plates displayed slight
anisotropy. The initial acoustic axes (pure-mode directions) in
the plane of the plates were found to be parallel to and
perpendicular to the rolling direction for both the thin and thick
plates.

The details of Okada's theory have been discussed in a number of papers, [8,10,11,14] therefore only the results and their application to this work will be presented here. In order to discuss the results of the theory, a (local) coordinate system is constructed in the plane of the plate, as shown in Fig. 2. The Y_0 and X_0 axes are the initial acoustic axes and remain fixed with respect to the rolling direction (material symmetry axes for our aluminum specimens). The phase velocity is greater along the Y_0 axis and is therefore denoted the 'fast' axis. Correspondingly, the X_0 axis is called the 'slow' axis. For the case of biaxial stress, the principal stress directions are displayed in Fig. 3 as σ_1 and σ_2. The principal stress directions are referenced to the slow acoustic axis by the angle Θ. It has been found that the application of stress causes the acoustic axes (polarization directions for propagation of pure-mode SH-wave) to rotate with respect to the initial acoustic axes. [8-11,14] The rotated acoustic axes, Y and X, are referenced to the initial slow axis by the angle ϕ. The Y and X axes are the rotated 'fast' and 'slow' axes, respectively.

The initial acoustic birefringence B_0 (in the unstressed state) is defined as

$$B_0 = \frac{V_{y_0} - V_{x_0}}{\langle V_0 \rangle} \tag{1}$$

where V_{y_0} and V_{x_0} are phase velocities of waves polarized along the initial fast and slow axes, respectively, and $\langle V_0 \rangle \equiv \frac{1}{2}(V_{y_0} + V_{x_0})$. Similarly, the acoustic birefringence B (in the stressed state) is defined as

$$B = \frac{V_y - V_x}{\langle V \rangle}. \tag{2}$$

V_y and V_x are the phase velocities of SH waves polarized along the rotated 'fast' and 'slow' acoustic axes, respectively. For a state of plane stress, the birefringence equation derived by Okada is of the form

$$B = \left\{ [B_0 + M_1(\sigma_1 + \sigma_2) + M_2(\sigma_1 - \sigma_2)\cos 2\Theta]^2 + [2M_3\sigma_{xy}]^2 \right\}^{1/2} \tag{3}$$

where B_0, B, σ_1, σ_2, and Θ are as previously defined, M_1, M_2 and M_3 are acousto-elastic constants, and $\sigma_{xy} = [(\sigma_1 - \sigma_2)\sin 2\Theta]/2$. The angle ϕ between the initial and stressed acoustic axes is calculated from

$$\tan 2\phi = \frac{2 M_3 \sigma_{xy}}{B_0 = M_1(\sigma_1 + \sigma_2) + M_2(\sigma_1 - \sigma_2)\cos 2\Theta} \tag{4}$$

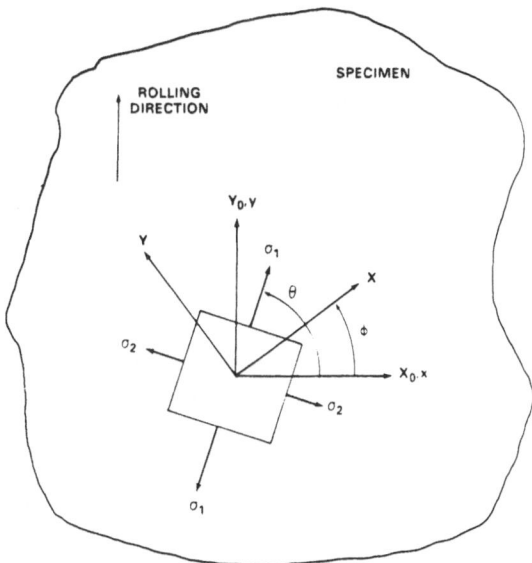

Fig. 2 Coordinate systems used in the analysis of SH wave
    propagation through rolled plate. X_0 and Y_0 denote the initial
    slow and fast acoustic axes, respectively. X and Y are the
    slow and fast axes in the stressed state. x and y are axes
    parallel to the initial acoustic axes and the principal
    stresses are labelled σ_1 and $\sigma_2 (\sigma_1 > \sigma_2)$.

From (3) and (4), the shear stress is given by

$$\sigma_{xy} = \frac{B \sin 2\phi}{2 M_3},\tag{5}$$

thus, the birefringence measurements can be used to calculate the
shear stress.

Equation (5) indicates that differences in the acousto-elastic
constant M_3, from one thickness plate to another thickness plate
will result in errors of the calculated shear stress which are
proportional to errors only in M_3. However, in the case
where Θ (the angle between the principal stress directions and
initial acoustic axes) is 0° or 90°, the shear stress will be zero
and Eq. (5) will be of no use. Only the birefringence Eq. (3) will
yield any information about the stresses. In these cases not only
will differences in the acousto-elastic constants M_1 and M_2 be of
concern, but differences in the initial birefringence, B_0, must
also be considered for various plate thicknesses.

For uniaxial stress, the birefringence equation (3) becomes

$$B = [(B_0 + M_1 \sigma + M_2 \sigma \cos 2\Theta)^2 + (M_3 \sigma \sin 2\Theta)^2]^{1/2}\tag{6}$$

where $\sigma_2 = 0$ and $\sigma_1 = \sigma$. Further simplification of the birefringence equation occurs for certain values of the angle Θ (the angle between the initial 'slow' acoustic axis and the principal stress direction). For $\Theta = 0°$ and $90°$ the birefringence is a linear function of stress:

$$(\Theta = 0°) \qquad B = B_0 + (M_1 + M_2)\sigma \tag{7}$$

$$(\Theta = 90°) \qquad B = B_0 + (M_1 - M_2)\sigma . \tag{8}$$

As discussed in the introduction, some earlier workers[6,7] found it was possible to subtract the initial birefringence from the total birefringence yielding good results in correlating the remaining birefringence with stress. Those results are supported by Eqs. (7) and (8) for the special cases when $\Theta = 0°$ or $90°$. However, in general, the birefringence is a nonlinear function of stress, as illustrated for $\Theta = 45°$ and $\Theta = 60°$, and the initial birefringence cannot be subtracted out.

$$(\Theta = 45°) \qquad B = [B_0^2 + 2M_1 B_0 \sigma + (M_1^2 + M_3^2)\sigma^2]^{1/2} \tag{9}$$

$$(\Theta = 60°) \qquad B = [(B_0 + M_1\sigma - 0.5\,M_2\sigma)^2 + 0.75\,M_3^2\sigma^2]^{1/2}. \tag{10}$$

BIREFRINGENCE MEASUREMENTS

A modified version of the pulse-echo-overlap (P.E.O.) method [15-17] was used to determine the orientation of acoustic axes and the acoustic birefringence. A 10 MHz, 1.27 mm (0.5 in) diameter AC-cut quartz shear wave transducer was mounted in a spring-loaded goniometer device. A viscous fluid acoustically coupled the transducer to the specimen and also permitted rotation of the transducer. A more detailed discussion of the apparatus and measurement technique have been reported elsewhere.[11]

Since SH waves polarized along both the fast and slow axes travel through the same material at each location using the P.E.O. method, the birefringence is given by

$$B = \frac{f_f - f_s}{0.5\,(f_f + f_s)}. \tag{11}$$

where f_f and f_s are the overlap frequencies associated with the fast and slow axes, respectively.

Four uniaxial tension specimens were cut from the thin plate at angles of $0°$, $45°$, $62°$, and $90°$ to the X_0 - axis and used to determine the acousto-elastic constants. A linear regression fit was used for the experimentally obtained birefringence as a function of stress for the specimens with Θ equal to $0°$ and $90°$. From the slopes of these lines and Eqs. (7) and (8), the acousto-elastic constants M_1 and M_2 were determined. The acousto-elastic constant M_3 was then determined from the $\Theta = 45°$ and $62°$ specimens by using a nonlinear least squares fit for each specimen. A final value of M_3 was obtained from the average of the two values. Similarly, four uniaxial tension specimens were cut from the thick plate at angles of $0°$, $45°$, $60°$ and $90°$ to the X_0 - axis. The values of M_1, M_2 and M_3 for the thick specimens were determined by the same procedure used for the thin specimens.

The experimental data and fitted curves were plotted as a function of stress and are shown in Fig. 3 for the thin plate and Fig. 4 for the thick plate.

It should be noted that the acousto-elastic constants are temperature-dependent. Consequently, all our acousto-elastic measurements were performed at a constant temperature of $20°$ C.

The measured values of the acousto-elastic constants (in units of $10^{-5}MPa^{-1}$) and average initial birefringence for the aluminum 'thin' plate were: $M_1 = 0.01$, $M_2 = 3.64$, $M_3 = 3.32$ and $\langle B_0 \rangle = 3.48 \times 10^{-3}$. The measured values for the 'thick' plate were: $M_1 = 0.23$, $M_2 = 3.51$, $M_3 = 3.68$ and $\langle B_0 \rangle = 6.52 \times 10^{-3}$, where $\langle B_0 \rangle$ is the average initial birefringence.

Only about a 10% difference was found in the acousto-elastic constants for the thick plate in comparison to the thin plate. If the equation relating birefringence to shear stress, Eq.(5), were used to calculate stress from birefringence data, the error in the calculated shear stress would be proportional to the error in M_3. In this case that error would be about 10%. However, if it were necessary to use the birefringence equation, Eq. (3), the initial birefringence would have to be used. As can be seen from the data, the initial birefringence of the thick plate is almost twice that of the thin plate. An 80% error would be involved in calculations using the initial birefringence of the thin plate for the initial birefringence of the thick plate.

X-RAY EXAMINATION OF TEXTURE

X-ray 'pole figures' were constructed in order to determine if there was any difference in the texture (preferred grain orientation) of the two plates. A pole figure is constructed from the x-ray diffraction patterns from a particular type of crystallographic plane and shows the distribution of grain orientations relative to axes of the sample.

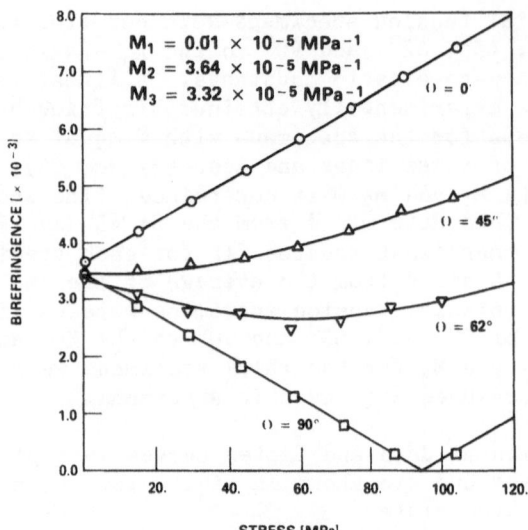

Fig. 3 Acoustic birefringence as a function of stress for
6.3 mm (0.25 in) thick 2024-T351 aluminum uniaxial tensile
specimens for $\Theta = 0^{\circ}$, 45°, 62° and 90°. Theory shown
as (———) and data by various symbols.

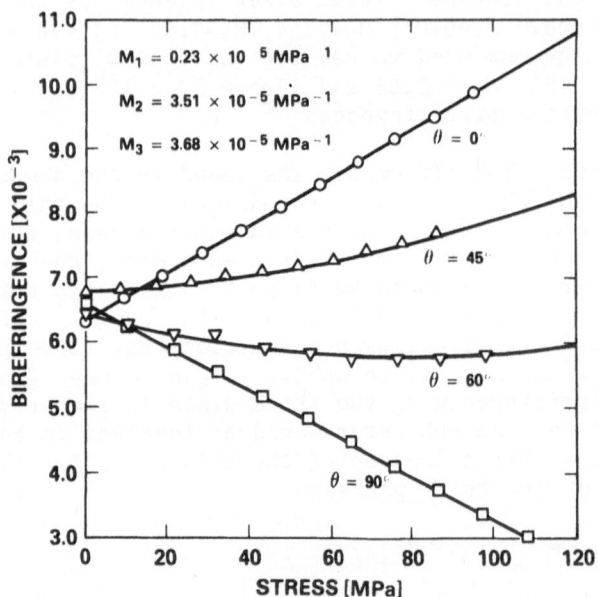

Fig.4 Acoustic birefringence as a function of stress for
19.0 mm (0.75 in) thick 2024-T351 aluminum uniaxial tensile
specimens for $\Theta = 0^{\circ}$, 45°, 60°, and 90°. Theory shown
as (———) and data by various symbols.

If a pole figure for an isotopic polycrystalline material is
constructed, the entire figure would be a 'gray' region of equal
x-ray intensity. The pole figure from a single crystal would have
a few sharp 'spots' of high x-ray intensity and the rest of the
figure would be of zero intensity. If a polycrystalline material
had some texture (preferred grain orientation), the pole figure
would show patterns of equal x-ray intensity that would be
displayed in a manner similar to a contour map. The contours, in
this case, are of equal x-ray intensity. For more information on
pole figures, refer to a text such as that by Barrett and
Massalski.[18]

Four {111} pole figures are shown in Fig. 5. The pole figures
on the left are of the thin plate and the ones on the right are of
the thick plate. The plane of the pole figures is parallel to the
rolling plane. The rolling and transverse direction are indicated
by R.D. and T.D., respectively. The top two pole figures are of
the surface of both plates and the bottom two are of the center
plane of the plates. The orientation of the crystallographic plane
parallel to the rolling plane is indicated by the {hkl} brackets
and the orientation of the crystallographic direction parallel to
the rolling direction is indcated by the ⟨hkl⟩ brackets.

The surface pole figures show similar but very weak texture,
the thick plate showing a relatively weaker texture than the thin
plate. These patterns indicate a {110} ⟨112⟩ as a possible
texture. The pole figures of the center plane of the plates show a
substantially stronger relative texture than the surface, but they
still have a weak texture. A significant difference also exists
between the amount and type of texture displayed at the center
plane of the two plates. The thin plate still exhibits a possible
{110}⟨112⟩ texture but the thick plate appears to be a combination
of {211}⟨110⟩ and a {110}⟨112⟩ texture. The {211}⟨110⟩ is the
predominant texture in the thick plate. The x-ray intensities
indicate that the amount of texture is also greater in the thick
plate than in the thin plate. Figure 6 illustrates these textures.

SUMMARY AND CONCLUSIONS

The differences found in the acousto-elastic constants and
initial birefringence between the thin plate and the thick plate
can be explained by the x-ray pole figures.

Plate materials are polycrystalline aggregates having elastic
moduli which depend upon the single crystal elastic moduli and the
average crystallographic orientation of the grains that make up the
aggregate. The pole figures give a measure of the average
orientation of the grains. The existence of a texture in the pole
figures indicates that the average grain orientation is not random.

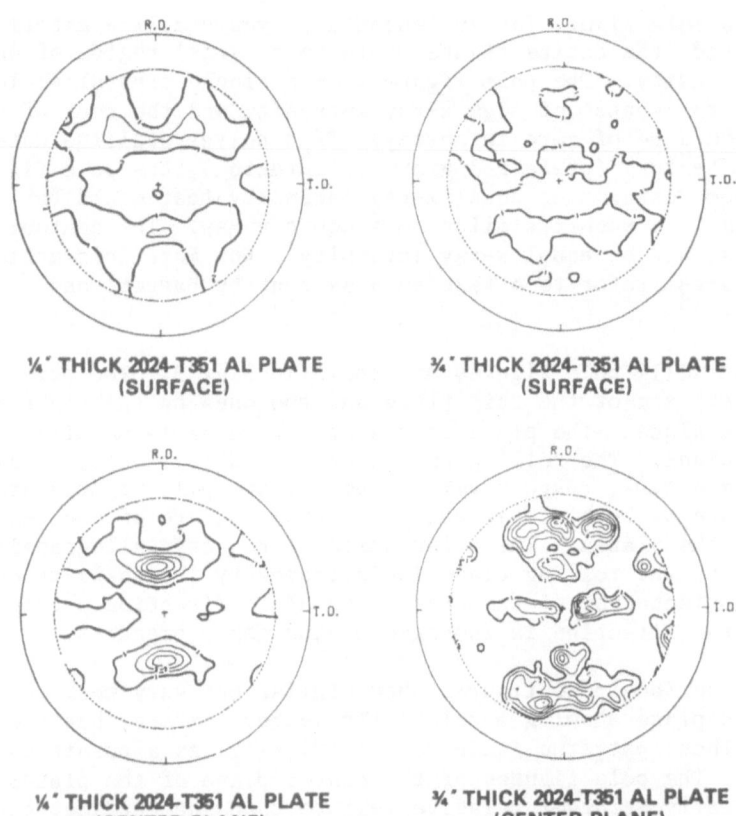

Fig. 5 {111} pole figures of surface and center plane in rolled
    2024-T351 aluminum plate, thin plate on left, thick plate on
    right.

The type of texture was found to be different in the thin
plate as compared to the thick plate, a {110}<112> for the thin
plate and a combination {211}<110> and {110}<112> for the thick
plate.

A difference in texture between the two plates implies a
difference in the number of grains having a particular orientation
and/or a different average crystallographic orientation of the
grains. Thus some difference in the aggregate elastic moduli
between the two plate thicknesses is to be expected. The acousto-
elastic constants are combinations of second- and third-order
moduli[9,10]. Consequently, differences in plate texture will also
give rise to differences in acousto-elastic constants.

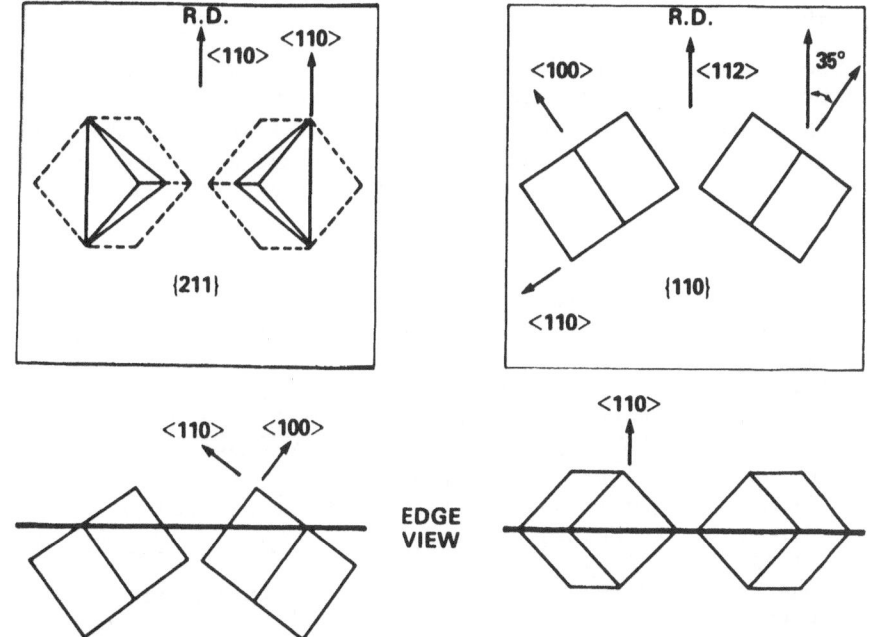

Fig. 6 Preferred grain orientations (represented by cubes) found
in the aluminum plates. Thick plate is a combination of both
textures; thin plate consists only of texture on right. The
plane of the top two illustrations is parallel to the rolling
plane; rolling direction is indicated by R.D. Bottom views are
of the edge of the plate looking into the rolling direction.

However, only a 10% difference (10% of the dominant constants
M_1 and M_2) in the acousto-elastic constants M_1, M_2, and M_3 was
found. From an engineering standpoint of measuring stress, the 10%
difference is not significant, but it is measurable. The most
significant acoustic difference was in the initial birefringence.
The initial birefringence was greater in the thick plate than in
the thin plate (6.52×10^{-3} as compared to 3.68×10^{-3}). This is
supported by the stronger texture observed for the thick plate.

As a result of the small (10%) difference in the acousto-
elastic constants for 2024-T351 rolled aluminum, of different plate
thicknesses, it may be possible to use the acousto-elastic
constants measured from one plate for various plate thicknesses.
However, if the initial birefringence is needed, significant errors
may be encountered in attempting to use the initial birefringence
measured for one plate thickness for the initial birefringence of
another plate thickness.

ACKNOWLEDGEMENTS

    The research reported in this paper was conducted in the
Nondestructive Evaluation Section of the Naval Research Laboratory
under the sponsorship of the Office of Naval Research. The authors
are indebted to Mr. J. P. Waskey for assistance in making specimens
and loading fixtures for the acousto-elastic experiments.

REFERENCES

1. Hughes, D.S. and Kelly, J.L. "Second-Order Elastic
        Deformation of Solid", Physical Rev., Vol. 92, (1953) 1145-
        1149.
2. Rollins, F.R., "Study of Methods for Non-Destructive
        Measurement Residual Stress," WADC Tech. Report 1959
3. Rollins, F.R., "Ultrasonic Methods for Non-Destructive
        Measurement of Residual Stress", WADD Tech. Report 1961.
4. Sullivan, P.F.and Papadakis, E.P., "Ultrasonic Double
        Refraction in Worked Metals", J. Acoust. Soc. Am. Vol. 33,
        (1961) 1622-1624.
5. Mahadevan, P., "Effect of Frequency on Texture-Induced
        Wave Birefringence in Metals," Nature, Vol. 211, (1966) 621-
        622.
6. Crecraft, D.I., "The Measurement of Applied and
        Residual Stressed in Metals Using Ultrasonic Waves," J.
        Sound Vib., Vol. 5, No. 1, (1967) 173-192.
7. Hsu, N.N., "Acoustical Birefringence and Use of
        Ultrasonic Waves for Experimental Stress Analysis," Exp.
        Mech., Vol. 14, No. 5, (1974)` 169-176.
8. Okada, K., "Stress-Acoustic Relations for Stress
        Measurement by Ultrasonic Technique," J. Acoust. Soc. Jpn.
        (E), Vol. 1, No. 3, (1980) 193-200.
9. Iwashimizu, Y. and Kubomure, K., "Stress-Induced
        Rotation of Polarization Directions of Elastic Waves in
        Slightly Anisotropic Materials," Int. J. Solids Structures,
        Vol. 9, (1973) 99-114.
10. Clark, A.V. and Mignogna, R.B., "A Comparison of Two
        Theories of Acousto-Elasticity," Ultrasonics, Vol. 21, No.
        5 (1983) 217-225.
11. Clark, A.V. and Mignogna, R.B., "Acousto-Elastic
        Measurement of Stress and Stress Intensity Factors Around
        Crack Tips," Ultrasonics, Vol. 21, No. 2, (1983) 57-64.
12. Sanford, R.J., "A Critical Re-examination of the
        Westergaard Method for Solving Opening Mode Crack Problems,"
        Mechanics Research Communications, Vol. 6, (1979) 289-294.
13. Sanford, R.J., et.al., "A Photoelastic Study of the
        Influence of Non-Singular Stresses in Fracture Test
        Specimens," NUREG/CR-2179, University of MD, College Park,
        MD, August 1981.

14. Okada, K., "Acousto-elastic Determination of Stress in Slightly Orthotropic Materials," Exp. Mech., Vol. 21, (1981) 461-466.

15. Papadakis, E.P., "Ultrasonic Attenuation and Velocity in Three Transformation Products in Steel," J. Appl. Phys., Vol. 35, (1964)   1474-1482.

16. Papadakis, E.P., "Ultrasonic Phase Velocity by the Method Incorporating Diffraction Phase Corrections," J. Acoust. Soc. Am., Vol. 42, (1967) 1045-1057.

17. Chung, D.H., Silversmith, D.J. and Chick, B.B., "A Ultrasonis Pulse-Echo-Overlap Method for Determining Sound Velocities and Attenuation of Solids," Rev. Sci. Instrum., Vol. 40, (1969) 718-720.

18. Barrett, C.S., and Massalski, T.B. "Structures of Metals," McGraw-Hill Book Co., New York (1966).

ULTRASONIC SHEAR WAVE MEASUREMENTS OF

KNOWN RESIDUAL STRESS IN ALUMINUM

G.V. Blessing, N.N. Hsu and T.M. Proctor

Ultrasonic Standards Group
National Bureau of Standards
Washington, DC   20234

ABSTRACT

Ultrasonic shear wave time-of-flight measurements were made at a nominal frequency of 4 MHz on a shrink-fit disk sample of 2024 aluminum alloy. The stress state of the sample was produced by shrink-fitting a plug and ring to produce a calculated 130 MPa region of uniform compression in the plug, and a concomitant nonuniform tension and compression in the ring. Time-of-flight measurement scans across sample diameters were made using a piezoelectric shear transducer with a viscous couplant, and repeated using a contactless electromagnetic acoustic transducer. The ultrasonic results were then compared with elasticity theory, assuming the acoustoelastic relationship between sound velocity and material strain.

INTRODUCTION

A reliable means to nondestructively evaluate the residual stress state of an object is of much interest to materials scientists, design engineers, and safety officials [1]. While substantial research effort has been expended to apply ultrasonic techniques to this problem [2], practical difficulties of measurement precision and unknown sample stress have limited its application. This paper addresses both of these problems by comparing the results using two distinctive transducers (one piezoelectric and the other electromagnetic) applied to an aluminum sample possessing a known stress state.

In previous work [3], we presented a candidate reference stress standard, wherein a shrink-fit ring-plug specimen was fabricated with a known level and distribution of residual stress. Ultrasonic longitudinal waves, generated with a piezoelectric ceramic (PZT) transducer, were used in that work to assess the residual stress pattern of the aluminum sample via the acoustoelastic effect. The relative velocity values were found to be in good qualitative agreement with elasticity theory. In this work, we present results using shear waves generated with a PZT shear transducer, requiring a viscous couplant between transducer and sample. These studies were extended to also include the results on a non-stressed reference sample of like geometry and material. Furthermore, we repeated these measurements using a non-contacting electromagnetic acoustic transducer (EMAT), and compared the results.

THEORY

Consider a shrink-fit disk-shaped sample consisting of a plug of radius R_1 within a ring of radius R_2 (see Fig. 1), where cylindrical coordinates appropriate to the sample symmetry are used. We propagate bulk shear waves in the z direction, normal to the (θ, r) plane of the sample which is in a state of plane stress. Identifying the principal plane stresses as σ_θ and σ_r, their acoustoelastic relationship with the shear sound speed V_s and the transit time t through the sample thickness may be written as [4]:

$$\frac{\Delta(V_{sr} - V_{s\theta})}{V_{s\theta}} = \frac{\Delta(t_{s\theta} - t_{sr})}{t_{so}} = \beta(\sigma_\theta - \sigma_r) , \qquad (1)$$

where subscripts θ and r refer respectively to the tangential and radial directions, and subscript o refers to the unstrained material assumed to be isotropic. The proportionality constant β, related to the acoustoelastic or stress-acoustic constants found in the literature, assumes a linear relationship between the stress and the change in velocity, as is observed in the case of applied stresses [5].

In order to relate the principal stress difference to sample fabrication parameters, we refer to elasticity theory as before [6]. From that analysis, we obtain the principal stress difference in terms of the contact pressure P between plug and ring. In the plug region, since it is in a state of uniform compression, the principal stress difference is simply:

$$\sigma_\theta - \sigma_r = -P-(-P) = 0, \quad r<R_1 . \qquad (2)$$

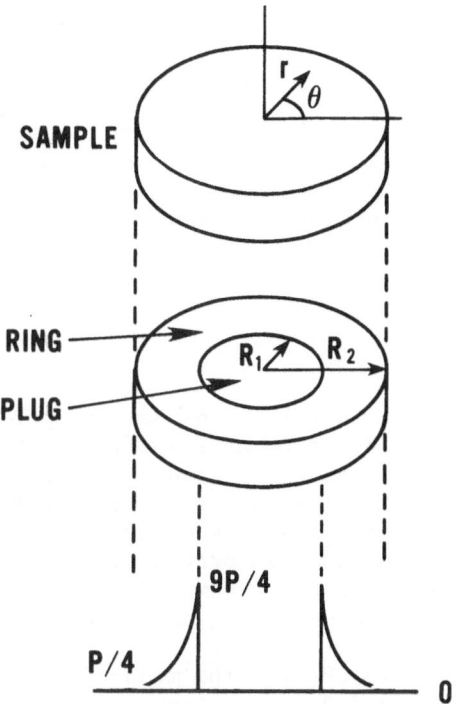

Fig. 1. Sample geometry illustrating the ring-plug configuration,
and the residual stress profile according to elasticity
theory.

In the ring region, there results an inverse r-squared dependence:

$$\sigma_\theta - \sigma_r = \frac{2P \ R_1^2 \ R_2^2}{R_2^2 - R_1^2} \left(\frac{1}{r^2}\right), \quad r > R_1 . \tag{3}$$

Here, the pressure P may be obtained from Young's modulus E for the specimen material, the diametral interference fit δ between ring and plug, and the disk component dimensions R_1 and R_2:

$$P = E \, \frac{(R_2^2 - R_1^2)}{2R_2^2} \, \delta \quad . \tag{4}$$

For a radii ratio R_2/R_1 equal to three, Eq.(3) shows that the ring experiences a maximum stress difference equal to $9P/4$ at the ring-plug interface, decreasing to $P/4$ at the outer ring perimeter. The resulting shape of the stress distribution is shown at the bottom of Fig. 1.

EXPERIMENTAL

Sample

As described previously [3], the samples were fabricated from a 2024-T351 aluminum rod, using an experimentally determined region of elastic homogeneity. Aluminum rather than steel was chosen in these initial studies for its larger acoustoelastic coupling coefficient. Three adjacent slices of material nominally 2.5 cm thick were cut from the rod, two providing the ring and plug, and the third the reference or blank sample. The diametral inter-ference δ (see Eq.(4)) between ring and plug corresponded to a shrink-fit residual contact pressure P of 130 MPa, based on a Metals Hankbook [7] value of 73 GPa for the modulus E, and respective ring and plug radii of 7.5 and 2.5 cm. Assemblage of the shrink-fit disk was achieved by heating the ring and cooling the plug. Final polishing yielded both the stressed and the non-stressed samples flat with parallel faces, and equal in thickness (nominally 2.2 cm), to better than one part in 10^4.

A photoelastic analog of the ring-plug sample is shown in Fig. 2, where polarized monochromatic light reveals the stress pattern in a ring-plug sample of similar geometry. In photoelastic theory, the isochromatic fringe density is proportional to the principal stress difference, in direct analogy with Eq.(1) [8]. The inverse r-squared spacing observed for the concentric fringes in the ring therefore indicates a like dependence for the radial stress difference. The lack of circular symmetry in the outermost fringes indicates an asymmetry for the residual stress. (A similar asymmetry was noted in the ultrasonic scans of the aluminum sample-- see RESULTS). The nondescript region within the plug includes the zeroth order fringe which corresponds to a zero principal stress difference.

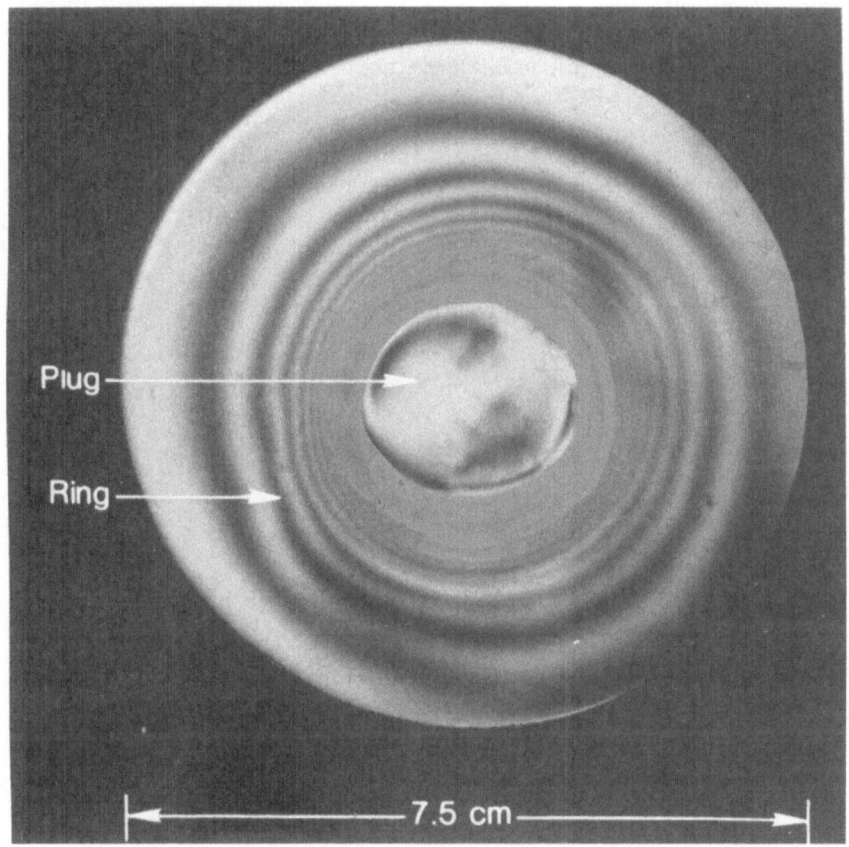

Fig. 2. The isochromatic fringes observed in a photoelastic ring-plug disk specimen.

Ultrasonic

A pulse-echo overlap technique was used to measure the shear
wave transit time (t_{ab} in Eq.(1)) through the sample thickness.
Multi-cycle RF tonebursts at a nominal frequency of 4 MHz were
used to drive the respective transducers. A viscous borax-
glycerine mixture [9] was used to couple the 1.2 cm diameter PZT
transducer to the samples. To facilitate sample scanning with the
EMAT, a thin sheet of teflon was taped across the entire surface
of its aperture (nominally 1 cm square of exposed coil) and ground
plane (see Fig. 3). Scans across respective orthogonal diameters
of the two samples were then made at increments of 1 to 2 mm
depending on the rate-of-change of the ultrasonic wave transit
time. The shear wave polarization was oriented first parallel to
a diameter for t_{sr} and then perpendicular for $t_{s\theta}$ (see Eq.(1)). The
transit times were measured using the delta-time feature of a
Hewlett-Packard 1743A oscilloscope, with a time averaging precision
capability of $\pm$ 1 ns.

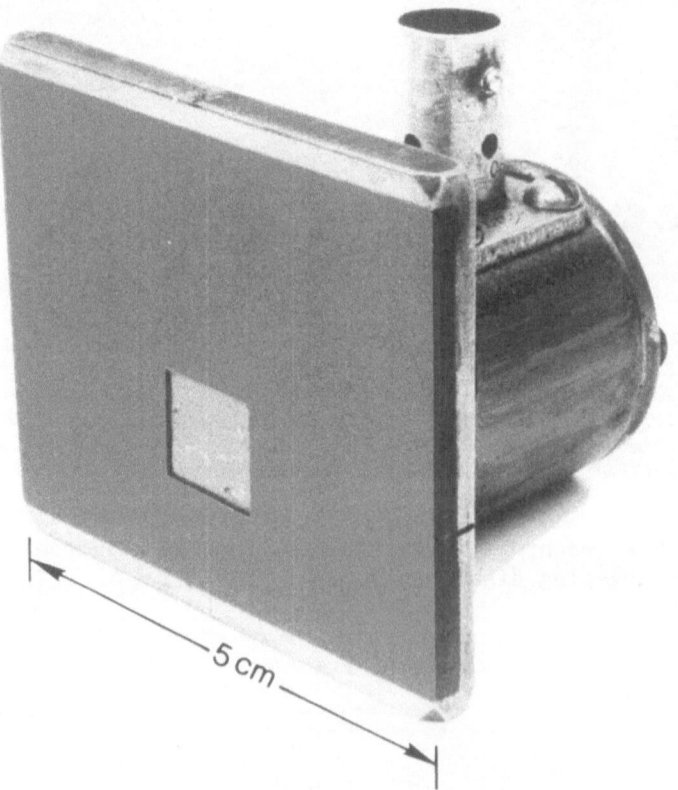

Fig. 3. Photo of the electromagnetic acoustic transducer (EMAT) [10].

RESULTS

PZT

Figure 4 illustrates the pattern obtained for the time difference when scanning with a PZT transducer across an arbitrarily chosen diameter of the ring-plug (asterisks) and blank (solid line) samples. The time difference relates to the sound velocity difference between orthogonal wave polarizations for two round trips of the ultrasonic wave measured from the output driving pulse to the second received echo. (The results of scanning across a perpendicular diameter were similar, but with a reduced time difference at the regions of peak stress in the ring.) The agreement with theory (Eqs.(2) and (3)) is encouraging in several respects. In the plug region of the stressed sample, the time difference is relatively featureless and close to zero, which is expected for uniform compression. In the ring, an inverse r-squared dependence may be observed, with the time difference becoming small at the ring perimeter. At the interface, a rapid change in the time difference follows the expected discontinuity in stress from the plug to the ring.

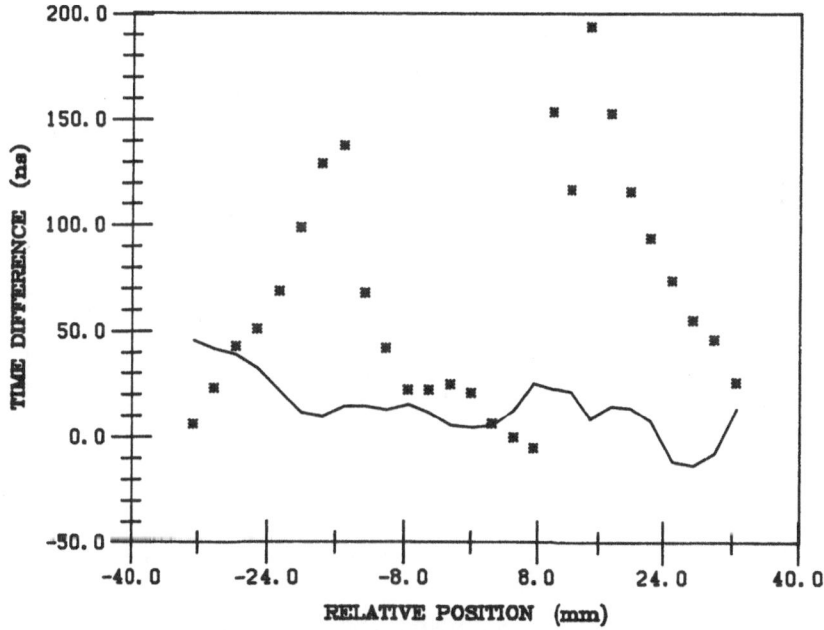

Fig. 4. Time difference profiles for two orthogonal shear wave polarizations as a function of position in the stressed (asterisks) and non-stressed (solid line) sample, using a PZT transducer.

The irregular pattern observed for the time difference in the blank sample (solid line in Fig. 4) was principally due to coupling variations between sample and transducer while scanning. Whereas the repeatability at a given sample position was approximately ± 5 ns for the time difference (involving two separate time measurements), the result of couplant build-up and/or squeeze-out during specimen scanning gave rise to variations as great as ± 30 ns.

EMAT

Figure 5 illustrates the pattern obtained using an EMAT transducer to traverse the same specimen diameters in the stressed (asterisks) and non-stressed (solid line) samples. In this case, the time difference relates to the sound velocity difference between orthogonal wave polarizations for one round trip of the ultrasonic wave pulse measured from the first to the second echo. Of special note is the dramatic improvement, by about a factor of ten as compared with the PZT transducer scans, that was obtained in measure-

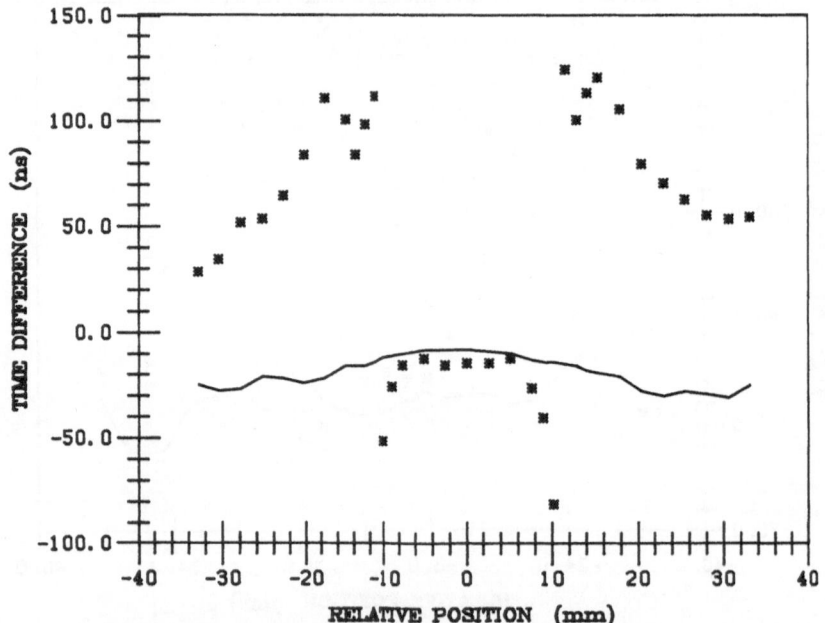

Fig. 5. Time difference profiles for two orthogonal shear wave polarizations as a function of position in the stressed (asterisks) and non-stressed (solid line) samples, using an EMA'

ment precision due to the absence of coupling effects between transducer and sample. Measurement repeatability of the time difference was nominally $\pm$ 1 ns, limited principally by a low signal to noise level for the observed echoes.

The increased precision revealed a noticeable improvement in the diametral symmetry of the time difference for both the stressed and non-stressed samples, as may be clearly observed in Fig. 5. A measurable sensitivity to sample temperature also became apparent, causing a change in the sample round trip time on the order of a few nanoseconds per degree Celsius. As a result, it was imperative that successive scans, for shear polarization respectively parallel and perpendicular to the diameter, be performed at the same temperature.

While the data pattern in Fig. 5 is basically the same as that in Fig. 4, there are some noticeable differences in the region of the ring-plug interface which are not fully understood. · The negative values at the plug boundary, and the dips at the regions of peak tension in the ring, are distinctive. Furthermore, we note that whereas the same basic pattern reoccurred for the perpendicular scans (and as previously observed with the PZT transducer, with a reduced time difference in the region of peak ring stress), the negative values in the plug did not reoccur. We conjecture that conductivity differences at the ring-plug interface ·play a significant role since EMAT performance depends on the generation and detection of eddy currents. This is in addition to real stress gradient differences (as a function of θ) that may result from slight geometrical asymmetries in the ring or plug at the interface during sample fabrication.

Another distinctive feature observed with the EMAT was a perturbation of the echo wave form as the transducer (with its copper ground plane surrounding the EMAT coil aperture) approached the sample perimeter. Since the aperture is not centrally located in the ground plane, this "edge effect" was not uniform at both ends of the diametral scan. This may explain the slight asymmetry observed for the time difference at the ring perimeters in Fig. 5.

Two limitations associated with present EMAT technology are revealed in Fig. 6: initial receiver saturation at resonance due to the high power RF output pulse, and a low signal to noise of the received echoes. Care was required to avoid interference, observed in the top trace of Fig. 6, between the RF output pulse and the first received echo by electrically damping the EMAT coil. The low signal to noise is unfortunately an intrinsic feature of EMAT performance [11]. Improvement will require innovative transducer designs to compensate for its low transduction efficiency, in particular as a transmitter at higher ultrasonic frequencies.

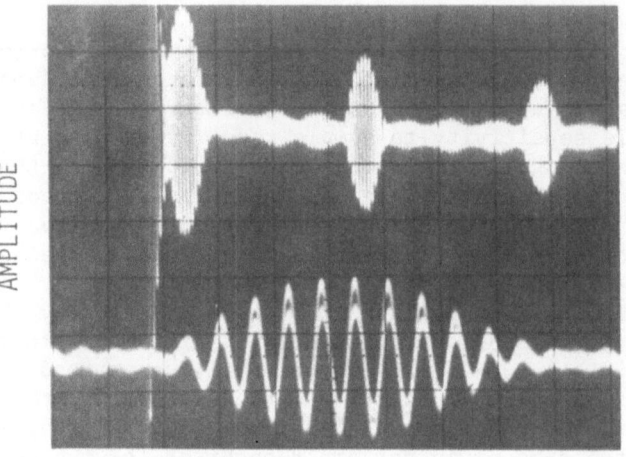

TIME: 5.0, 0.5 μs/div.

Fig. 6. (a) Pulse-echo pattern obtained on a disk sample using an
         EMAT; and (b) the resulting overlap of the second and third
         echoes.

CONCLUSIONS

    We have extended the experimental ultrasonic analysis of a ring-
plug shrink-fit residual stress specimen by using shear waves. The
experimental results are in good qualitative agreement with elasticity
theory, and are amenable to future quantitative analysis.

    In addition, we have applied the relatively new electromagnetic
acoustic transducer technology to the same task as a piezoelectric
transducer was applied, in order to improve measurement precision by
eliminating ultrasonic coupling effects which are unavoidable with
piezoelectric transducers. The improvement in repeatability was very
significant, but resulted in a reduced signal to noise ratio.
Furthermore, sample boundary effects which will require further
investigation were observed.

ACKNOWLEDGEMENT

The laboratory help of Mr. Carl Thompson, an NBS Cooperative Student from the University of California at Berkeley, is gratefully acknowledged.

REFERENCES

[1] See e.g., "Residual Stress and Stress Relaxation," E. Kula, V. Weiss, eds., Plenum Press, New York (1981).
[2] See, e.g., Vol. 10, No. 5, J. of Testing and Eval. (1982).
[3] N.N. Hsu, T.M. Proctor, and G.V. Blessing, An Analytical Approach to Reference Samples for Ultrasonic Residual Stress Measurement, J. of Testing and Eval. 10:230 (1982).

[4] R.E. Green, Nonlinear Elastic Waves in: "Ultrasonic Investigation of Mechanical Properties," Treatise on Materials Science and Technology, Vol. 3, Academic Press, New York (1973).

[5] See, e.g., G.S. Kino, D.M. Barnett, N. Grayel, G. Herrman, J.B. Hunter, D.B. Ilie, G.C. Johnson, R.B. King, M.P. Scott, J.C. Shyne, and C.R. Steele, Acoustic Measurements of Stress Fields and Microstructure, J. of Nondestructive Eval. 1:67 (1980).
[6] M.F. Spotts, Miscellaneous Machine Elements in: "Design of Machine Elements," Prentice Hall, New Jersey (1978).
[7] Anon., "Metals Handbook," Vol. 2, Am. Soc. for Metals, Metals Park, OH (1965).
[8] J.W. Dally and W.F. Riley, Analysis Techniques in "Experimental Stress Analysis," McGraw Hill, New York (1965)
[9] Couplant manufactured by Automation Industries, Inc., Sperry Division, Danbury, CT.*
[10] EMAT manufactured by MEA (Calif.) Ltd., Oakland, CA.
[11] H.M. Frost, Electromagnetic-Ultrasound Transducers: Principles, Practice, and Application, in: "Physical Acoustics," Vol. XIV, W.P. Mason, ed., Academic Press, New York (1979).

---

* Commercial laboratory equipment is identified in this report to accurately describe the experimental procedures used. In no case does such identification imply recommendation or endorsement by the National Bureau of Standards, nor does it imply that this equipment was necessarily the best available for the purpose.

RAYLEIGH WAVE MEASUREMENT OF SURFACE STRESSES IN

STAINLESS STEEL PIPING

D. Husson, S. D. Bennett and G. S. Kino

Edward L. Ginzton Laboratory
W. W. Hansen Laboratories of Physics
Stanford University
Stanford, California

ABSTRACT

The acoustoelastic effect for surface waves is used to measure the residual stress field on the inside and outside surfaces of a stainless steel pipe, near a circumferential weld. Rayleigh waves are launched on the pipe with electromagnetic transducers. An in-situ calibration technique is used to calculate the proportionality coefficients between stress and variation of velocity of the Rayleigh wave. The stress profiles obtained are compared with theoretical and experimental results published by other authors.

INTRODUCTION

A great deal of effort has been expended in the development of nondestructive methods for stress measurement in structural components. One of the main areas of interest has centered on the acoustoelastic effect, which relates the velocity of an acoustic wave to the state of strain of the medium in which it is propagating. Although the effect is weak, it has a major advantage in that it is totally nondestructive. Much of the work that has been described in the literature[1,2] has been confined to the use of laboratory specimens with well-machined surfaces and uniform thickness. The work described in this paper addresses the problem of making such acoustoelastic measurements in more practical situations. For the purpose of the study, we have chosen to look at the stresses near a circumferential weld in a section of stainless steel piping, a topic of considerable practical importance in the electrical power generation industry.

We have developed a measurement scheme based on the use of
Rayleigh waves which propagate along the surface of the specimen,
with their energy confined to a region of the order of a wavelength
deep. The advantage of this mode is that thickness variations have
no effect on the measurement; further, as we will show, sensitivity
to surface roughness can be largely overcome, and stresses on both
surfaces of the sample can be measured separately.

An objection to the use of acoustoelasticity which is
frequently raised is the fact that the presence of stress is not
the only physical variable which can result in a change in the
propagation velocity of the acoustic wave. The presence of
microstructural changes can frequently cause velocity changes of
the same order as or larger than the stresses one wishes to
measure. A further complication is that for some materials the
variation in microstructure from sample to sample is sufficient
that a one-time calibration of the material type is inadequate.
Rather, one must calibrate each sample of the material (or at least
maintain records of material batches). We demonstrate here an
in-situ calibration technique which could readily be adapted to
field use.

A detailed theory has been developed to predict the variation
of acoustic velocity with applied stress.[3] It may be shown that
the fractional change in Rayleigh wave velocity $\Delta V/V$ depends on
stress in the following manner

$$\frac{\Delta V}{V} = B_1 \sigma_{11} + B_2 \sigma_{22} \qquad (1)$$

where σ_{11} and σ_{22} are the surface stresses parallel and
perpendicular to the direction of propagation. There are thus two
constants which must be determined in order to calculate the value
of the material stresses from the measured changes in acoustic
velocity. If the two stress components are to be distinguished,
velocity measurements in two, preferably orthogonal, directions are
necessary.

EXPERIMENTAL SYSTEM

Sample

We have received from the EPRI Research Center, North
Carolina, a piece of 33mm (13") diameter 304 stainless steel pipe
with a circumferential weld. This sample is 46mm (18") long, the
weld being 30mm (12") from one of the ends of the pipe. We performed
all our measurements on the longest side of the weld, at points between 2 and 28 cm from its center, on the outside surface. On the

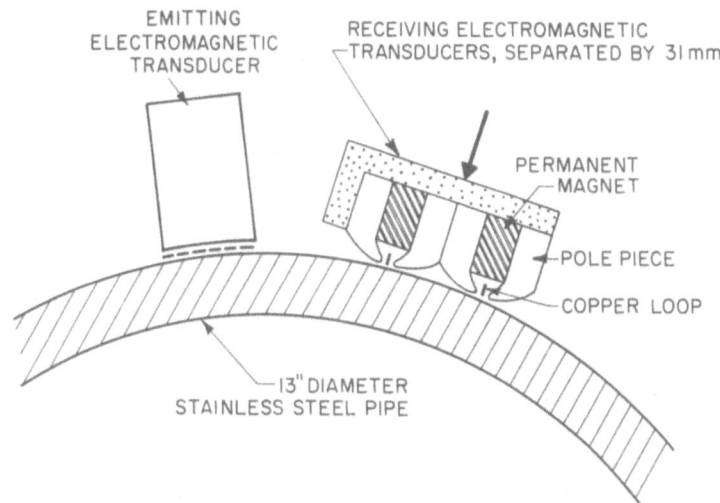

Fig. 1. Transmitting and receiving electromagnetic transducers.

inside surface, the machining done in preparation for the weld
limited our measurements to a region between 6 and 26 cm from
the weld.

Electrical System

The use of Rayleigh waves to measure acoustoelastic effects
has involved the development of special purpose transducers. The
most straightforward methods of launching surface waves (mode
conversion from a liquid or through a wedge) are not satisfactory
because they tend to launch spurious bulk waves which propagate
with a velocity very close to that of the Rayleigh wave and thus
interfere with the signals we wish to measure. A second, and
possibly more important, weakness of these methods of launching
surface waves is that they are dependent on the surface condition
of the pipe. This is particularly true of wedge transducers and of
edge-bonded transducers.[4] For these reasons we have developed a
system of electromagnetically coupled transducers (EMATs) which,
although they are very inefficient, have the great advantage that
the signal is coupled through an evanescent electromagnetic field
and therefore its phase is not affected by small changes in the
specimen-to-transducer separation.[5]

The configuration we have evolved is shown in Fig. 1. A
single transducer is used to launch two tone bursts which propagate

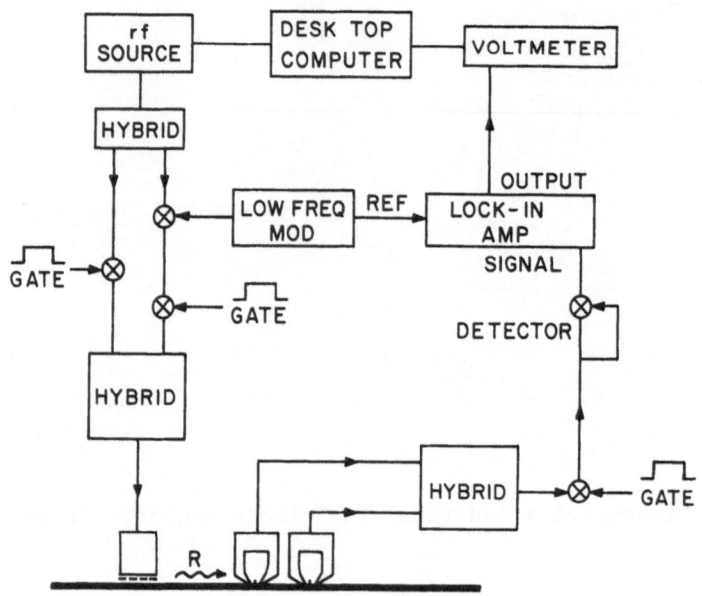

Fig. 2. Digitally-controlled phase measurement system.

toward a pair of receiving transducers. The separation of the tone
bursts in time is arranged to correspond to the spatial separation
of the receivers so that the two output signals occur
synchronously. The phase difference φ between these output
signals is given by

$$\phi = \frac{2\pi f L}{V} \tag{2}$$

where f is the excitation frequency, L is the separation of the
transducers, and V is the Rayleigh wave velocity. A closed-loop
phase measurement system is used (Fig. 2) to measure the phase
difference φ and adjust the frequency of the source f in such a
way that cos φ = 0 . Now, if V changes and L is kept
constant, the frequency must be changed by an amount Δf to
maintain the condition cos φ = 0 . We find, that to first order

$$\frac{\Delta f}{f} = \frac{\Delta V}{V} . \tag{3}$$

Combining Eqs. (1) and (3), we are able to relate changes in the
lock frequency of the measurement loop to changes in the stress
state of the material.

The operating frequency of the transducers we have constructed
is about 1.5 MHz and we observe an equivalent electronic noise
level of a few parts in 10^6. Experimental scatter is somewhat
larger than this because of scratches and pits in the surface of
the pipe. It is here that the value of the EMAT technology becomes
apparent. Surface discontinuities give rise to some scatter of the
measured data about some smoothly varying curve, but the
repeatability of the measured data is very much better than the
scatter. Our earlier experiments with edge-bonded transducers for
transmission[6] gave acceptable results, but we were unable to
achieve a very high degree of repeatability due to variations in
the coupling conditions for the transmitting transducer.

The overall system insertion loss is of the order of 120 dB
and it is necessary to use a rather high power drive amplifier
(about 300 W) and to take great care to provide a well-shielded
electrical environment for the receivers. The receivers are
mounted in a specially designed brass housing which accommodates
the magnets, pole pieces and the matching circuits. The output
signal is fed directly to low noise, high gain amplifiers through
short sections of rigid coaxial cable.

Calibration System

The determination of the material constants B_1 and B_2 was
achieved by making measurements of $\Delta V/V$ and strain while a line
load was applied. The arrangement is shown in Fig. 3. The line
load was applied with a hydraulic jack on the inside surface of the
pipe and strain gauges were positioned on the outside surface in a
pattern symmetrically opposite the acoustic transducers. As the
load was increased, the strain was recorded using conventional
bridge techniques. At the same time, the variation in the lock
frequency of the phase measurement system was recorded. It is
clear that a loading technique which involves access to the inside
surface of the pipe is of limited practical value. However, the
load could equally well be applied to the outside surface, the only
difference being that more sophisticated jigs and fixtures would be
required.

EXPERIMENTAL RESULTS

Calibration

The calibration technique we have described was used to
determine the coefficients required to calculate the stresses from

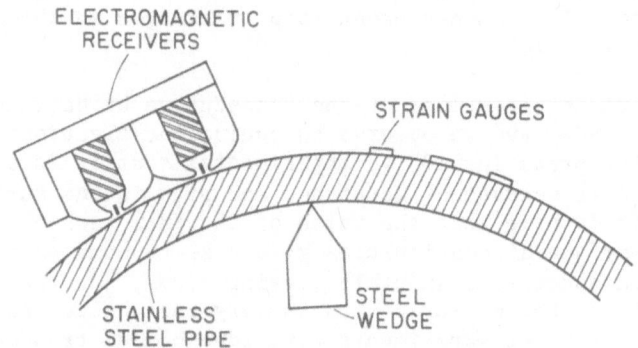

Fig. 3.   Configuration of the calibration experiment.

the measured $\Delta f/f$ data. We found that $B_1 = 8.72 \times 10^{-6}$/MPa and $B_2 = -12.7 \times 10^{-5}$/MPa .

Residual Stress Measurement

The experimental apparatus we have constructed allows the EMAT transducer assembly to be lifted clear of the pipe surface, translated a known distance by means of a stepper motor driven stage, and then pressed back into contact with the surface using compressed air filled metal bellows. All of these operations, and the subsequent phase measurement, are controlled by an HP-9826 desk top calculator.

The data logging procedure consisted of making measurements of velocity in the circumferential direction at points separated by 5 mm along the length of the pipe for four different azimuthal positions on the outside surface. For each position, the measurements were repeated three times to reduce the (small) experimental noise. The typical repeatability for a given point was of the order of 5×10^{-5} in $\Delta V/V$. This is an order of magnitude smaller than the scattering due to the surface condition of the pipe. The transducers were then rotated through 90° and the velocity in the axial direction was measured at the same set of positions on the pipe as the circumferential data set. Figures 4 and 5 show, at a particular azimuthal position on the pipe, the variation of the circumferential and axial velocities as a function of distance from the center of the weld. We assumed that the stress field at a point 20 cm away from the weld was zero, and added a constant offset to our data to obtain a zero of the

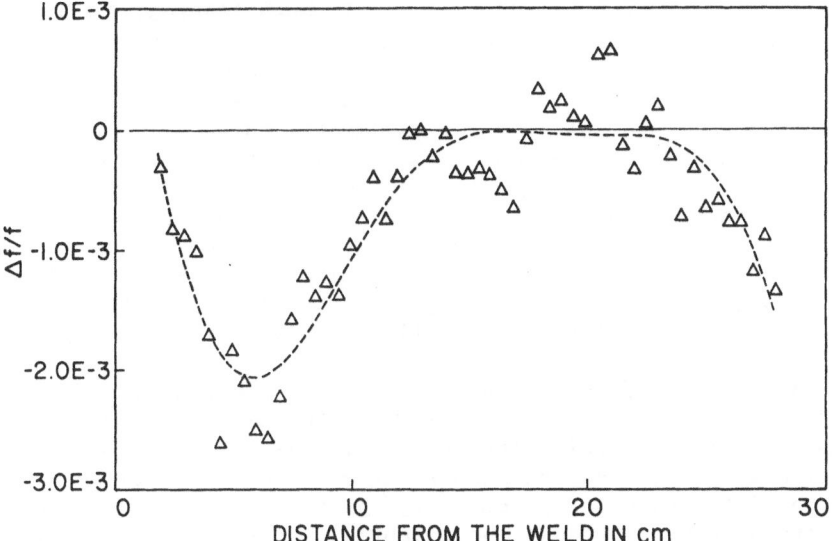

Fig. 4. Profile of the velocity of Rayleigh waves propagating
along the circumference of the pipe, on the outside
surface.

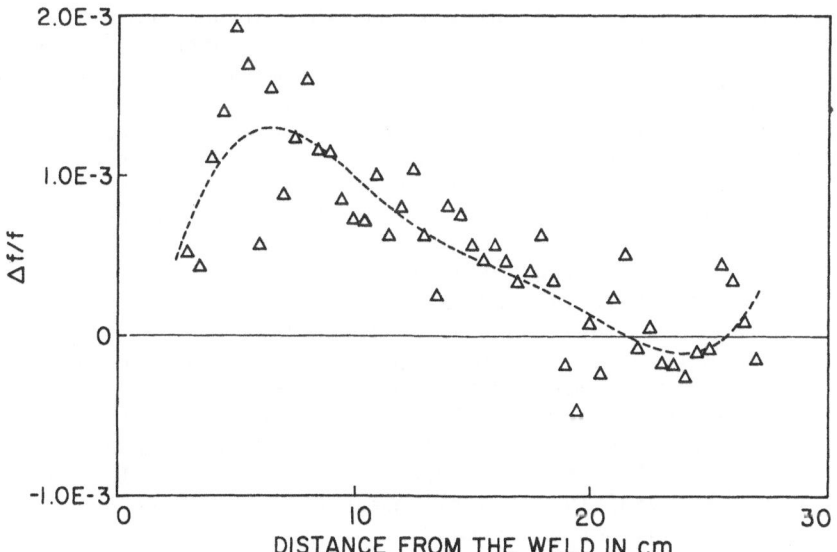

Fig. 5. Profile of the velocity of Rayleigh waves propagating
along the axis of the pipe, on the outside surface.

velocity perturbation at this point. In both figures, the dotted line represents a mean square fitting of the data with a Fourier series of 7 terms. Once again, the high-frequency spatial components of the data are due to the scratches and defects of the pipe's surface, and not to the experimental noise. In order to attenuate these high-frequency components, the data was digitally filtered and smoothed. Then Eq. (1) and the coefficients B_1 and B_2 obtained in our calibration experiments were used to calculate the variations of stress on the outside surface of the pipe. These variations are plotted in Fig. 6. The dotted lines were obtained by solving Eq. (1), using the Fourier series approximation of the velocity profiles as data.

The entire procedure was repeated with the transducers mounted inside the pipe. The spatial extent of the measurement was slightly smaller in this case because it was not possible to bring the transducers closer than about 60 mm from the weld line because of the shoulder produced by the counterboring operation when the pipe was welded. The measured data was processed in the same way as before, and is plotted in Fig. 7. The results obtained at the three other azimuthal positions were qualitatively and quantitatively similar to those presented here.

CONCLUSIONS

We believe that the results presented here show that the use of acousto-elastic Rayleigh wave measurements should prove to be a viable field applicable method for the determination of residual stresses in stainless steel piping and other structures of practical importance. The effect of surface scratches and other discontinuities is to give a scatter in the measured data that is much larger than the intrinsic repeatability of the technique. It appears, however, that with relatively simple data processing, the underlying stress distributions can be readily and unambiguously extracted. The large increase in the stress levels at the open end of the pipe sample is apparently due to the counterboring process that had been carried out on our sample in preparation for another test weld. It was observed that this large increase was consistent at all azimuthal positions and was not therefore due to any random surface scratches. The possibility of spurious reflections of the acoustic waves from the end of the pipe was also considered and can be eliminated for a number of reasons, such as the fact that where end reflections did exist (in the axial measurements), they occurred at different times from the required signals and could be rejected, and the fact that interference from other reflections would be highly sensitive to small changes in alignment and position. This was not observed.

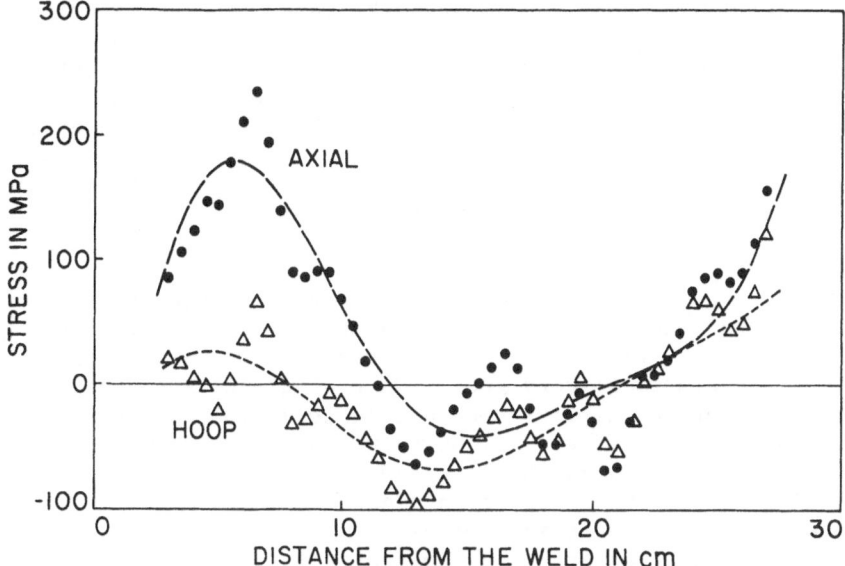

Fig. 6. Stress distribution measured on the outside surface.

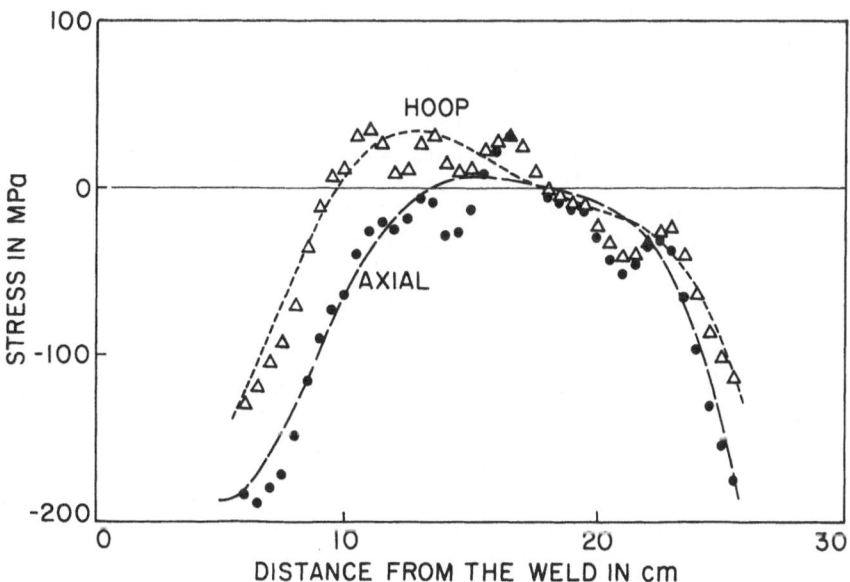

Fig. 7. Stress distribution measured on the inside surface.

The general form of the stress distributions that we predict compare well with other published work, based on both experimental measurements and theoretical models.[7,8,9] It will be observed that the axial distributions for both the inside and outside surface peak at a distance of about 6 cm from the weld (approximately three wall thicknesses). The outside peak is tensile and of the same order of magnitude as the yield stress. The inside peak is of opposite sign and of almost equal magnitude. All of these characteristics are consistent with expectation.[9]

The hoop stress on the outside surface is smaller than the axial stress and its peak on the inside surface is of the same order of magnitude as the axial stress peak. The qualitative agreement between the inside hoop stress and the work by Finnie[7], Fidler,[9] and others is very good. On the outside surface the hoop stress distribution differs somewhat in form from some of the other published material. There is, however, considerable variation in the results presented in the literature and frequently, for these particular components, there is a significant difference between measured data and theoretical prediction.[8] We are in the process of carrying out both x-ray and semidestructive (trepanning) measurements to provide independent data for comparision with the acousto-elastic measurements.

One would expect that any localized microstructural changes, due to the welding process, would cause serious perturbations of the data we measured. There are two basic reasons why this should not be a problem here. First, it is reasonable to assume that Rayleigh waves propagating in the circumferential and axial directions would be similarly affected by heat-induced microstructural changes. We observe, however, that the changes in these two components have opposite signs near the weld. Secondly, direct microstructural measurements have been reported[10] that show that the extent of the heat affected zone for material of this type and welded in this fashion is only of the order of 2 cm , somewhat closer to the weld than any of the data we are able to collect. For these reasons, we believe that it is fair to assume that the velocity variations we observe are mainly due to residual stresses.

To our knowledge, the data presented here represents the most complete set of measurements carried out on this particular problem. The results are in good agreement with independent theoretical and experimental data and we are in the process of establishing further verification. We have overcome the problem of variations in material properties from sample to sample by making in-situ calibration measurements. The final piece of information which is lacking at the present time is accurate zero stress value. This is difficult to obtain on a laboratory specimen of the pipe material. In practice one would be able to make measurements at a much greater distance from the weld and well away from the influence of machining or other end effects. One alternative solution would be to make a single trepanning measurement to establish this base line, providing this level of destructive testing is acceptable on a structure in service.

ACKNOWLEDGEMENT

This work was supported by the Electric Power Research Institute on Contract No. T107-5-5.

REFERENCES

1. M. Hirao, H. Fukuoka, and K. Hori, "Acoustoelastic Effect of Rayleigh Surface Wave in Isotropic Material," J. Appl. Mech. 103 (1), 119-124 (March 1981).
2. K. Jassby, and D. Saltoun, "Use of Ultrasonic Rayleigh Waves for the Measurement of Applied Biaxial Surface Stresses in Aluminum 2024 T351 Alloy," J. Mat. Eval. 40, 198 (February 1982).
3. D. Husson, and G. S. Kino, "A Perturbation Theory for Acoustoelastic Effects," J. Appl. Phys.
4. J. E. Bowers, B. T. Khuri-Yakub, G. S. Kino, and K. H. Hu, "Design and Application of High Efficiency Wideband SAW Edge Bonded Transducers," Proc. IEEE Ultrasonics Symp., 744 (1978).
5. H. M. Frost, "Electromagnetic Ultrasound Transducers: Principles, Practice and Applications," Phys. Acoust. Vol. XIV, Academic Press, New York (1979).
6. D. Husson, S. D. Bennett, and G. S. Kino, "Measurement of Surface Stresses Using Rayleigh Waves," Proc. IEEE Ultrasonics Symp. (1982).
7. S. Vaidyanathan, A. F. Todaro, I. Finnie, "Residual Stresses Due to Circumferential Welds," J. Eng. Mat. Tech. 95 (4), 233-237 (October 1973).
8. E. F. Rybicki, D. W. Schueser, R. W. Stonesifer, J. J. Groom, and H. W. Mishler, "A Finite Element Model for Residual Stresses and Deflection in Girth-Butt Welded Pipes," J. Press. Vess. Tech. 100, 256 (August 1978).
9. R. Fidler, "The Complete Distribution of Residual Stress in a 1/2 Cr, 1/2 Mo, 1/4 V/2CrMo Main Steam Pipe Weld in the As-Welded Condition," GEGB Report R/M/R261 (December 1977).
10. "Studies on AISI Type 304 Stainless Steel Piping Weldments for Use in BWR Applications," EPRI Report NP-944 (December 1978).

ACOUSTO-ELASTIC EFFECTS IN TWO STRUCTURAL STEELS

G. Mott
Metal Systems Department
Westinghouse Research Laboratories
Pittsburgh, PA 15235

and

M. C. Tsao
(Formerly at the same address as G. Mott above)
Electric Boat Division
General Dynamics Corp.
Groton, CT 06340

ABSTRACT

The change in the velocity of elastic waves caused by an applied uni-axial stress have been measured for two steels (ASTM A533B and A471). A novel tensile specimen, permitting the accurate measurement of absolute velocity, was designed and used for measuring the change in velocity of longitudinal and two orthogonally polarized shear waves for applied stresses ranging near to the yield point. Also measured was the velocity of Rayleigh surface waves, propagating in the large surface of a plate specimen of A533B steel. The variation of this surface wave velocity as a function of the angle, θ, between the propagation and the applied plate extensional stress directions was determined over the range $0° \leqslant \theta \leqslant 90°$. In all experiments, contact transducers were employed and the frequency ranged from 1.0 to 7.5 MHZ. The influence of these measurements on the quantification of internal residual stresses in solids is discussed.

1. INTRODUCTION

Whenever solid materials are formed into the shapes that are required for their operational use as a component part, one can say that, quite commonly, although not always, residual mechanical stresses will everywhere be present within the volume of the piece after its manufacture is complete. When the component's volume is small and if operational requirements render it necessary, these residual stresses can easily be reduced to very small magnitudes by a variety of annealing processes. However, when the component's volume is large, as it is in the case of heavy forgings or castings of metals, the time and cost of performing such annealing may be considerable becoming, thereby, a significant enough economic factor to warrant its omission.

If the surface residual stresses alone are thought to be of significance (e.g. in promoting more rapid fatigue failure, or stress enhanced surface corrosion cracking (SCC) caused by the residual extensional stresses in surface layers) these can be neutralized, for example, by ball-peening which leaves behind compressive residual stresses which act to impede SCC. In large components, however, both the surface and interior bulk stresses are important, not only, in the first instance, because of SCC, but also because the sum of operational load mechanical stresses and residual bulk stresses may exceed the yield stress of the material. In addition to this aspect of operational use, the designer of large components, having no measured values of residual stress to rely on, may have to assume that they are substantial, modifying his design safety factors to take them into account leading ultimately to unnecessary waste of material and space.

There is a variety of methods, both destructive and nonde-structive, by which residual surface stresses can be quantified such as strain gauges combined with hole drilling, acoustic surface waves, X-ray etc., and the reader is referred to a recent article by RUUD[1] for a review of them. However, recent literature on the subject indicates that, at the moment, there is only one viable nondestructive testing (NDT) technique potentially capable of quantifying bulk residual stresses in metals and this employs acoustic (ultrasonic) waves as the means of interrogating the material volume. The reason for this is due to the simple fact that only acoustic waves potentially have a sufficiently long enough range and resolution capability that are pre-requisite to minutely examining the whole of the interior of a large volume of material.

The physical property that makes it possible to investigate the volume distribution of bulk residual stresses by means of acoustic waves is the acousto-elastic effect. This is a property possessed by all homogeneous solids and it is the basic cause of the experimental observation of changes, admittedly very small, of all three bulk acoustic velocities (the longitudinal and the two orthogonally polarized shear waves) whenever the local stress state is changed. The relative change in any velocity, $\delta v/v$, is of the order of 10^{-4} for an applied stress of 69 MPa (10 KSI) and it is somewhat chagrinning to the experimenter to realize that this is about the same magnitude as the relative change in velocity caused by a one degree change in temperature. See SALAMA.[2]

Further, if the acousto-elastic relation is, for the moment, linearized in the form $\delta v/v = K\sigma$ where σ is a stress increment and K is an acousto-elastic constant, it has been found by KINO[3] that K can vary substantially for materials having the same nominal composition but which were acquired from different suppliers. This may be partially explained using the calculations of JOHNSON,[4] on the basis of differences in spatial averages between random and less-than-random orientations of crystallites forming the material, but is most probably due to the combination of small variations in thermal, compositional and forming histories.

The answer to the question why is $\delta v/v$ affected by changes in so many state variables lies in the form of the most general thermo-dynamic internal energy equation where second order effects lead to a coupling between the main energy storage variables. As a result of such coupling, one may expect that, for example, for a magnetic material, the effective permeability will be a slight function of stress and, reciprocally, that elastic constants will be a slight function of magnetic field etc. The expected inter-relationship between temperature, magnetic (electric) field, elastic stress, carrier densities (in semi-conductors) etc. makes it extremely difficult to isolate one of them as causing a given change, $\delta v/v$, if the others are not kept in strict control. In spite of the increased complication of having to deal with many state variables at one time, there is, nevertheless, optimism by a number of workers in the field that the acousto-elastic effect can be used to find the volume distribution of bulk residual stresses by means of measurements, at a large number of external points, of $\delta v/v$ along many internal paths.

This article represents our attempt to obtain a data base on the acousto-elastic properties of two structural steels in order to provide information for possible future work in calculating the three-dimensional distributions of residual stresses within a solid body.

2. TEST SPECIMEN DETAILS

The specimens that were used in this investigation were of a simple plate and dumb-bell geometries. The plate had a cross-section of 25.4 cm X 1.27 cm and was 45 cm in length. It was made from ASTM A533B steel and was used only for surface wave measurements. The dumb-bell specimens were typical of tensile test types and were made from ASTM A533B and A471 steels, but were so unusual in cross-sectional shape they require some additional background discussion. In the past, specimens having a rectangular or circular cross-section were typical of those whose third order elastic constants were being examined. In these earlier investigations, especially in the late 1960's, there were difficulties associated with phase effects in the liquid (or solid) couplant, necessary to provide transmission of elastic waves, together with phase effects due to diffraction and reflection of pulses at a fluid-specimen boundary. See TRUELL ,et al.[5] Most of the phase effects due to couplant difficulties can be overcome by using water immersion techniques, although the experimental test arrangement becomes somewhat more complicated with the potential side effect of corrosion of surfaces looming as a source of further difficulty.

To circumvent all of these actual and potential problems, we have designed a unique specimen for use in tests where uni-axial stresses are applied. Dimensional details of the tensile test specimen are shown in Figure 1 where it can be seen that the cross-section in the gauge region has an unusual squat T shape. The T-shaped section gives the specimen its unique quality in that it permits, in solid materials, the use of an ultrasonic technique that is virtually equivalent to the split-beam interferometer commonly used in optics.

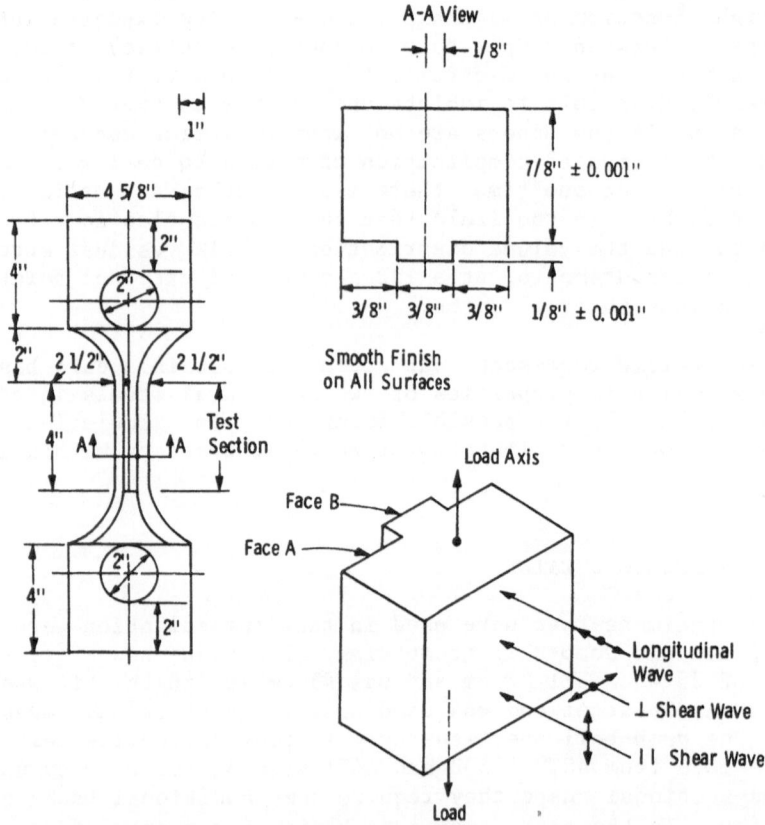

Fig. 1 - Design detail of novel tensile specimen showing load axis direction, wave propagation vectors and particle motions of the three bulk wave types.

The sketch in Figure 2 explains the physical basis underlying the T design. With a transducer transmitting elastic waves across the gauge section in the position shown, the vertical leg of the T and a half of the horizontal arm become equivalent to two specimens intimately adjoined with one another. By adjusting the position of the transducer, the amplitudes of ultrasonic pulses reflected from the plane surfaces opposite the transducer in the leg and arm segments can be brought to equality. When equal pulse-echo amplitudes have been achieved by small adjustments of the transducer, the split-beam equivalence is assured. This arrangement nullifies the effects of uncertainties of pulse phase due to transducer coupling coefficient, reflection coefficient of the free surfaces and diffraction effects since both "halves" of the split beam share these equally provided that the individual path lengths are not very different. In practice, the path length difference was chosen to be 0.317 cm and is thought to be sufficiently small enough to permit perfect equality in both paths of all the phase effects mentioned.

All surfaces in the gauge section were given a No. 4 finish (4 micro.cm/cm) and were everywhere accurately parallel to the main axis of the specimen.

3. APPARATUS AND METHODS FOR MEASURING ULTRASONIC VELOCITY

All specimens were provided with strain gauges of which some were mounted parallel to and others perpendicular to the direction of the applied uni-axial stress. The specimens were placed in hydraulically driven material test machines capable of providing sufficient loads so that stresses up to the yield stress could easily be applied.

Starting at zero applied stress, the three bulk wave phase velocities were measured for the tensile specimens. Increasing axial loads were impressed and the three bulk velocities were again

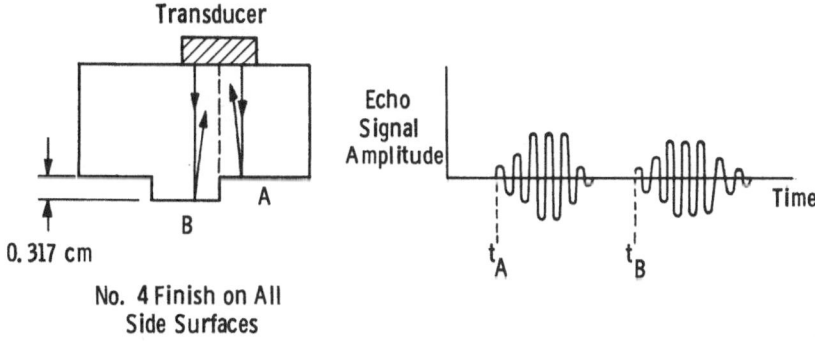

Fig. 2 - Sketch of T-section of tensile specimen showing ray paths and typical oscilloscope traces.

measured at increments of 34.4 MPa (5 KSI) until the applied stress
was close to the yield point. The load was then decreased and all
velocities were again measured at decrements of 34.4 MPa until the
no load condition was reached.

The A533B plate specimen was treated somewhat differently. A
special holder was constructed which supported the surface wave
launcher/receiver transducer pair. The surface was cleaned with a
solvent, allowed to dry and then given a very light spray of Krylon
mold release after which the transducer was put into place and
maintained in light contact with the surface of the plate. A plate
extensional load was applied until a stress of 309.6 MPa (45 KSI)
was reached. This stress was maintained constant and the surface
wave velocity was measured for fixed values of, θ, the angle
between the propagation and applied stress directions. The angle θ
was varied, by 10° increments, between 0° and 90°. See Figure 3
for details.

The heart of the apparatus that was used for making velocity
measurements is the Ultrasonic Stress Analyser (USA) instrument
purchased from Ultrasonic Analysis Inc., Stamford, CT. It is
basically a low power ultrasonic pulser/receiver incorporating a
timing clock that possesses the capability of measuring time delays

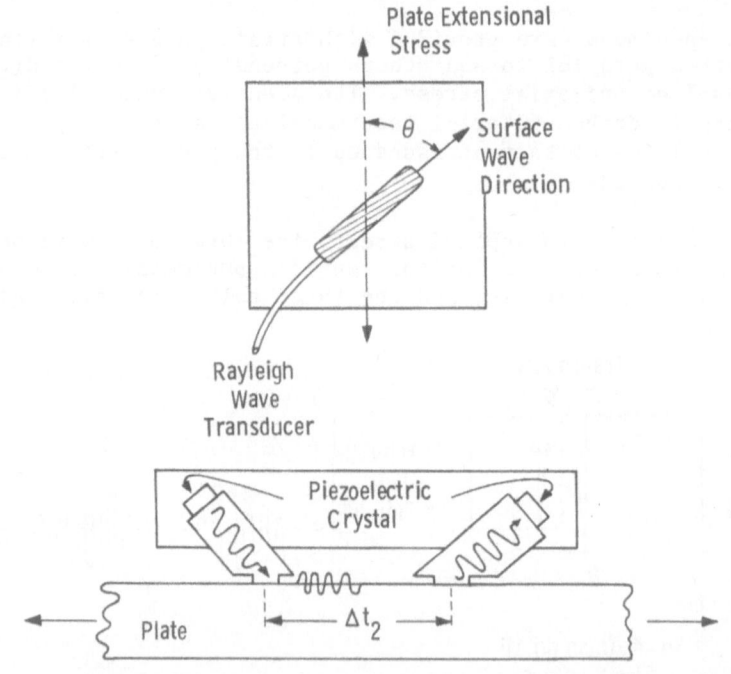

Fig. 3 - Diagram of Rayleigh wave specimen set-up and transducer
         detail showing typical time spent in the surface wave
         path segment.

of pulse arrivals over the range 0-100 μsecs with a nominal accuracy of 10^{-10} secs.

The instrument is provided with a set of transducers for generating longitudinal, shear and Rayleigh surface waves covering the frequency range 1.0-7.5 MHZ. The bulk wave transducers were used in tests on the tensile specimens while the surface wave transducers were used only in plate tests.

The relative time of arrival of a particular ultrasonic pulse is ascertained by means of a signal brightening technique that permits the brightness level within the pulse to be intensified at a selected number of consecutive points on the time axis each point separated from the next neighbor by a half period of the main frequency. The brightening signal is derived from a reference train of coherent rectangular pulses with which the ultrasonic signal of interest can be synchronised by adjusting its triggering time delay by any amount up to 100 μsecs. When the brightened points are all at the zero crossings well within the ultrasonic pulse signal, a straight line of bright dots can be seen on the CRT screen. Time delay errors of multiples of half a period of the main frequency can easily be avoided because the instrument has the capability of viewing the ultrasonic pulse signal and the reference rectangular pulse train simultaneously by making use of the available double beam oscilloscope traces. Approximate synchronism of ultrasonic and reference pulses can be obtained by first using coarse time delay adjustment and judging by eye. Then, by use of a 'mixing' switch, brightened dots can be seen on the CRT screen and made to form a straight line at zero crossings within the ultrasonic pulse by using the fine adjustment switch on the time delay. In this way, the relative time of arrival of a pulse can easily be measured to within 10^{-10} secs. See Figure 4 for details of pulse trains and brightening dots.

3.1 Measurement of Bulk Wave Phase Velocity by Time Differences

The relative times of arrival of pulse echoes being reflected from the faces at the end of the leg and cross arm regions of the T-section at A and B are t_A and t_B. Then the equation

$$v(t_A - t_B) = 2L = v(\Delta t) \tag{1}$$

accurately describes the relationship between velocity, distance and arrival time difference for these pulses.

When the total differential of this equation is formed, one obtains

$$dv/v + d(\Delta t)/(\Delta t) = dL/L \tag{2}$$

and when the applied uni-axial stress, σ, is the only variable of interest,

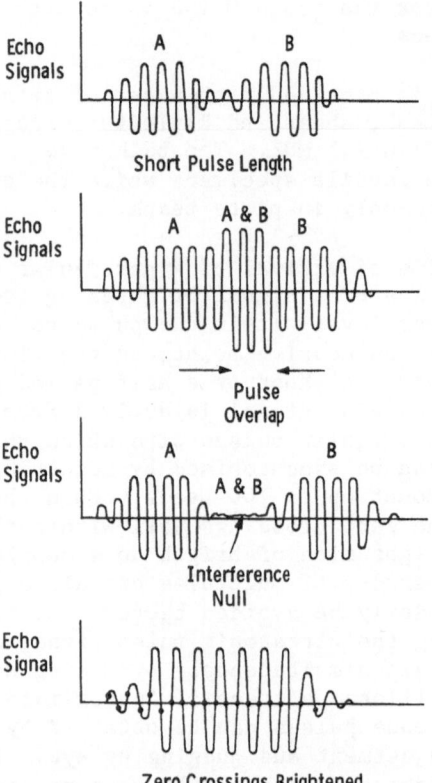

Fig. 4 – Sketch of typical oscilloscope pulses showing brightening
        dots, overlap and interference null.

$$dL/L = -\nu\sigma \qquad\qquad\qquad\qquad\qquad\qquad (3)$$

where ν is Poisson's ratio. From this, one obtains

$$dv/v = -\nu\sigma - d(\Delta t)/(\Delta t) \qquad\qquad\qquad\qquad (4)$$

and, when $d(\Delta t)/(\Delta t)$ has been measured for a number of σ values,
one is able to write

$$dv/v = K\sigma \qquad\qquad\qquad\qquad\qquad\qquad\qquad (5)$$

where K is an acousto-elastic constant.

3.2   Measurement of Bulk Wave Phase Velocity by Pulse Overlap

   The two pulses received from A and B (see Figure 1) will
overlap when the physical length of a pulse within the specimen
(when measured normal to the planes at A and B) exceeds twice the
distance between A and B; i.e. when the physical length of the
pulse exceeds 0.635 cm (0.25 inch). The USA equipment possesses

pulse length variability and when this length is well in excess of
0.635 cm, the pulse amplitude in the region of pulse overlap can,
by destructive interference, be made extremely small by varying the
central frequency of the pulse. See Figure 4. This central, or
main, frequency can be measured by counting many time periods
within the pulse by using the brightened zero crossings as
reference points or by using an ancillary pulse frequency counter.
Once the average periodic time, or frequency, has been ascertained
in this way, the phase velocity of the wave can be calculated from

$$v = f_n \lambda_n \tag{6}$$

where λ_n is the wavelength and f_n is the frequency so that

$$(2n - 1)\lambda_n = 4L \ (n = 1,2,3....) \tag{7}$$

whenever the pulse overlap destructive interference minimum has
been found for the frequency f_n. Thus, in this way, v can be found
for a range of frequencies at any given state of stress,
temperature etc.

3.3 Method for Finding the Surface Wave Velocity

The Rayleigh wave device is composed of a surface wave
transmitter/receiver pair. The two transducers are kept at a fixed
distance apart and, for making relative velocity measurements, it
is not necessary to know the separation distance exactly. The
relevant velocity equation that governs the surface wave
propagation is simply

$$v(\Delta t) = \text{constant} \tag{8}$$

where (Δt) is the travel time taken by the surface wave in getting
from the transmitter to the receiver sensor.

The total travel time contains a part (Δt_1) due to travel in
the coupling wedges as well as a part (Δt_2) due to the travel in
the surface of the specimen. See Figure 3. By forming the total
differential of equation 8, one obtains

$$dv/v + d(\Delta t)/(\Delta t) = 0 \tag{9}$$

and since

$$(\Delta t) = (\Delta t_1) + (\Delta t_2) \tag{10}$$

equation 9 can be written as

$$dv/v = - d(\Delta t_2)/(\Delta t_2) \tag{11}$$

for, because $(\Delta t1)$ is constant, $d(\Delta t1)$ is zero.

The change, $d(\Delta t_2)$, in (Δt_2) is measured by means of the USA instrument while the initial value of (Δt_2) is calculated accurately enough from the mean separation distance of surface transducer coupling wedges, Young's modulus and Poisson's ratio (obtained from static load strain gauge measurements).

4. Test Results and Discussion

Preliminary calibration runs allowed Young's modulus and Poisson's ratio to be calculated from strain gauge readings and engineering (approximate) values for the longitudinal and shear wave velocities were calculated from these elastic parameters. These approximate velocities are shown in Table 1 and have been calculated by using an acceptably accurate value (obtained from materials handbook data) for the mean density.

Extremely accurate absolute values for these velocities were also obtained at zero applied stress by employing the USA test equipment and using both the pulse overlap and time difference techniques. The values obtained for these velocities are also shown in Table 1 and may be compared with the engineering values there.

Table 1 - Approximate and accurate wave velocity measurements

Material	Longitudinal wave Velocity $C_p = [(\lambda+2\mu)/\rho]^{1/2}$ m/sec	Shear Wave Velocity $C_s = (\mu/\rho)^{1/2}$ m/sec
(A533B)	Eng. Value 5972	3150
	Acc. Value 5913	3249
(A471)	Eng. Value 5847	3180
	Acc. Value 5796	3185

In Figures 5 and 6 are shown the measured results of the relative velocity change, $\delta v/v$, of all three bulk waves for A533B and A471 steels as a function of the applied uni-axial stress. The propagation direction was perpendicular to the load axis and, for the two shear waves, the particle velocities were either parallel or perpendicular to the load axis as shown in Figure 1. The measurements were done at several frequencies in the 1.0 to 7.5 MHZ range and, occasionally, slight variations with frequency of the acousto-elastic constant were noticed and an example is shown in Figure 6. There, the values for K at 2.25 and 5.0 MHZ are very slightly different due, possibly, only to experimental error because the limit of accuracy is being approached.

The Murnaghan third order elastic constants have been calculated from our experimental data using the equations of Egle and Bray[6] and these, along with the calculated second order elastic constants, are shown in Table 2 and compared there with similar results published by Egle and Bray. Bearing in mind the different steels represented in Table 2, agreement between these selected results is felt to be very good.

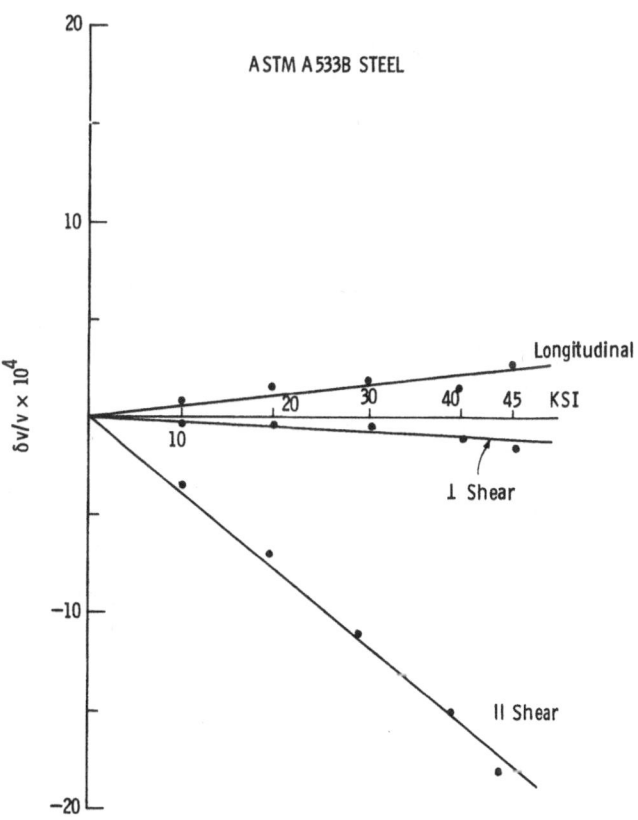

Fig. 5 – Experimental measurements of $\delta v/v$ as a function of applied stress for bulk waves in ASTM A533B steel.

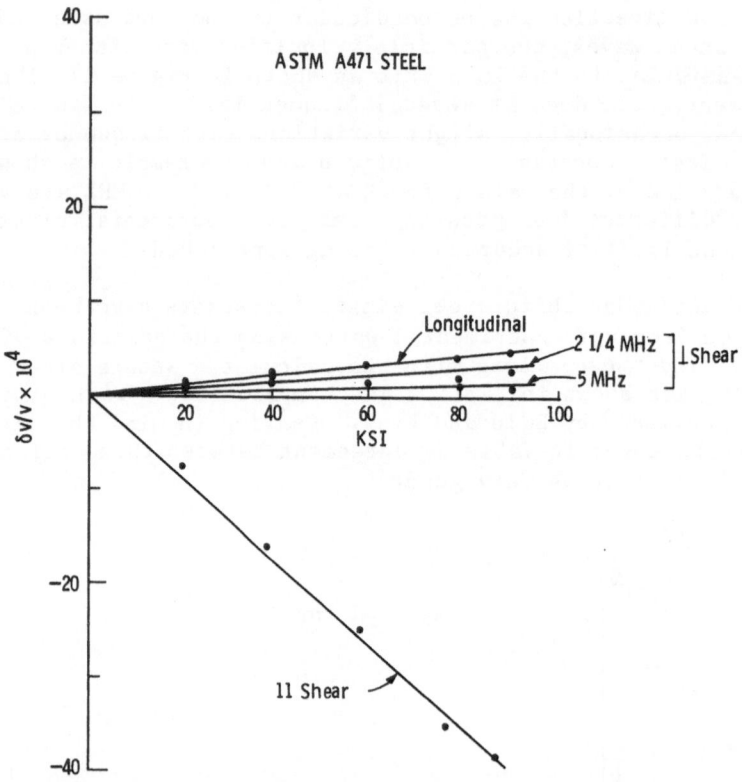

Fig. 6 - Experimental measurements of $\delta v/v$ as a function of applied
stress for bulk waves in ASTM A471 steel.

Table 2 - Comparison of the present results with those of Egle and
Bray (6)

Material	Reference (6)		Present Results	
	Rail 1	Rail 2	A533B	A471
Density Kg/m^3	7800	7799	7799	7800
Lamé constant N/m^2	11.58	11.07	11.90	12.0
Shear modulus N/m^2	7.99	8.24	7.91	7.94
Murnaghan l	−24.8	−30.2	−21.8	−17.9
Third order m	−62.3	−61.6	−48.6	−49.6
Constants n N/m^2	−71.4	−72.4	−56.4	−62.8

The surface wave velocity measurements were troublesome due to the unexpected presence of residual plate stresses. Because of the residual stresses, the measured values of δv/v as a function of θ showed a fairly large scatter and the results shown in Figure 7 are a mean of the many values observed. Individual sets of values of δv/v for a given θ were collected from many small areas of the plate's surface all of which were within the central region of the specimen. The angular variation of velocity for the fixed plate stress of 309 MPa (45 KSI) depicted in Figure 7 shows that the surface wave velocity continually increases as θ changes from 0° to 90°. The results in Figure 7 indicate that the direction of applied principal plate stresses can be established from the angular variation of surface wave velocity and that, therefore, the in situ measurement of surface wave velocity might be used to find the direction of principal residual surface stresses.

While all of the preceding completes the tally of the positive aspects of this work, this discussion would be incomplete if it were not to include some mention of the experimental difficulties that caused us some setbacks. The first difficulty arose from the use of liquid couplants for the tensile specimen work. The

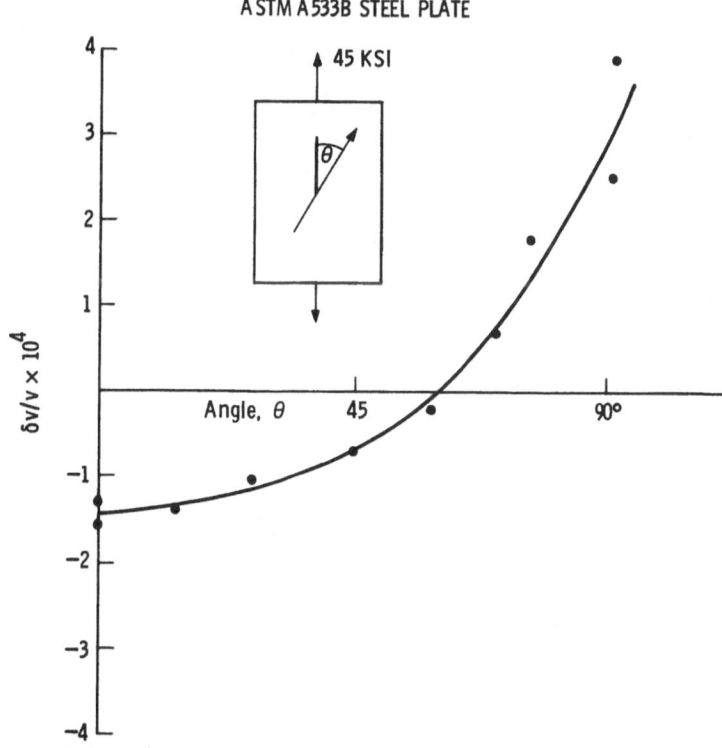

Fig. 7 – Experimental measurements of the variation of Rayleigh wave velocity as a function of angle, θ, between plate stress and wave propagation vectors.

permanent seating of the transducer required up to an hour of
waiting for the couplant film to settle down enough to be
reliable. Using the pulse overlap destructive interference method
detailed in Section 3.2, the variability in the coupling can easily
be observed by the drifting of the specific frequency required for
the maintenance of the interference null. Long waits were
necessary before drifting became minimal.

The second difficulty was due to the residual stresses in the
plate. We had hoped to show results that included the variation of
$\delta v/v$ at constant θ as a function of applied plate stress but the
experiments were troubled by both stress/velocity hysteresis and
drifting of the measured velocity at zero applied stress. There
was such a large scatter in our measurements that we felt them to
be too unreliable to be reported here.

The third difficulty arose when we attempted to measure $\delta v/v$
as a function of applied stress using a specially designed variant
of the tensile specimen shown in Figure 1. This variant was
equipped with unusual extensions to the side surfaces in the gauge
region that were meant to permit propagation of bulk waves in a
direction at $45°$ to the load axis. Even though great care was taken
in the experiments, we very quickly found that a severe loss of echo
amplitude occurred at quite low values (207 MPa) of applied
stress. We believe that this was possibly due to the fact that the
stress induced anisotropy, even though it is weak, leads to a
sufficient divergence between the phase velocity vector and the wave
(energy, or ray) vector that some of the energy of the ultrasonic
pulse was slightly diverted away from and did not return to the
source transducer. This effect is shown graphically in the sketch
in Figure 8 and the interested reader can find excellent discussions
of the theory of the phenomenon by HEARMON.[7] The phase velocity-
wave vector divergence phenomenon may produce little or no effect
along the principal stress directions, but it may be expected that,
when the propagation direction is well away from the principal axes
of stress, the resulting phase-wave deviations will become more
pronounced. This marked deviation between the two vectors may pose
an interesting difficulty for those workers who wish to use shear
wave bi-refringence as a basis for measuring internal stress because
the difference in the phase-wave divergences for two orthogonally
polarized shear waves can be expected to lead to a more pronounced
(possibly doubled) effect than we have observed.

During the course of all measurements reported here, great
care was necessary in order to maintain constancy of specimen
temperature because, as was earlier mentioned, a one degree change
in temperature can sometimes lead to the same relative velocity
change as can a change in applied stress of 69 MPa (10 KSI).

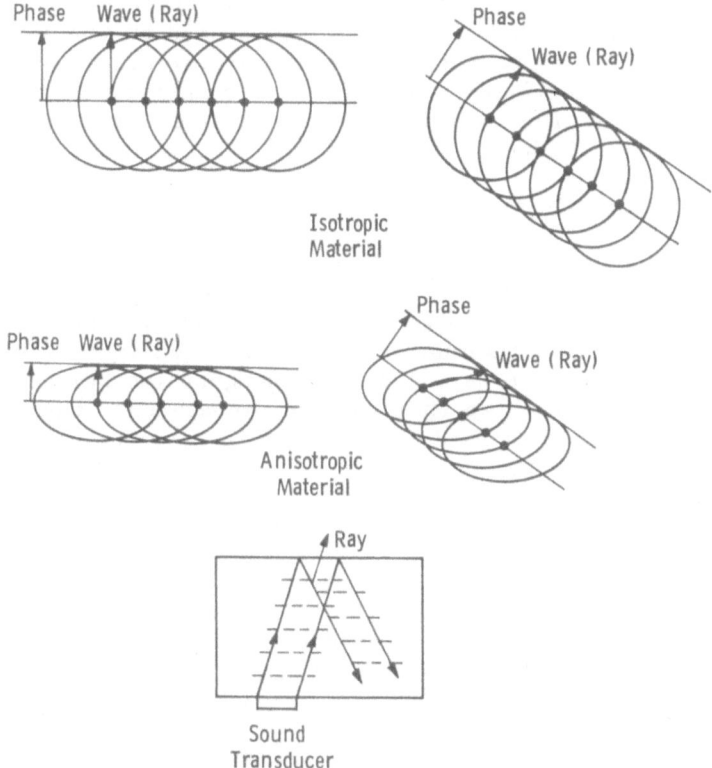

Fig. 8 - Sketch of wave and velocity surfaces for an anisotropic
material showing deviation between phase- and wave-vector
directions.

5. CONCLUSIONS

The main objective of this work, which was to accumulate a data
base of acousto-elastic coefficients for two steels of interest to
us, has been achieved. These coefficients have been measured for
the three independent bulk elastic wave types using applied uni-
axial stresses ranging to near the yield stress. In order to avoid
the necessity of using immersion methods, which have been employed
by other workers specifically to overcome phase effect errors in
accurate velocity determination, a second objective was established
and successfully realized by the development and use of a unique
tensile specimen design. The special T-shaped cross-section of
this specimen makes it possible to use contacting transducers whilst,
at the same time, eradicating all errors due to phase effects (caused
by diffraction and reflection coefficients) which potentially always
occur during measurements using contacting transducers.

Our results, displayed in Figures 5 & 6 agree extremely well
with those obtained by other workers in the field and this

agreement shows that the novel specimen design can confidently be used in high sensitivity measurements of this type. A variant of this specimen design has successfully been used by the first author (G.M.) to measure the temperature dependence of ultrasonic parameters in solids. An article on that work has been accepted for publication at Review of Progress in Quantitative NDE, Santa Cruz, CA., Aug., 1983.

The coupling between state variables (referred to in Section 1) involved in the internal energy equation and the experimental difficulties encountered in the present work (discussed briefly in the previous section) pose potentially serious experimental difficulties which might delay the development of a successful technique to quantify residual internal stresses in solids. It is clear that considerable care and more thoughtful experiments are required in future work so that the individual contributions to the combination effect that leads to ultrasonic velocity variations can separately be estimated.

6. ACKNOWLEDGEMENTS

This work was supported wholly by the Nuclear Technology Division, Westinghouse Electric Corporation, Monroeville, PA, and its sponsorship is gratefully acknowledged.

7. REFERENCES

1. C. O. Ruud, "A Review of Nondestructive Methods for Residual Stress Measurement", Journal of Metals, pp. 35-40, July, 1981.
2. K. Salama, "Development of an Ultrasonic Method for the Nondestructive Evaluation of Residual Stress in Steels", Nondestructive Evaluation Program Progress in 1980. E.P.R.I. NP-1690-SR, Dec., 1980.
3. G. S. Kino & J. H. Shaw, "Acoustic Techniques for Measuring Stress Regions in Materials", E.P.R.I. Report NP-1043, Research Report 609-1 (Interim Report) pp. 2-13, April, 1979.
4. G. C. Johnson, "The Effect of Texture on Acousto-elasticity", Review of Progress in Quantitative NDE, University of California, San Diego, Aug., 1982.
5. R. Truell., C. Elbaum, & B. B. Chick, "Ultrasonic Methods in Solid State Physics", pp. 53-158, Academic Press, New York, 1969.
6. D. M. Egle & D. E. Bray, "Measurement of Acousto-elastic and Third-order Elastic Constants for Rail Steel", J.A.S.A. Vol. 60, No. 3, pp. 741-744, Sept., 1976.
7. R. F. S. Hearmon, "Applied Anisotropic Elasticity", pp. 68-88, Oxford University Press, 1961.

RELATIONSHIP BETWEEN STRESS AND TEMPERATURE

DEPENDENCE OF ULTRASONIC SHEAR VELOCITY

N. Chandrasekaran and K. Salama

Mechanical Engineering Department
University of Houston
Houston, Texas   77004

ABSTRACT

The effects of applied elastic stress on the temperature dependence of 10 MHz ultrasonic shear velocity have been studied in
A533B steel. The measurements were performed with stress perpendicular to the propagation direction of ultrasonic shear waves.
The polarization direction was either parallel to or perpendicular
to that of stress. In all these measurements, the ultrasonic
velocity is found to decrease linearly with temperature, and the
slopes of the lines of best fit of ultrasonic velocity versus temperature vary considerably when the specimen is subjected to stress.
The results obtained when the stress is applied in a direction perpendicular to ultrasonic propagation and parallel to polarization
show that the temperature dependence of ultrasonic shear velocity
increases linearly with applied tensile stress, and decreases when
the stress is compressive. When the stress is applied in a direction perpendicular to both propagation and polarization, the temperature dependence decreases linearly with applied tensile stress
and increases with compressive stress. Calibration curves relating
relative changes in the temperature dependence of ultrasonic velocity to applied stress in A533B steel are constructed. Using these
curves, the sensitivity in determining unknown applied stress in
steel is estimated to be ±32 MPa, when the stress is parallel to
polarization and ±35 MPa when the stress is applied perpendicular
to polarization.

INTRODUCTION

Earlier studies have demonstrated[1-4] that the method of the
temperature dependence of ultrasonic velocity can be used to measure

applied as well as residual stresses. Experiments performed by
Salama et al on copper[1], aluminum[2,3] and steel[4] have shown that the
relative change in temperature dependence is a linear function of
applied elastic stress. In steel, this relationship is found to be
represented by,

$$\frac{(dV/dT)_\sigma - (dV/dT)_o}{(dV/dT)_o} = \pm K\sigma \tag{1}$$

where $(dV/dT)_o$ is the temperature dependence of ultrasonic velocity
at zero stress, $(dV/dT)_\sigma$, is the temperature dependence of ultrasonic velocity at an applied stress σ and K is a material constant.
The value of $(dV/dT)_o$ in stress-free steel is found to be -0.645
m/sec/K, while K for the same steel is equal to $5.88 - 10^{-4}$ per MPa.
All the previous temperature dependence work in steel as well as
other materials, however, has been performed using ultrasonic
longitudinal waves. In practice, the magnitude as well as the
direction of stress in a specimen needs to be determined. Also the
stress in the specimen may be biaxial or triaxial. In order to
apply the temperature dependence method to these stress measurements,
the relationships between the temperature dependence of shear waves
polarized parallel to or perpendicular to stress are needed.

In this paper, we investigated the effects of externally applied elastic stresses on the temperature dependence of ultrasonic
shear velocity in A533B steel. The investigation is made when the
stress is applied in a direction perpendicular to propagation and
either parallel to or perpendicular to polarization. The result
of this investigation is also combined with the temperature dependence of ultrasonic longitudinal velocity to form a complete
set of calibration curves for A533B steel.

EXPERIMENTAL PROCEDURE

Carbon steel ASTM 533B steel was used in the present research.
This steel is generally used in the manufacturing of pressure vessels in the nuclear industry. The specimen was machined in the
form shown in Fig. 1.   The two faces of the specimen were made
flat and parallel to within ±0.001 inch in order to avoid wedge
effects of ultrasonic waves during propagation. The specimen
shown in Fig. 1 was used when tensile stresses were applied. For
compression tests, the tensile specimen was cut at the dotted lines
to avoid possible buckling and bending stresses. The application
of external stress was carried out with a model 1125 floor type
Instron machine of 20,000 Kg maximum load capacity. Two different
types of loading arrangements were used in the present investigation. These arrangements are described in detail elsewhere[5,6].
To help insure the uniformity of stress in the specimens, special

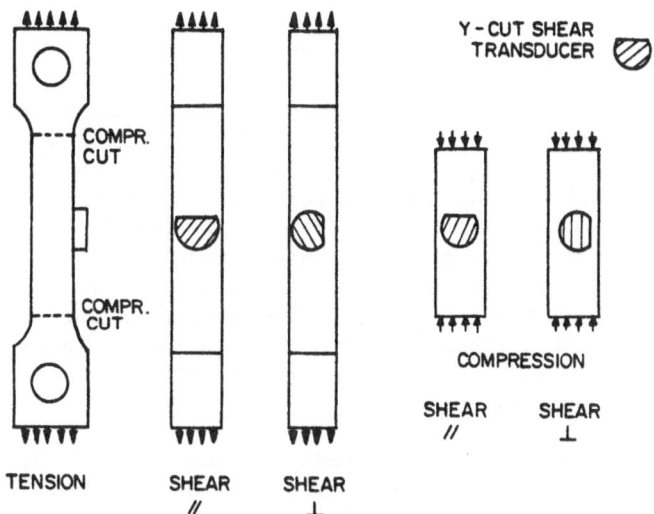

Fig. 1:   Specimens used to study the effect of
          applied stress on ultrasonic velocity
          measurements.

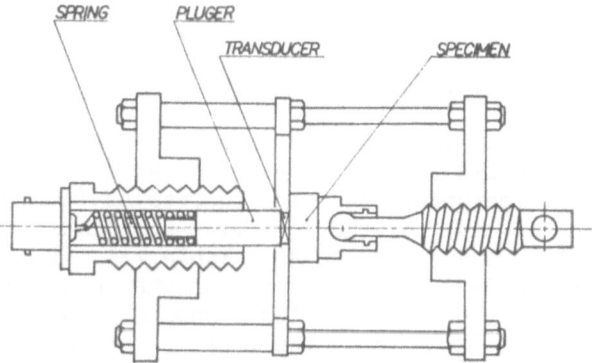

Fig. 2:   Transducer holder used for the measurements
          of ultrasonic velocity.

effort was made in designing these stress application systems to
minimize any effects of misalignment between the axis of the specimen and the loading frame.

The ultrasonic velocity was measured using the pulse-echo-overlap method which has been fully described elsewhere[7]. Y-cut transducers are used for the generation of the shear waves, which are
used in all measurements. The transducer is placed on the specimen
by means of a special holder which is designed to clamp to the
specimen as shown in Fig. 2. The spring-supported plunger serves
as the inner conductor of the coaxial cable which carried out the
10 MHz electrical signals to and from the transducer which is bonded to the specimen. These signals are transmitted from the plunger
by a piece of teflon-coated wire to a BNC receptacle which is mounted
directly to the transducer holder. The spring is used to produce
a pressure-coupled transducer which is required to insure uniformity
of the thin layer bond. A small clamping force of 30 to 50 Newtons
was used in the present work.

The temperature control systems are designed to enclose the
specimen and the measuring accessories to insure stabilized temperature during the time required for the velocity measurements.
The desired subambient temperature is obtained by controlling the
flow rate of liquid nitrogen to the insulated enclosure. After the
specimen is cooled to the desired low temperature, the supply of
nitrogen is stopped. The warming rate of the system is achieved
by using a low heat input light source within the enclosure. The
warming rate is adjusted so that the rise in temperature is steady
at about 0.25°K/minute. All the measurements were made in the
range of 220°K to 260°K.

RESULTS

Accuracy of Measurements

The measurements of utrasonic velocity as a function of temperature were repeated for the cases of longitudinal, shear (parallel) and shear (perpendicular) in order to estimate the repeatability and the accuracy of the measuring technique. The results
of these experiments are tabulated in Table (1). The table also
shows the standard deviation, δ, based on the results of the four
sets of measurements performed. The standard error, S, is used to
obtain the error in determining the temperature dependence of ultrasonic velocity, dV/dT[8]. This is given by,

$$S = \delta/\sqrt{n} \qquad\qquad\qquad (2)$$

where n is the number of measurements and δ is the standard deviation of an infinite number of measurements.

Table 1: Variations of measured dV/dT using different types of
          waves.

Type of Wave	dV/dT	% Dev.
Longitudinal	-0.640	1.84
Longitudinal	-0.652	0.0
Longitudinal	-0.613	0.0
Longitudinal	-0.629	2.54
Shear (Par.)	-0.401	0.0
Shear (Per.)	-0.407	1.47
Shear (Par.)	-0.474	0.0
Shear (Per.)	-0.483	1.86
Shear (Par.)	-0.443	0.0
Shear (Per.)	-0.445	0.44

Table 2: Variations of temperature dependence of ultrasonic
          shear velocity with applied tensile and compressive
          stresses in A533B steel. Stress is applied in a direc-
          tion perpendicular to the direction of propagation
          and in a direction parallel to polarization.

Stress MPa	dV/dT m/sec/MPa	$(dV/dT)_\sigma - (dV/dT)_0$ m/sec/MPa	$\Delta\%$	Dev. in %
181.8	-0.460	-0.042	10.0	0.35
151.5	-0.454	-0.036	8.6	0.17
121.2	-0.454	-0.036	8.6	-1.4
90.9	-0.440	-0.022	5.2	0.41
60.6	-0.435	-0.017	4.1	0.05
0.0	-1.418	0.0	0.0	0.88
0.0	-0.452	0.0	0.0	0.88
-60.6	-0.442	0.01	-2.2	-0.07
-90.9	-0.438	0.014	-3.1	-0.74
-121.2	-0.424	0.028	-6.2	0.77
-151.5	-0.419	0.033	-7.3	0.29
-169.7	-0.416	0.036	-8.0	0.04

In this investigation, dV/dT was measured only once for a given condition of the specimen using a particular type of wave propagation. Using the results in Table (1) and equation 2, S is found to be 1.7%. This accuracy in the temperature dependence measurements agrees well with the accuracy of ±1.6% determined by Barber[9] in his investigation on Aluminum. It can be noted from Table 1 that the values of dV/dT for shear-parallel and shear-perpendicular at zero applied stress are equal within the accuracy of measurements. This means that at zero applied stress, dV/dT of shear waves in the steel investigated is texture insensitive. This is in contrast to that observed from measurements of the absolute values of shear velocities which vary considerably with polarization direction.

Temperature Dependence of Ultrasonic Velocity-Shear Waves Polarized Parallel to Stress

The velocity of shear waves propagating in a direction perpendicular to applied stress and polarized parallel to stress was measured as a function of temperature on type A533B steel. Fig. 1 shows the orientation of the Y-cut transducer used in the velocity measurements with respect to the direction in which the stress is applied to the specimen. All the measurements are made at temperatures ranging from 220 K to 270 K, while the specimen is subjected to various amounts of tensile and compressive stresses ranging between 0 and 190 MPa. Typical examples of the results of the variations of ultrasonic velocity with temperature at applied tensile stresses of 0, 60, 120, 180 MPa are shown in Fig. 3. Because of the small change in velocity with stress, the velocity axes are shifted for clarity. From Fig. 3, it can be seen that, in the temperature range used in these measurements, the velocity is found to decrease linearly with temperature. A computer program was used to process the velocity-temperature data to determine the temperature dependence of ultrasonic velocity, dV/dT. Table 2 lists the results obtained on A533B steel for both applied tensile and compressive stresses. The table includes the values of the temperature dependence of shear velocity (polarization parallel to stress) determined at stresses ranging from 0 to 180 MPa, both in tension and in compression. These results indicate that the temperature dependence of shear waves polarized parallel to stress decreases as the amount of applied tensile stress is increased, and increases when the applied compressive stress is increased.

The values of dV/dT at zero stress in Table 2 do not match, possibly because of the variation of internal stress introduced when the tensile specimen was cut to be used for compression measurements. Hence the relative changes in the temperature dependence Δ, was calculated and its values are also listed in Table 2. Δ is calculated using the relationship,

$$\Delta = \frac{(dV/dT)_\sigma - (dV/dT)_o}{(dV/dT)_o} \tag{3}$$

Figure 4 shows the effect of applied tensile and compressive stresses on the relative change in the temperature dependence, $\Delta\%$. The plot shows that all the points lie on a straight line and that Δ increases with the increase of applied tensile stress, and decreases with the increase of applied compressive stress. The slope of this linear relationship is 5.2×10^{-4}/MPa, which yields a maximum change of 10% at a stress of 180 MPa. From Fig. 4, it can also be seen that the line of best fit of the relative changes in the temperature dependence versus applied stress does not pass through the origin and the value of intercept is 0.88%. This intercept could be due to misalignment in the load application system, where bending stress could be introduced.

Temperature Dependence of Ultrasonic Velocity Shear Waves Polarized Perpendicular to Stress

The velocity of shear waves propagating in a direction perpendicular to applied stress and polarized perpendicular to stress was measured as a function of applied tensile and compressive stresses. The experiments were performed in the temperature range between 220 K and 260 K, while the specimen was subjected to various amounts of tensile and compressive stresses. The results of the relative changes in the temperature dependence, dV/dT, at various stresses ranging from 0 to 180 MPa in tension and in compression are shown in Fig. 5 and tabulated in Table 3. Also listed in the table are the relative percentage changes of temperature dependence. The plots of Fig. 5 show that Δ decreases linearly with the increase of applied stress and increases with the increase of compressive stress. The line of best fit for both tension and compression is plotted in Fig. 5, and found to have a slope, K, equal to -4.79×10^{-4} per MPa. Here also it can be seen that the line of best fit does not pass through the origin and the Y-intercent is 0.7%. The intercepts in both shear (parallel) and shear (perpendicular) are almost equal which supports the idea that this intercept could be due to misalignment. Eliminating the effect of the bending stress results in a straight line which is parallel to the line of best fit and which passes through the origin.

DISCUSSION

The relative changes in the temperature dependence, Δ, of ultrasonic shear velocities (parallel and perpendicular) in A533B steel are plotted in Fig. 6 along with the published data for longitudinal velocities. For both shear and longitudinal measurements shown in Fig. 6, the propagation direction was perpendicular to

Table 3: Variations of temperature dependence of ultrasonic
          shear velocity with applied tensile and compressive
          stresses in A533B steel. Stress is applied in a direc-
          tion perpendicular to both the directions of propagation
          and polarization.

Stress MPa	dV/dT m/sec/MPa	(dV/dT) - (dV/dT) m/sec/MPa	%	Dev. in %
181.8	-0.409	0.027	-6.2	-0.54
145.4	-0.411	0.025	-5.7	0.18
90.9	-0.418	0.018	-4.1	0.42
60.6	-0.422	0.014	-3.1	0.54
0.0	-0.436	0.0	0.0	-0.61
0.0	-0.442	0.0	0.0	-0.83
-36.36	-0.447	-0.005	1.1	-0.35
-72.72	-0.458	-0.016	3.6	-0.13
-96.96	-0.461	-0.019	4.3	-0.96
-151.5	-0.480	-0.038	8.6	0.08
-181.8	-0.492	-0.05	11.3	0.71

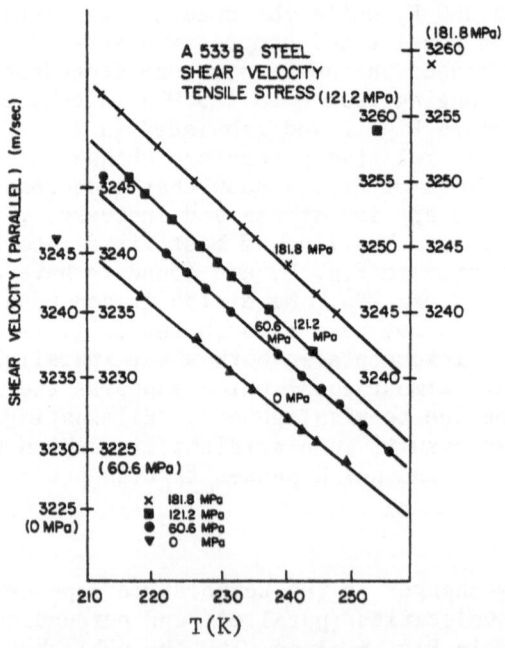

Fig. 3: Shear velocity variations as a function
         of temperature at different applied tensile
         stresses in A533B steel.

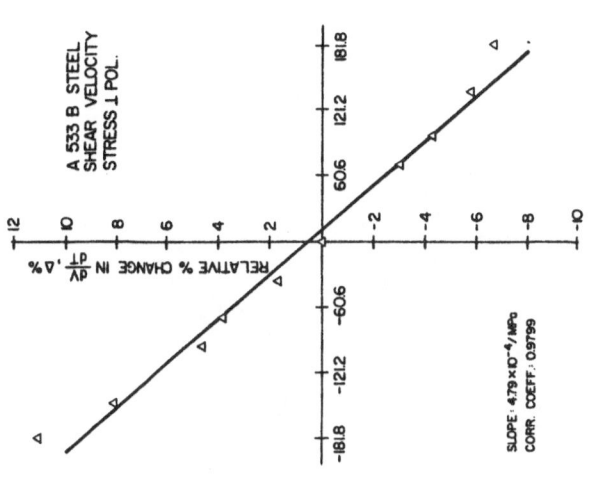

Fig. 5: Relative percentage change in the temperature dependence of ultrasonic shear velocity as a function of applied stress in A533B steel. Stress is perpendicular to polarization direction.

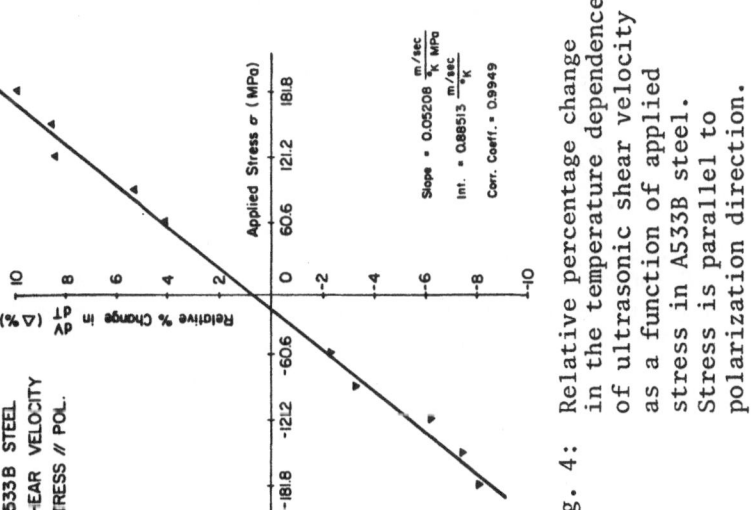

Fig. 4: Relative percentage change in the temperature dependence of ultrasonic shear velocity as a function of applied stress in A533B steel. Stress is parallel to polarization direction.

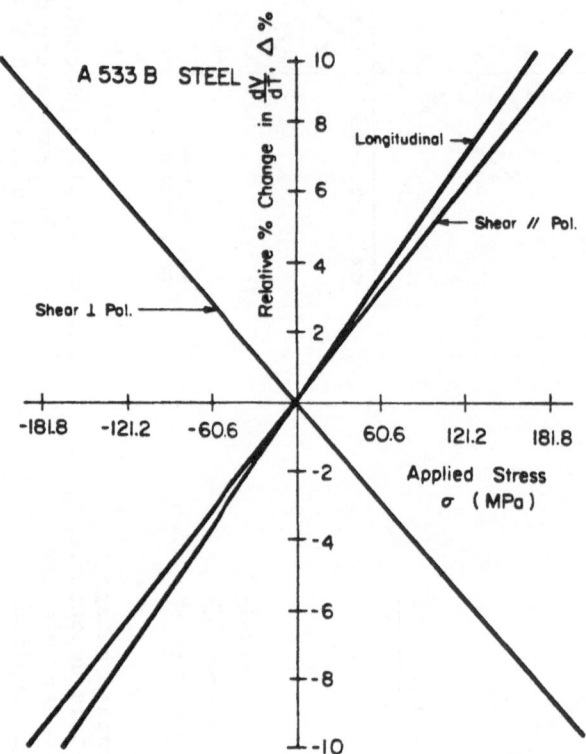

Fig. 6:   Relative percentage changes in the temperature dependences
          of longitudinal and shear velocities as a function of
          applied stress in A533B steel. Stress is perpendicular
          to propagation direction.

that of stress. From the figure, it can be seen that in the cases
of ultrasonic longitudinal velocity, and shear velocity of waves
polarized in a direction parallel to the direction of the applied
stress, the relative change in the temperature dependence of ultrasonic velocity, Δ, increases with the increase of applied tensile
stress and decreases as the applied compressive stress is increased.
In both measurements, the stress is applied in a direction perpendicular to that of wave propagation. The slope of the calibration
curve for longitudinal velocity is positive and is equal to
5.88×10^{-4} per MPa, and is slightly greater than the slope of the
calibration curve for the shear velocity (parallel), which is equal
to 5.208×10^{-4} per MPa. However, when the stress is applied in a
direction perpendicular to both directions of propagation and polarization (shear-perpendicular), then the relative change in the temperature dependence of ultrasonic velocity, Δ, decreases with the
increase of applied tensile stress and increases with the increase
of applied compressive stress. The slope of the calibration curve
is negative and is equal to -4.79×10^{-4} per MPa. Figure 6 represents
the complete set of calibration curves in A533B steel for uniaxial
stresses applied perpendicular to wave propagation.

ACKNOWLEDGEMENT

This work is supported by the Electric Power Research Institute
under Contract T107-2.

REFERENCES

1. Salama, K. and Ling, C. K., J. Appl. Phys. 51, 1505 (1980).

2. Salama, K. and Ling, C. K., Proc. ARPA/AFML Review of Progress
    in Quantitative NDE, p. 96 (1979).

3. Salama, J., Ling, C. K. and Wang, J. J., Experimental Technique
    5, 14, (1981).

4. Salama, K., Collins, A. L. W. and Wang, J. J., Proc. DARPA/AF
    Review of Progress in Quantitative NDE, p. 256, (1980).

5. Salama, K., Barber, G. C. and N. Chandrasekaran, Proc. 14th
    Symposium on NDE (1983).

6. Salama, K., Barber, G. C. and N. Chandrasekaran, Ultrasonic
    Symposium, p. 877, (1982).

7. Papadakis, E. P., J. Acoust. Soc. Am. 42, 1045 (1967).

8. Spiegel, M. R., Schaum's Outline of Propability and Statistics
    McGraw-Hill Book Company, New York, (1975).

AUTHOR INDEX

A

Adler, L., 179
Alreja, V., 39
Altpeter, I., 115

B

Bennett, S.D., 365
Bertram, A.L., 327
Blessing, G.V., 353
Borgonovi, G.M., 47
Buck, O., 161
Bussière, J.F., 289
Buxbaum, S.R., 271

C

Chandrasekaran, N., 393
Chihoski, R.A., 81
Clark, A.V., 339
Clark, W.G. Jr., 89

D

Deka, M., 135
Droney, B.E., 237

E

Eberhardt, N., 135

F

Fiedler, C.J., 161
Friant, C.L., 301

G

Good, M.S., 189
Green, R.E. Jr., 271

H

Hahn, H.T., 315
Hsu, N.N., 353
Husson, D., 365

J

Junker, W.R., 89

K

Kino, G.W., 365

L

Liu, J.M., 39

M

Masubuchi, K., 123
Mehrabian, R., 207
Mignogna, R.B., 339
Molian, P.A., 71
Monchalin, J.-P., 289
Mott, G., 377

O

Oien, C.T., 59
Oravecz, M.G., 259

SUBJECT INDEX